Nuclear Debates in Asia

Nuclear Debates in Asia

The Role of Geopolitics and Domestic Processes

Edited by Mike M. Mochizuki
and Deepa M. Ollapally

ROWMAN & LITTLEFIELD
Lanham • Boulder • New York • London

Published by Rowman & Littlefield
A wholly owned subsidiary of The Rowman & Littlefield Publishing Group, Inc.
4501 Forbes Boulevard, Suite 200, Lanham, Maryland 20706
www.rowman.com

Unit A, Whitacre Mews, 26-34 Stannary Street, London SE11 4AB, United Kingdom

British Library Cataloguing in Publication Information Available

Library of Congress Cataloging-in-Publication Data

Names: Mochizuki, Mike, editor. | Ollapally, Deepa Mary, editor.
Title: Nuclear debates in Asia : the role of geopolitics and domestic processes / edited by Mike M. Mochizuki and Deepa M. Ollapally.
Description: Lanham : Rowman & Littlefield, [2016] | Includes bibliographical references and index.
Identifiers: LCCN 2016005008 (print) | LCCN 2016005442 (ebook) | ISBN 9781442246997 (cloth : alk. paper) | ISBN 9781442247000 (electronic)
Subjects: LCSH: Nuclear energy—Political aspects—Asia. | Nuclear weapons—Political aspects—Asia.
Classification: LCC HD9698.A762 N824 2016 (print) | LCC HD9698.A762 (ebook) | DDC 333.792/4095—dc23
LC record available at http://lccn.loc.gov/2016005008

∞™ The paper used in this publication meets the minimum requirements of American National Standard for Information Sciences—Permanence of Paper for Printed Library Materials, ANSI/NISO Z39.48-1992.

Printed in the United States of America

To our children—
Miles and Ella Mochizuki and
Kavitha and Siddharth Anandalingam

Contents

Acknowledgments

In contrast to most multiauthor works, we can happily say that this book was done in record time with minimum fuss. This is not entirely our doing. We were lucky enough to recruit an extraordinarily conscientious and collegial group of authors. We also received generous funding from the MacArthur Foundation, making the book possible in the first place—a special thanks to Emma Belcher and John Fei for the confidence reposed in us.

Having a convivial setting to conduct our work helped, too. Two successive directors of The George Washington University's Sigur Center for Asian Studies, Ed McCord and Bruce Dickson, and the Center's staff, could not have done more to support us. The Dean's office of the Elliott School of International Affairs and Dean Michael Brown, were consistent backers, along with internal Strategic Opportunity for the Advancement of Research (SOAR) funding for this project (and other endeavors) undertaken as part of the Sigur Center's Rising Powers Initiative (RPI). With great ingenuity and aplomb, Matt Grieger, as RPI's Assistant Director, handled every budgetary and programmatic hurdle that cropped up and repeatedly saved us time and effort. In addition to coauthoring the chapter on Southeast Asia, the project's research and program associate Timothy Westmyer provided excellent administrative, research, and editing support throughout this project. Without his dedication and contribution, successful completion of this volume would have been impossible. As research manager for the Rising Powers Initiative, Winnie Nham enabled the project to proceed smoothly and managed the project website.

We would also like to thank our colleague Janne Nolan, who chairs the Nuclear Security Working Group, for organizing a series of study group meetings. These discussions gave us valuable insights into the policy and economic circumstances regarding nuclear energy and proliferation. Joyce Connery, then Director of Nuclear Energy Policy within the Office of International Economics in the U.S. National

Security Council, and Seth Grae, President and CEO of Lightbridge Corporation, were especially gracious in sharing their expertise.

One of the highlights of writing this book was the "Nuclear Perspectives in Asia" workshop we jointly organized with Vietnam National University (VNU) in 2014 in Hanoi, bringing together scholars and top policymakers. It gave many of us from the United States a firsthand look at the perspectives of perhaps the important nuclear aspirant country in Asia right now. We want to thank VNU's Rector Professor Nguyen Van Khanh for his interest and support, and single out our Vietnam-based contributor, Professor Nguyễn Thị Thanh Thủy, for making the workshop a reality. Our colleague Linda Yarr, who coauthored the chapter on Vietnam with Professor Thanh Thủy helped immensely in making the arrangements with VNU.

We organized the workshop at an opportune moment as the annual Nuclear Power Asia conference was being held in Hanoi in 2014 for the first time. Our group was able to attend this event that convened regulators, nuclear energy agencies, and regional and global operators. As a research team, it gave us a rare chance to hear viewpoints of individuals at the ground level of nuclear energy—and we might add, a forum for us to question and debate nuclear policies, plans, and propositions these experts expressed.

In addition to Scott Snyder who authored the South Korea chapter and Linda Yarr, our colleagues Marcus King, Janne Nolan, and Douglas Shaw joined us in participating in both the VNU workshop and the Nuclear Power Asia conference. Their presentations enriched the workshop and assisted the contributors of this volume in establishing the international context of domestic nuclear debates in the Asian countries examined here.

Each of us enjoyed great personal advantages, making it much easier to get the book done. Deepa's husband Anand, daughter Kavi, and son Siddhu, kept letting her off the hook for familial duties and instead provided just the right amount of entertaining diversions. Despite her grueling first year in medical school, Kavi found time to send me a steady stream of engrossing literary books way off best-seller lists, giving a much-welcomed balance to nuclear reading. Siddhu provided an exceptionally agreeable work environment for me with his original and virtuosic music compositions and performances. For this and much more, Deepa dedicates this book to her children.

During the course of this project, Mike and his wife Clare had the opportunity to take their son Miles and daughter Ella to Hiroshima. The memorable day spent at Hiroshima Peace Memorial Park reminded us that the international community must work harder to prevent another Hiroshima, Nagasaki, Chernobyl, or Fukushima. With the hope that the world can prevent future nuclear catastrophes, Mike also dedicates this book to his children.

Deepa Ollapally and Mike Mochizuki

1

Introduction

Uncovering Nuclear Thinking in Asia

Deepa M. Ollapally

The rise in global demand for nuclear energy is heavily concentrated in emerging and aspiring Asian powers. While nuclear power may alleviate energy shortages and climate change concerns, the promotion of nuclear energy compounds Asia's nonproliferation problem with the spread of nuclear technology and safety risks. Moreover, Asia remains at risk for high stakes nuclear weapons competition, nuclear conflict, and nuclear terrorism. All this is exacerbated by rising geopolitical tensions in Asia, with more assertive policies in the region testing regional stability. Against this setting, our book questions the extent to which we can infer nuclear thinking simply from external conditions and considers the notion that policy thinking on nuclear power and proliferation in Asia is more complex and variegated than often posited.

Most analysts in the nuclear field have viewed state policies as inexorably driven by purposive unitary actors responding to external stimuli.[1] Nuclear weapons development is said to be based on interstate competition; and the pursuit of nuclear energy is often viewed as a pathway to gaining sensitive dual-use technology or energy security writ large. Not surprisingly then, nuclear trends in Asia are mostly taken for granted and seen to be determined by exogenous factors. More recent scholarship has questioned the assumption of traditional neorealist frameworks that see a state's nuclear choices as directly determined by external security forces, with little consideration given to domestic contexts. Scholars in this vein argue that even nuclear policy, uppermost on the "high politics" ladder, is often shaped by domestic factors ranging from political survival and bureaucratic compulsions, to social norms and identity politics. We share the skepticism that strategic circumstances are sufficient to explain the choices that states make on nuclear weapons and nuclear energy questions, and our framework argues for the necessity of a domestic lens of analysis.

Specifically, our book addresses a key question: How do we understand the trajectory of nuclear energy, nuclear security, and nuclear nonproliferation discourse and debate in seven key Asian countries (China, India, Japan, Pakistan, South Korea, Taiwan, and Vietnam) as well as countries in the Association of Southeast Asian Nations (ASEAN)? We seek to understand how nuclear debates in Asia change over time and why they are changing the way they are. We explore what drives the debates and how decisions are framed, what the interplay is between domestic dynamics and geopolitical calculations in the discourse, where the center of gravity of debates lies in the countries, and what this means for regional cooperation or competition, and U.S. nuclear energy, nonproliferation, and deterrence policy in Asia. We propose that domestic debates are structured (not random) along what we term "Nationalist," "Realist," and "Globalist" foreign policy schools of thought (to be elaborated later in the chapter).[2]

CONVENTIONAL WISDOM ON NUCLEAR POLICY: THE INTERNATIONAL LENS

Dominant neorealist theory holds that nuclear policy is a byproduct of competition in the international system.[3] Variations of realism such as defensive realism (which in particular strongly recognizes the idea of the "security dilemma"), in the end still do not fully see the need to go beyond those state preferences that can be more or less derived from the external environment. Charles Glaser describes his strategic choice theory as "integrating defensive realism with neoclassical realism," wherein both the international environment and a state's motives matter.[4] Still, for Glaser, fundamental state interests are constant and his focus remains on the strategic interaction between states. He sparingly collapses state motivations into essentially two types: greedy and security seeking, leaving the door tantalizingly ajar for ascertaining motives from domestic political and identity urges.[5]

Muthiah Alagappa in his book *The Long Shadow: Nuclear Weapons and Security in 21st Century Asia* concludes that "security interaction in Asia increasingly approximates behavior associated with defensive realism."[6] Alagappa is one of the first to conduct a comprehensive study of nuclear strategies of new and old nuclear states in post–Cold War era Asia, broadly defined, with his work squarely located in the neorealist framework. A major objective of Alagappa is to counter deterrence pessimists on Asia by arguing that unlike the "grim scenarios associated with nuclear weapons in Asia," the preferences and behavior of new and aspiring entrants to the nuclear club are consistent with neorealist deterrence logic and indeed, "on net, they [nuclear weapons] have reinforced national security and regional stability in Asia." Indeed, Alagappa suggests that arguments by nonproliferation adherents, especially in the West, opposing developing countries acquiring nuclear weapons capabilities because of higher perceived "risks" are not based on substantive grounds. He bluntly states, "Intellectual (and possibly racial) biases and national strategic interests are often cloaked in this argument, which is usually advanced on behalf of the international community."[7]

The type of bias that Alagappa alleges is entirely plausible. If bias of this sort exists, it ironically undercuts a foundational assumption of neorealism—that is, the likeness of states. Of course it is not possible to easily prove the existence of such biases.

On the other hand, what would be useful in disputing value judgments regarding issues such as nuclear risk-taking are more even-handed analyses of domestic thinking on nuclear questions across both developed and developing countries. Jacques Hyman's work with his cross-regional case studies is instructive in this regard.[8] This bias issue is beyond our scope here, but it is worth noting that our analysis of competing worldviews in Asian states can serve as a sort of internal check on such a bias in our own work by our consideration of what different in-country elites and opinion shapers are themselves saying. Of course this does not eliminate another potential bias—the unfortunate reality that national leaders across the globe tend to shroud their nuclear preferences with realist principles or "reasons of state," disguising other motivations. As Leon Sigal has insightfully noted, "Realism seems to be the secular religion of the foreign policy establishment."[9] All those who work anywhere on foreign policy using official state sources are vulnerable to this evidentiary bias. It is practically an occupational hazard. However, if nothing else, it is clearly worth acknowledging it rather than the usual practice of simply ignoring it. In this connection, we would note that our attention to contending viewpoints goes some way to alleviate this particular bias.

Like Alagappa, the National Bureau of Asian Research's (NBR) *Strategic Asia 2013–14: Asia in the Second Nuclear Age*, also ends up ascribing mostly realist security motivations to states. This is not surprising since the authors of the seven comprehensive country case studies in the book take a more or less state-centric view of nuclear weapons and strategy despite some attention to societal factors.[10] The NBR volume is generally less optimistic than Alagappa on regional deterrence stability. The lead editor, Ashley Tellis, in particular cites what he views as the inevitable and dangerous growth of nuclear multipolarity in Asia. For Tellis, the link between a deteriorating security environment and affirmation of nuclear weapons is pretty direct: "Given the contested geopolitics of Asia which is defined by several enduring rivalries, many unresolved territorial disputes, significant local power transitions, and now the continent-wide anxiety provoked by the rise of China, it is not surprising that nuclear weapons have retained their critical importance."[11] A major policy recommendation from the NBR study is for continuing American nuclear supremacy (if not nuclear hegemony) in Asia and shoring up America's extended deterrence commitments to allies against China in no uncertain terms. What is not adequately elaborated is the domestic nationalist reaction that this policy might produce in China (as well as domestic dilemmas in allied countries it might even create regarding the right amount of "hugging" of the United States).

Meanwhile, more fine-tuned work homing in on the nuclear postures and strategies of six regional powers—China, India, Pakistan, France, Israel, and South

Africa—by Vipin Narang, shares to a significant extent Alagappa's deterrence optimism on China and India, but not so on Pakistan. Taking inspiration from neorealism but going beyond it (as he puts it), Narang aims to explain why and how a regional power might "optimize" its nuclear posture, taking into account both external security and internal political and financial constraints. At the domestic level, Narang identifies two critical variables that influence a state's choice of nuclear posture—civil-military relations and resource constraints.[12] His explanation of India's and China's stability-inducing "assured retaliation" nuclear posture and its future continuity rests significantly on the assertive civil-military setup in the two countries in which civilian leadership presumably keeps the more aggressive military in check. Conversely, Pakistan's alarming "asymmetric escalation" nuclear posture is premised to rest on its perceived dire security conditions and accentuated by the powerful hold of the military.

To anticipate—Pakistan, as described by Christopher Clary in this volume, emerges as an exception to the general Globalist/Realist nuclear outlooks of other states. Given Pakistan's special status, it merits greater scrutiny here. In Narang's book, Pakistan's nuclear posture is described as "the most aggressive option available to nuclear states and involves fielding both tactical and strategic capabilities that can be mobilized and launched quickly, often through pre-delegated authority in the event of conventional conflict."[13] As Clary noted in the NBR volume, Pakistan "is increasingly reliant on its nuclear arsenal for deterrence, despite the fact that India's conventional force-advantages are not yet dire."[14]

The work by Narang and authors of the NBR volume tends to support the view of Pakistan as an exception to the rule of deterrence stability promoting behavior among Asian states (save North Korea), and it hints that security factors are underdetermining. Instead, the civil-military asymmetry in Pakistan takes on special significance to understand Pakistan's nuclear stance.[15] For example, it turned out that Pakistan's surprise war against India in 1999, so soon after the two countries went nuclear and apparently included nuclear posturing by Pakistan, was a military enterprise locking out the civilian leadership who were engaged in peace talks with India. It was widely seen as an attempt to "internationalize" the Kashmir dispute by linking it with a nuclear flashpoint. Instead, the result was to reinforce international opposition to changing territorial status quo by any sort of force.[16] Without the particular type of hyper-nuclear Nationalism including nuclear Islamism that Clary lays out in this volume, and that the military in Pakistan encourages and fans, we would be hard-pressed to understand how Islamabad's highly risky (and internationally roundly condemned) nuclear postures and actions are possible. Pakistan's most recent pursuit of battlefield nuclear weapons, its evolving nuclear war–fighting doctrine, and its willingness to buck international opinion and suffer the associated costs simply underline this point. Most ominously, what we see is how nuclear Nationalism in Pakistan is able to successfully rationalize even the most egregious behavior of proliferators like A. Q. Khan. As a final point here, it should be noted that China's checkered record on assisting Pakistan's nuclear and missile program casts a strong shadow on China's otherwise deterrence-promoting practices that most authors see prevailing.[17]

To sum up and to put nuclear outlooks in a broader context: In a region where China's neighbors view its rise and assertiveness as a growing threat to their national security interests and energy sea lines of communication, we would expect a particularly high salience for external security factors. After all, at least since 2010, China has been widely seen as the source of spiking tensions from East Asia to South Asia: the standoff with Japan in the East China Sea; butting up against Vietnam and Philippines with increasing sovereignty claims and build-up in the South China Sea; and encroachment into the disputed border with India. Meanwhile, leaders of the two other major Asian countries—Shinzō Abe of Japan and Narendra Modi of India—have displayed a greater willingness to respond more forcefully.

Partly in anticipation and partly in response to this rise of China, many observers are evoking power-transition theory as the most applicable general international relations theory to contemporary Asia, and in particular China-U.S. relations in Asia. Given the high profile of power transition theory popularized, for example, as the "Thucydides Trap" by Graham Allison, and its antecedents in neorealism discussed above, it is worthwhile to specifically consider this theory's value to understanding nuclear debates in the region.[18]

Power transition theory harks back to A. F. K. Organski and Jacek Kugler's status disequilibrium model, which pits an ascending, dissatisfied, revisionist state against a relatively declining, status quo power—in our case China's challenge to U.S. dominance in Asia.[19] The most unstable and conflict-prone time frame is identified as when the declining hegemon is about to be overtaken by the ascending challenger. For structural realists like Aaron Friedberg and John Mearsheimer, rivalry between China and the United States (and Japan for that matter) is set to intensify as China grows richer and stronger.[20] For Friedberg and Mearsheimer, this type of competition will inevitably be settled through war (or "hegemonic war," to use Robert Gilpin's term).[21] Power transition theorists of recent vintage tend to see the challenger as initiating war.

The competing maritime and territorial claims in the region that impinge on a number of China's neighboring states and the growing intensity of these claims since 2011 at the same time that American capability and will are seen to be sapped by Afghanistan, Iraq, and the Middle East, fuels this conflictual scenario. Thus the nuclear debates and outlooks of countries like Japan, India, and Vietnam—all which have ongoing territorial or maritime disputes with China, could be expected to reflect these geopolitical tensions and uncertainties. Taken to its logical conclusion, the so-called "second nuclear age" could even be viewed as wrapped up in these geopolitical anxieties.[22] As Ashley Tellis states, "continuing interstate competition, along with the expectations of many states that nuclear weapons will enhance their security and offer deterrence value, ensures that regional arsenals will grow."[23]

The strong claims of inevitable conflict that power transition theory posits may be questioned. Like critics of neorealism, several scholars looking at power transition theory have argued that shifting material capabilities are not enough to predict conflictual outcomes, and that the distribution of benefits from the prevailing system needs to be taken into account.[24] Others point out that lest we forget, it is state officials who make strategic choices and even "construct realities" that cannot be predetermined.[25] A power shift to Asian states is also seen by some as part of a more general diffusion of global power having to do with technological, economic, and institutional changes. Referring to such fragmentation of power and the Asia-Pacific region, Rosemary Foot explains, "That diffusion of power and the hybrid nature of world order complicates decision-making for all political actors and suggests that we need to investigate the full spectrum of concerns—domestic as well as external—that governments face when they determine their policies."[26] In contemporary Asia, a critical question then becomes whether the dual or hybrid nature of economic interdependence and strategic rivalry is firmly tipped in favor of the former or not for China, power transition notwithstanding. The shifts in domestic viewpoints and orientations during this period of power transition thus deserve more attention than they get.

If we take the criticism of power transition theory seriously and extend it to nuclear policy, we can conclude that geopolitical competition—even under the more acute power transition version in Asia—need not inevitably result in the most extreme outcome of war or a ratcheting up of the nuclear arsenal as part of competitive strategic capabilities. This of course also has implications for how America should be thinking about its rebalancing strategy and deterrence strategy in Asia.

From a critical reading of power transition theory in Asia, we can infer for our own work that geopolitical calculations need not drive domestic debates about nuclear policy to the degree that we might expect (consistent with our reading of neorealism earlier). This is not to suggest that external conditions and geopolitical pressures do not matter for nuclear debates. We recognize that the international system sets up constraints and incentives for a country's nuclear choices. In the nuclear choices made by Japan, South Korea, and Taiwan for example, we are not downplaying the importance of alliance with the United States and extended deterrence. However, we concur with Etel Solingen's point that to understand these countries' "relative receptivity to persuasive and coercive aspects of the U.S. alliance requires us to delve into their domestic politics."[27] What we are putting forward is the need to factor in interactive dynamics between external and internal contexts—more specifically, that nuclear policies are mediated by domestic debates. In other words, even nuclear issues are often a source of domestic contestation and policy alternatives between competing foreign policy worldviews, something that tends to be passed over by many nuclear analysts.

We now turn from the international security literature to the growing studies of nuclear issues that give weight to domestic factors and we situate our own work within it.

BRINGING IN THE DOMESTIC LENS

As already alluded to, some important work has emerged arguing for the importance of domestic level variables to the understanding of nuclear policymaking in Asia and beyond. One early critique of neorealism proposed techno-military elitism as a driver of nuclear policies. Scott Sagan and Itty Abraham are two proponents of this thinking on nuclear developments in Asia, though Sagan also gives weight to purely domestic political considerations at the leadership level.[28] We would argue that these so-called "strategic enclaves" (composed especially of nuclear science and technology bureaucracies) do not operate in a vacuum and that their ability to hijack nuclear policymaking is unlikely to remain impervious to nuclear debates and the competing schools of thought articulated by other political and bureaucratic groups. We agree that these groups have tended to wield disproportionate influence over nuclear decisions in the past, especially in India, Japan, and South Korea. A key question for our own work is the extent to which such elitist strategic enclaves are operating in the country case studies and their historic versus contemporary level of influence on nuclear debates and vice versa.

Another type of argument comes from Etel Solingen's innovative work on domestic political coalitions (liberalizing versus nationalist-confessional coalitions) and their impact on the decision to nuclearize. In *Nuclear Logics: Contrasting Paths in East Asia and the Middle East,* Solingen contrasts East Asia (liberalizing) with the Middle East (with many nationalist-confessional coalitions) and argues that those states that aim to integrate into the global economy are willing to give up the pursuit of nuclear weapons, while states controlled by the inward coalitions do the opposite. This is an important work in a field that is still surprisingly short on investigations of why states go after nuclear weapons—that is, what their preferences are—as opposed to the copious material on the operation of deterrence stability or instability.[29] In Solingen's domestic "political survival" model,

> Leaders and ruling coalitions interpret security issues through the prism of their own efforts to accumulate and retain power at home. Internationalizing leaders define economic growth and global access as crucial for advancing state security, rejecting nuclear weapons if the latter endanger those core objectives. Conversely, inward-oriented leaders thrive by defining security as "self-help" while protecting and promoting constituencies that variously favor economic, political, strategic, religious or cultural autonomy.[30]

Our structured nuclear debates approach captures the divergent domestic opinions and coalitions that we too argue populate the nuclear arena. Like Solingen's model, our fine-grained depiction of contrasting schools of nuclear thought seeks to "avert the frequent overestimation of *state* [emphasis in original] security as a source of nuclear behavior."[31] But Solingen's liberalizing coalition argument for eschewing nuclear weapons runs up against a counter case when we turn to South Asia (not included in her book) and away from the paradigmatic East Asia and the Middle East.

India's decision to openly test nuclear weapons and "go nuclear" after twenty-four years of nuclear limbo came precisely during the height of its historic economic liberalization begun in 1991. By doing so, India courted economic sanctions and diplomatic isolation just when Solingen's model would have sent Indian leaders in the opposite direction, decisively against weaponization—for example, by supporting the Comprehensive Test Ban Treaty (CTBT) over which international discussions were in full swing. By tracking domestic nuclear discourse, our work shows how a critical surge in what we term Realist thinking on security matters in India over the 1990s, along with long-standing Nationalist preference for strategic autonomy, came together to support the nuclear choice India finally made. This is despite the serious inroad that a liberalizing Globalist model had made during the 1990s.

Another domestic take on why states make a decision to go or not go nuclear comes from Jacques Hymans in *The Psychology of Nuclear Proliferation: Identity, Emotions, and Foreign Policy*. As the title indicates, the key variable determining why some states say yes to nuclear weapons and others say no, is the nature of the decision-making leader's so-called "national identity conception."[32] Again, this work is very useful in determining leadership preferences regarding nuclear choices that allow us to transcend state preferences narrowly defined in neorealist security terms. Hymans argues that "any full explanation of decisions to acquire the bomb must account not for their ultimate valence but also their timing and the manner in which they are made."[33] Even if we agree on this general point, a micro-focus on a single decision-making point and leader is apt to leave out the richness and twists and turns of the debates and decision trajectories that preceded it. This is especially true of his study of India, the one case that overlaps in our respective books. Hymans himself concedes that of his four cases, "the Indian one is the most complex," and that "it does not perfectly fit" his framework.[34] Our book's aim is to unpack such complexities in the domestic nuclear discourse, and in the process, we will be able to better determine whether and how well Hyman's propositions operated in the Indian case to go nuclear. Generally speaking, though, Hyman's theoretical framework built on identity-based nuclear preferences is entirely consistent with our book's approach.

Maria Rublee's book *Nonproliferation Norms: Why States Choose Nuclear Restraint* shows how social norms against nuclear weapons emerged in Japan and provided the political foundations for Japan's policy not to possess, develop, or permit the deployment of nuclear weapons.[35] However, this work does not analyze explicitly how and why Japan decided to vigorously embrace nuclear energy while eschewing nuclear weapons. Rublee's "social norms" argument also needs to be updated to consider the impact of the unprecedented 2011 Fukushima nuclear disaster and the changing East Asian security environment on the Japanese nuclear policy debate.

In contrast to the voluminous literature on nuclear weapons, there is much less scholarly discussion of the link between nuclear weapons and nuclear energy, and what the drivers are toward nuclear energy acquisition. It is to this question that I now briefly turn.

NUCLEAR ENERGY AND SECURITY

Broadly speaking, the assumption that nuclear energy is often a back door to nuclear weapons capability is fairly commonplace. Alagappa asserts that one reason for increased international concern with nuclear weapons is linked to the renewed interest in nuclear energy and potential for diversion.[36] Ever since North Korea's withdrawal from the Nuclear Nonproliferation Treaty (NPT) in 2003 to openly pursue nuclear weapons and Iran's alleged violation of the NPT to build nuclear weapons capability, nuclear energy programs have become more suspect. Both North Korea and Iran were able to construct a nuclear infrastructure based on peaceful use of nuclear energy allowed under the NPT.[37]

Miller and Sagan point out that the link between nuclear power and nuclear weapons has been a serious worry throughout the nuclear age. In their view, states are likely to maintain a "latent nuclear weapons capability" with enrichment and reprocessing capabilities if they cannot be assured access to nuclear fuel. This is especially true if "nuclear newcomers acquire the full panoply of technology associated with nuclear power, including fuel cycle technologies that have direct weapons applications." [38] How can this risk be reduced? According to them, "only if the acquisition of uranium enrichment or plutonium reprocessing technologies is discouraged or prohibited." However, they concede that these very technologies are indispensable for producing nuclear fuel for nuclear power reactors. For these authors, a key condition for states to give up national fuel cycle capabilities that keep the nuclear weapon option open is reliable outside fuel supplies.

How might states be persuaded to give up the more expansive nuclear route? Is it simply a matter of having access to fuel, or do the preferences and dispositions, incentives and disincentives of state actors come into the picture? Using the domestic lens, we see how factors beyond the nuclear technical level—for example, norms, coalitions, identities, globalization, national ambitions—and security perceptions, could provide alternative explanations. The structure of domestic thinking can thus give us greater insight into the various options and complexities underlying nuclear choices internally in states, and why some states are willing to give up the nuclear fuel cycle capabilities even in a context containing a degree of uncertainty. Authors in this book explore whether debates on nuclear energy are a proxy or precursor for weapons capabilities or not. The limited research that does exist on potential versus actual nuclear proliferation shows there is no automatic link. As Hyman notes, "This yawning gap between technical potential and military reality should have led to widespread rethinking of the phenomenon of nuclear proliferation. To a surprising extent it has not."[39]

One of our objectives is to assess the extent to which positions on different nuclear issues cluster—in particular, on nuclear weapons versus nuclear energy—and whether these clusters reflect Nationalist, Realist, or Globalist orientations. This would be a unique contribution to the nuclear literature. Even when there has been a direct attempt to understand the complexities of nuclear energy debates in Asia, like Rajesh Basrur and Koh Swee Lean Collin's *Nuclear Power and Energy Security in Asia*, the link or interconnectedness between energy and weapons debates is not fully explored. [40] Basrur and Collin do find that domestic debates on nuclear energy are especially important in Asia, a finding that provides some evidence for the validity of our approach. It becomes all the more urgent to track such debate after Fukushima, when the perils of nuclear energy became all too obvious.

Apart from pursuing nuclear energy for gaining the technical capabilities for weapons, another motivation may be attaining energy security—indeed even energy independence.[41] When we look at the galloping energy needs of aspiring Asian powers outstripping their domestic resources, in combination with maritime and transit vulnerabilities of supplies, the geopolitical argument for nuclear energy as a way to address energy insecurity seems quite plausible.

Emerging economies, led by Asian countries, accounted for 80 percent of growth in energy consumption in 2013.[42] During the decade from 1995 to 2005, one third of the entire growth in global demand for oil came from China and India alone.[43] In 2013, China became the world's largest net importer of petroleum, after having already overtaken the United States as the world's largest energy consumer three years earlier.[44] By 2014, India was the fourth largest oil consumer in the world behind China, the United States, and Japan.

Along with this steep rise in energy demand is the growing external dependence for energy resources by key Asian states. In 2014, India overtook Japan as Asia's number two crude oil importer with nearly 85 percent of its oil requirements met through imports.[45] In 2015, China's oil imports rose nearly 9 percent compared to 2014, even with the country's economic slowdown.[46] India imported 62 percent of its oil from the unstable Middle East region in 2013, with the Middle East remaining the biggest supplier of crude for China as well.[47] One compilation of a so-called Energy Insecurity Index found that Asia has higher energy insecurity than Europe or the United States with especially high degrees of insecurity for Japan, South Korea, Taiwan, India, and China.[48] It is not a big leap from this situation of energy vulnerability to countries' possibly seeing great value in pursuing nuclear energy as a step toward energy security and autonomy.

However, global energy markets have been functioning fairly smoothly, even in times of crisis or war. So one question is the extent to which states are willing to trust the international market for their energy needs. Likewise, to what extent are states, especially states like China and India willing to rely solely on energy transit security provision by the United States? [49]

Some analysts have made the argument that the idea of "resource nationalism" can hold appeal for states, even when it goes against their economic interests according to market logic. I do not want to digress here, but the point is that "resource nationalism" and the so-called "cult of energy insecurity" could operate in countries that would then be predisposed toward nuclear energy for energy security reasons.[50]

In light of the above, a key question for our work is the significance of energy security in driving nuclear energy acquisition in Asia and how it shows up in domestic nuclear debates. Given the region's geopolitical environment and energy vulnerabilities, the internal energy production capability afforded by nuclear energy should be expected to make it particularly attractive. We would then expect energy security arguments to be high on the list of reasons governments in Asia turn to nuclear energy. The level of nuclear power in Asian countries' current energy portfolios presents a mixed picture, though generally on the lower side. In 2014, the percentage share of nuclear power in overall electricity generation was as follows: China (2.4 percent), India (3.5 percent), Japan (0 percent), Pakistan (4.3 percent), South Korea (30.4 percent), and Taiwan (18.9 percent).[51]

Nuclear energy plans and ambitions, however, are much greater almost across the board in Asia. Hence, it is important to determine how domestic opinion is arrayed on nuclear energy issues and how it interacts with security concerns: the importance of energy security, nuclear safety and regulatory aspects especially post-Fukushima, openness to international fuel cycle controls, and views on regional nuclear energy cooperation.

METHODOLOGICAL FRAMEWORK: NUCLEAR SCHOOLS OF THOUGHT

Our country case studies represent a cross section in terms of key variables: (1) geopolitical position and security environment, (2) type of political system, (3) level of economic development, and (4) energy endowments and energy demand trajectories. This variation should give greater confidence to our domestic debates approach that I have argued above makes an important contribution to the literature.

Additional support for our particular approach comes from some noted experts. In their high-profile study exploring whether the world has reached a nuclear "tipping point," Kurt Campbell, Robert Einhorn, and Mitchell Reiss conclude that a better understanding of the internal factors that could propel a country toward nuclear weapons is essential. According to them, "Given the importance of domestic factors in driving nuclear choices . . . there is a pressing need for a more finely tuned appreciation of the attitudes of key individuals, institutions, and political movements toward the nuclear issue."[52]

Accessing and Assessing Domestic Debates

Each chapter in this book examines change and continuity in the domestic debate on a range of nuclear issues in a given country: development of nuclear energy, nuclear safety and security, national security and nuclear weapons, nuclear nonproliferation, and nuclear disarmament. We address the debates systematically across the cases through a "schools of thought" approach developed in our earlier work. This framework offers a basic three-prong spectrum of foreign policy opinion or "schools of thought": Nationalist, Realist, and Globalist. Although Nuclear Nationalism,

Nuclear Realism, and Nuclear Globalism might at times have unique meanings across the cases, we offer a general understanding of the three orientations,[53] and then how they might play out more specifically on different nuclear questions.

Realists

The proponents of the Realist school of thought tend to emphasize the following set of ideas. First, they stress the importance of self-strengthening and self-reliance in the international arena: the internal balancing strategy. Second, Realists are relatively open to forming alliances with other states: the external balancing strategy. Third, Realists place an emphasis on great and rising powers, and prioritize relations with those states. Fourth, Realists stress the importance of hard power: military and/or economic. Fifth, Realism is often as much about power as prudence, so most Realists, especially defensive Realists, recognize weakness as much as strength; they accept the idea of security dilemmas and their dangers, and are willing to exercise self-restraint or be restrained by others. Realism is distinct from the other schools of thoughts due to the degree to which this perspective is embraced by members of the political and the military establishment. This school of thought influences debates and policymaking in a straightforward manner: it is especially likely to be embraced by powerful people.

Nationalists

Nationalists, like the Realists, emphasize self-reliance and self-strengthening. However, Nationalists may embrace these goals not only as a means to the end of meeting foreign threats, but also as a goal in and of itself. First, they may view the rise of their nation or attaining nuclear capability as a matter of pride and as a moral obligation. Both Nationalists and Realists tend to share an emphasis on protecting their country's sovereignty. Second, Nationalists are skeptical of international alliances and international regimes. They emphasize the downside of alliances and regimes: the loss of autonomy. Moreover, some Nationalists perceive a junior partner role as an affront to national pride. Third, other isolationist Nationalists are not interested in their country's rise internationally or in international relations in general. These Nationalists emphasize domestic over international concerns and prefer a foreign policy of limited scope.

Globalists

Proponents of the Globalist school of thought tend to favor international political and/or economic integration. First, they believe international institutions, as well as formal and informal regimes, can effectively resolve security and political issues in the region. Second, Globalists tend to favor democratic institutions. Some argue for emphasizing intergovernmental ties with other democracies. Third, Globalists are relatively skeptical about military power as a tool of statecraft.

Nuclear Realists, Nuclear Nationalists, and Nuclear Globalists

The discussion below is not to say that individual viewpoints will not be blended or manifested differently across specific issues. Nationalist and Realist frameworks, for example, may be especially permeable, but then Globalists and Realists may align more closely on specific nuclear policy questions. Nationalists and Globalists may be expected to occupy different ends of the spectrum on most nuclear issues. Individuals might also hold somewhat inconsistent views across issues—for instance, on the desirability of nuclear energy on the one hand, versus an aversion to nuclear weapons on the other.

Nuclear Realists would likely have the strongest views on nuclear weapons strategy in light of the priority they accord national security. For example, given how crucial American extended deterrence in Asia is for Japan, South Korea, and Taiwan, we should find Realists as the most active supporters in those countries (depending on the extent of doubt about American dependability).

Characteristically, Realists tend to be practical cost-benefits maximizers and believe that hard-nosed negotiations over arms control, nuclear technology trade, and nonproliferation can produce security-enhancing outcomes as well. They would be motivated by self-interest in greater safety and security of nuclear materials, and to varying degrees, fear of nuclear terrorism. Engaging in nuclear negotiations is also valued, since Realists see that as an important mechanism for gaining external legitimacy as a responsible power and offering the opportunity for setting the rules of the international game with the big powers. They would be more willing to trade off a level of transparency in exchange for nuclear technology or fuel and to restrain their country's nuclear program against perceived benefits.

On nuclear energy, the Realist school could comprise a mixed group. While all would favor the nuclear energy route to strengthen energy security, some may also see civil nuclear technology as a backdoor to the weapons option in the future and would be less willing to compromise on nonproliferation and transparency. Others may view the pursuit of nuclear energy as one of the important energy strategies to reduce their country's energy vulnerability and improve energy security. These individuals might go along with restrictive fuel-cycle measures as long as access to fuel supplies and disposal of nuclear waste is assured. Still others might also want to build up the country's nuclear energy capability to position it as a viable nuclear technology exporter. Realists holding the latter two sentiments would champion integration with global regimes to get the most out of the nuclear energy trade. Post-Fukushima, Realists may be more amenable to circumscribing the nuclear energy program, at least in the short run. They are also likely to have some concerns about climate change and rising pollution and its domestic impacts—especially in the energy-intensive, fast-growing economies—and consider nuclear energy as an alternative source of energy.

Nuclear Nationalists can be expected to be the most resistant to restraining their country's nuclear capabilities and to strive for nuclear self-reliance and autonomy, if not complete nuclear self-sufficiency. Though both Nationalists and Realists prioritize the protection of their state's sovereignty and see a role for nuclear weapons, the former are prone to take umbrage at the slightest perceived discrimination in nuclear affairs as an affront to sovereignty. Nationalists also give more symbolic weight to nuclear capabilities, seeing them not just as assets for their country's strategic power, but also as reflections of status, pride, and levers of influence in competitions with other states. While Nationalists tend to view the rest of the world as hostile, some countries are clearly viewed with more distrust than others. Nonproliferation efforts led by the United States will likely evoke different reactions from Nationalists in the various Asian states. Nationalists might at times be willing to engage in nuclear nonproliferation efforts and greater transparency in exchange for coveted nuclear technology to build up indigenous capabilities. But Nationalist suspicions are likely to keep such bargains difficult and to a minimum, and thus bring Nationalists at loggerheads with Realists who also see a hostile, but more manageable world. On U.S. extended deterrence, Nationalists are the likeliest to sharply question its adequacy and reliability against Realist convictions.

Nuclear Nationalists would be expected to champion nuclear energy for energy security and, perhaps less overtly, for retaining the nuclear weapons option. In their pursuit of energy security, Nationalists may have to face a tough choice of cooperating with nonproliferation regimes or losing access to required technology or markets. Some Nationalists might even want to attain energy self-sufficiency and see nuclear energy as desirable for that purpose, over and beyond a Realist objective to reduce the country's energy vulnerability. Nationalists would be most opposed to any moves seen as unfair restrictions on attaining full fuel-cycle capabilities, which they would view as their country's right. Nuclear Nationalists are unlikely to link nuclear energy with climate change; their country's growth takes precedence over an environmental crisis that many believe has been caused by the West, which is now making unfair demands to cut and cap greenhouse emissions to the detriment of other countries' development.

Nuclear nationalism may also take unique forms across states—Pakistan's Nuclear Islamists, for example. Nationalist ideas sometimes transcend and structure the debates themselves. For example, in many aspiring and rising powers, we might find a consensus on the importance of sovereignty or on the belief that the rise of one's country is just and desirable.

Nuclear Globalists can be expected to favor any measure that increases the integration of their country into international economic and security structures such as global nonproliferation efforts. Given their low priority for military means, they should be the most predisposed to nuclear weapons constraints and compromise. They see proliferation and nuclear terrorism as a serious challenge, would favor tight and intrusive safeguards, and argue against reprocessing. They support nuclear disarmament and promote measures such as nuclear weapons–free zones.

On nuclear energy, Globalists will be motivated by climate change concerns rather than energy security or energy autonomy objectives. They will want to ensure that nuclear energy is strictly for peaceful purposes. Some may even favor phasing out nuclear energy and adopting renewables—especially after Fukushima, but also because of their worry about proliferation risks from dual-use technology and nuclear material stockpiling. They are most likely to hold their governments accountable for nuclear power expansion and local safety.

Globalists would argue from an image perspective as well: the importance of being seen as a responsible nuclear power internationally and the reputational advantages to be gained. Some Realists would agree on this point, especially as a way of establishing credentials with the great powers, and even with the nuclear watchdog organization, the International Atomic Energy Agency.

Nuclear debates are complex and deserve careful screening. We suggest that it is possible to identify dominant strands in thinking. To assess the trajectory of debate, we consider the following aspects of the debate using the schools of thought:

- Cleavage Patterns: to what extent do the nuclear debates reflect or mix the Realist, Nationalist, and Globalist orientations?
- Clusters across Issues: to what extent do positions on different issues cluster (for example, nuclear weapons and nuclear energy) and do these clusters reflect the Nationalist, Realist, and Globalist orientations?
- Center of Gravity: what is the relative influence of different positions in the debate and where is the center of gravity?

We think that there are likely to be important crosscutting linkages as well as cleavages and thus various coalitional partnerships in the nuclear arena that may not be obvious at first sight. The center of gravity of domestic opinion and where it is trending will tell us which of these clusters and cleavages stand the best chance of prevailing in debates and the likeliest nuclear policies. Our findings will allow us to draw two types of conclusions: (1) a set of empirical generalizations on key nuclear trends and dispositions in Asia; and (2) an informal, if not formal, "test" of the validity of external security explanations on nuclear debates in Asia.

ORGANIZATION OF THE BOOK AND CHAPTER PRÉCIS

The book first takes up the countries in the region with the greatest geopolitical impact, biggest energy consumption, and in possession of the most advanced nuclear capabilities: China, India, and Japan. The next three chapters turn to South Korea, Taiwan, and Vietnam, all aspiring nuclear energy states at differing stages. Subsequently, we take up the special case of ASEAN. ASEAN has its own nuclear orientations, but also comprise possible nuclear aspirants besides Vietnam—especially politically turbulent Thailand and a rather enigmatic, but now fast-changing, Myanmar. Lastly, we consider Pakistan, which stands out for its uniqueness and follow with a concluding chapter.

For all the attention that China has received in recent years, the factors that go into its nuclear decision-making remain opaque. Analysis tends to be by preconceived notions or theory driven. Hui Zhang draws out the more fine-tuned thinking and divergences on nuclear energy and weapons that nonetheless underlie the rather impenetrable Chinese system. With his deep knowledge of the Chinese nuclear decision-making environment, Zhang provides extensive coverage of the numerous domestic actors and variables that drive Chinese policy in the nuclear realm and their orientations. Does the center of gravity in Chinese thinking conform to hard-nosed Realist thinking on energy security and nuclear energy? On stepping up its nuclear weapons capabilities and modifying its nuclear posture as tensions in the region rise? What is the dominant thinking on global nonproliferation regimes and what outcomes might we expect? The findings of this chapter will be critical in confirming or challenging conventional thinking on Chinese nuclear policy that sees significant buildup and more confrontational postures ahead along Nationalist lines. The China case is singularly important given its centrality to the region's shifting politics and geopolitics.

India's current special status as a nuclear state outside the Nonproliferation Treaty came after a long-running nuclear program subjected to international sanctions. Deepa Ollapally and Rajesh Rajagopalan explore the various domestic pressures and opinions that drove India's early ambiguous nuclear weapons strategy and its ambitious nuclear energy project, and clarify the new fault lines in nuclear thinking. How much change can we expect in thinking about India's nuclear weapons strategy as India rises to be a major power, and what will that mean for nonproliferation and regional stability? What are the prospects for India's nuclear energy program, and how does it fit into the country's concept of energy security and nuclear autonomy? How has the arrival of a self-avowed Nationalist government in Delhi affected India's orientations?

Post-2011 Japan has had to face the devastating consequences of the Fukushima/tsunami disaster, as well as rising geopolitical tensions with neighboring China. How will Japan handle the crisis in its large and established nuclear energy program? How stable is Japan's non-nuclear weapons commitment and policy of pacifism? Mike Mochizuki delves into the intricacies of Japanese thinking about its long-standing nuclear energy program and how domestic opinion is coalescing after the Fukushima shock as well as the extent of discernible shifts in the broader nuclear realm as the regional security climate deteriorates and Japan debates the future of American extended deterrence.

South Korea's grand nuclear bargain with the United States on the one hand, and nuclear fuel-cycle questions in Seoul's ambitious nuclear energy program on the other hand, is renewing strong interest in the country's nuclear thinking. Scott

Snyder provides an examination of deep-seated sentiments that have driven South Korea's approach to nuclear weapons and nuclear energy, against newer conceptions on nonproliferation and global integration. How are divergent views on U.S. extended deterrence faring, and what type of resolution can we expect, in light of both alliance politics and nuclearization of North Korea? How critical is the reprocessing issue in domestic debates and what is the array of opinions? In which direction are Nationalist-, Realist- or Globalist-leaning debates influencing broader South Korean worldviews as an aspiring nuclear power?

Taiwan has historically operated in precarious security environments, and it has at varying times held ambitions for both nuclear weapons and nuclear energy. Given China's resistance to its official participation, Taiwan is also in the peculiar position of having to work outside most international regimes. Robert Sutter and Yuan-Ming Alvin Yao offer a detailed account of the ebbs and flows of domestic thinking on the development of weapons, nonproliferation, and nuclear energy development before and after Fukushima. How is China's rise influencing the balance of domestic debate on approaches to nuclear nonproliferation, energy security and its special relationship with the United States? Are changes on the horizon or more continuity?

Southeast Asian countries, in particular Vietnam, exemplify those with aspirational goals of nuclear energy. Within a centralized political system however, nuclear debates are not obvious. Linda J. Yarr and Nguyễn Thị Thanh Thủy track the opinions of key government bureaucracies and officials, as well as a narrow, but important, band of commentators and civil society groups to ascertain various strands of thinking on developing and implementing a new nuclear energy program, nuclear security and safety, nonproliferation regimes, and global integration. Given Vietnam's long and costly nationalist wars, and now growing Chinese power and maritime disputes, what degree of Nationalist sentiment is found in the nuclear arena?

Given the rising importance assigned to nuclear energy by several Southeast Asian countries, we suggest that it is time to reassess the outlook of ASEAN and its key players. Catharin Dalpino and Timothy Westmyer offer a close reading of ASEAN's institutional positions on nuclear weapons and the long-awaited Southeast Asian Nuclear-Weapon-Free Zone (SEANWFZ) Treaty, the ups and downs of Thailand's nuclear energy ambitions, the evolution of Burma/Myanmar's historically cryptic nuclear decisions, as well as the orientations of other ASEAN states. Do Southeast Asian states and ASEAN provide a model of nuclear restraint and multilateral avenues along Globalist lines going forward, or are the security challenges in the region overwhelming? Do the nuclear energy ambitions of a growing Southeast Asia threaten nuclear safety and security, or are there constraints on unbridled nuclear energy development?

Pakistan's nuclear weapons have come under intense scrutiny, especially since 9/11 and the rising tide of violent extremism in the country, the exposure of the A.Q. Khan–led network of illicit nuclear technology transfers, the ongoing struggle for power between the civilian political parties and the military, Pakistan's apparent propensity for nuclear risk taking and its fraught relations with neighboring India, and the uneasy partnership with the United States. Christopher Clary opens up Pakistan's complex nuclear landscape and untangles the various lines of thinking on the possession and use of nuclear weapons, on how far to cooperate with international actors and regimes, and the desirability of nuclear energy. How do various groups assess costs and benefits of Pakistan's nuclear isolation over global integration? How ossified is Pakistan's nuclear Nationalism and are there cracks?

In the concluding chapter, Mike Mochizuki engages in a comparative analysis across the cases to determine the key patterns that emerge in nuclear energy, nuclear weapons, and nonproliferation debates. He identifies the main cleavages or fault lines in nuclear debates and the extent to which they reflect or mix the Realist, Nationalist, and Globalist orientations; whether orientations on different nuclear issues cluster or coalesce; and most importantly, the relative influence of the various viewpoints in the debate and where the center of gravity lies. He then considers the role of the United States in shaping the nuclear environment in Asia and how this affects the prospects for multilateral cooperation in the nuclear arena.

NOTES

1. For a classic work offering different views on why states acquire nuclear weapons, see Scott D. Sagan and Kenneth N. Waltz, *The Spread of Nuclear Weapons: A Debate Renewed* (New York: W. W. Norton, 2002).

2. This approach is laid out in Henry R. Nau and Deepa M. Ollapally, eds., *Worldviews of Aspiring Powers: Domestic Foreign Policy Debates in China, India, Iran, Japan, and Russia* (New York: Oxford University Press, 2012). It takes off from research showing that general foreign policy attitudes at both the elite and mass levels are structured—that is, they are organized along political attitudes more generally. See Jon Hurwitz and Mark Peffley, "How Are Foreign Policy Attitudes Structured? A Hierarchical Model," *American Political Science Review* 80 (1987).

3. Kenneth N. Waltz, *The Spread of Nuclear Weapons: Why More May Be Better*, Adelphi Papers, Number 171 (London: International Institute for Strategic Studies, 1981); Kenneth N. Waltz, *Theory of International Politics* (New York: Random House, 1979); John J. Mearsheimer, *The Tragedy of Great Power Politics* (New York: W. W. Norton, 2001).

4. Charles L. Glaser, *Rational Theory of International Politics: The Logic of Competition and Cooperation* (Princeton, N.J.: Princeton University Press, 2010), 14.

5. Ibid., 167.

6. Muthiah Alagappa, ed., *The Long Shadow: Nuclear Weapons and Security in 21st Century Asia* (Stanford, Calif.: Stanford University Press, 2008), 509.

7. Ibid., 508 and 513.

8. For more, see Jacques E. C. Hymans, *The Psychology of Nuclear Proliferation: Identity, Emotions, and Foreign Policy* (Cambridge: Cambridge University Press, 2006), 46. The cases he considers are France, India, Argentina, and Australia.

9. Quoted in Etel Solingen, *Nuclear Logics: Contrasting Paths in East Asia and the Middle East* (Princeton, N.J.: Princeton University Press, 2007), 258. For a good discussion of this dilemma for political science research, see Solingen, 258–259.

10. Ashley J. Tellis, Abraham M. Denmark, and Travis Tanner, eds., *Strategic Asia 2013–14: Asia in the Second Nuclear Age* (Washington, D.C.: National Bureau of Asian Research, 2013). The countries are Russia, China, India, Pakistan, Korean peninsula, Iran, and Japan.

11. Ashley J. Tellis, "No Escape: Managing the Enduring Reality of Nuclear Weapons," in *Strategic Asia 2013–14*, eds. Tellis, Denmark, and Tanner, 14.

12. Vipin Narang, *Nuclear Strategy in the Modern Era: Regional Powers and International Conflict* (Princeton, N.J.: Princeton University Press, 2014), 8–9.

13. Ibid., 78.

14. Christopher Clary, "The Future of Pakistan's Nuclear Weapons Program," in *Strategic Asia 2013–14*, eds. Tellis, Denmark, and Tanner, 160.

15. On the overweening role of the military in Pakistan's polity generally, see Ayesha Jalal's *The State of Martial Rule: The Origins of Pakistan's Political Economy of Defence* (New York: Cambridge University Press, 2008).

16. Bruce Riedel gives an insider account of this crisis and outcome. India's high-level *Kargil Review Committee Report* notes, "What Pakistan attempted at Kargil was a typical case of salami slicing . . . . Since India did not cross the LOC [Line of Control] and reacted strictly within its own territory, the effort to conjure up escalation of a kind that could lead to nuclear war did not succeed." Kargil Review Committee, *From Surprise to Reckoning: The Kargil Review Committee Report* (New Delhi: Sage Publications, 2000), 242. For more, see Bruce Riedel, *American Diplomacy and the 1999 Kargil Summit at Blair House Policy Paper Series* (Philadelphia, Pa.: Center for the Advanced Study of India, 2002).

17. See for example, Shirley A. Kan, who cites senior U.S. intelligence officials in *China and Proliferation of Weapons of Mass Destruction and Missiles: Policy Issues,* CRS Report for Congress (Washington, D.C.: Congressional Research Service, July 17, 2006), 5.

18. Graham Allison, "The Thucydides Trap: Are the U.S. and China Headed for War?" *The Atlantic,* September 24, 2015.

19. A. F. K. Organski and Jacek Kugler, *The War Ledger* (Chicago: Chicago University Press, 1980).

20. Aaron L. Friedberg, *A Contest for Supremacy: China, America, and the Struggle for Mastery in Asia* (New York: W. W. Norton, 2011); Mearsheimer, *Tragedy of Great Power Politics.*

21. Robert Gilpin, *War and Change in World Politics* (Cambridge: Cambridge University Press, 1981).

22. Paul Bracken posits a post–Cold War second nuclear age and nuclear multipolarity rife with instabilities. For more, see Paul Bracken, *The Second Nuclear Age: Strategy, Danger, and the New Power Politics* (New York: Times Books, 2012), 2.

23. Tellis, "No Escape," 3.

24. For a critical analysis of power transition theories including Organski and Friedberg, see Peter Harris, "Problems with Power-Transition Theory: Beyond the Vanishing Disparities Thesis," *Asian Security* 10, no. 3 (2014). See also Robert Powell, *In the Shadow of Power: States and Strategies in International Politics* (Princeton, N.J.: Princeton University Press, 1999); G. John Ikenberry, "The Rise of China, the United States, and the Future of the Liberal International Order," in *Tangled Titans: The United States and China,* ed. David Shambaugh (Lanham, Md.: Rowman & Littlefield, 2013).

25. Steve Chan, *China, the U.S., and the Power-Transition Theory: A Critique* (Abingdon, U.K.: Routledge, 2008), 4.

26. Rosemary Foot, "Constraints on Conflict in the Asia-Pacific: Balancing 'the War Ledger,'" *Political Science* 66 no. 2 (December 2014): 119–142.

27. Solingen, *Nuclear Logics*, 13.

28. Scott D. Sagan, "Why Do States Build Nuclear Weapons?: Three Models in Search of a Bomb," *International Security* 21 no. 3 (Winter 1996–1997): 54–86, http://iis-db.stanford.edu/pubs/20278/Why_Do_States_Build_Nuclear_Weapons.pdf; Scott D. Sagan, "The Causes of Nuclear Weapons Proliferation," *Annual Review of Political Science* 17 no. 4 (March 2011), 225–241; Steven Miller and Scott D. Sagan, "Alternative Nuclear Futures," *Daedalus* (Winter 2000): 132; and Itty Abraham, *The Making of the Indian Atomic Bomb: Science, Secrecy and the Postcolonial State* (New York: Zed Books, 1998). Abraham's work has been seminal in developing the idea of "strategic enclaves."

29. Solingen notes this shortcoming well in Solingen, *Nuclear Logics*, 8.

30. Ibid., 52–53.

31. Ibid., 53.

32. Hymans, *The Psychology of Nuclear Proliferation*, 46.

33. Ibid., 12.

34. Ibid., 172.

35. Maria Rost Rublee, *Nonproliferation Norms: Why States Choose Nuclear Restraint* (Athens, Ga.: University of Georgia Press, 2009).

36. Alagappa, *The Long Shadow*, 11.

37. For a critical review of writings on nuclear weapons proliferation as well as its links to nuclear energy technology, see Sagan, "The Causes of Nuclear Weapons Proliferation," 225–244.

38. Miller and Sagan, "Alternative Nuclear Futures," 128.

39. Hymans, *The Psychology of Nuclear Proliferation*, 4.

40. Rajesh Basrur and Koh Swee Lean Collin, *Nuclear Power and Energy Security in Asia* (New York: Routledge, 2012).

41. Japan's is a good case in point. For energy security compulsions generally, see Andrew Phillips, "A Dangerous Synergy: Energy Securitization, Great Power Rivalry and Strategic Stability in the Asian Century," *The Pacific Review* 26, no. 1 (2013): 17–38; and Daniel Moran and James A. Russell, "The Militarization of Energy Security," in *Energy Security and Global Politics: The Militarization of Resource Management*, eds. Daniel Moran and James A. Russell (New York: Routledge, 2009).

42. "BP Statistical Review of World Energy 2014," *BP*, June 2014, http://www.bp.com/content/dam/bp-country/de_de/PDFs/brochures/BP-statistical-review-of-world-energy-2014-full-report.pdf.

43. Kang Wu, Hassaan Vahidy, and Aarti Uplenchwar, "China and India," in *Asia's Energy Future: Regional Dynamics and Global Implications*, eds. Kang Wu and Fereidun Fesharaki (Honolulu, Hawaii: East-West Center, 2007), 63.

44. Candace Dunn, "China is Now the World's Largest Net Importer of Petroleum and Other Liquid Fuels," *U.S. Energy Information Administration—Today in Energy*, March 24, 2014, http://www.eia.gov/todayinenergy/detail.cfm?id=15531.

45. "How Can India Secure its Energy Needs?," *Business Wire India*, January 21, 2015, http://businesswireindia.com/news/news-details/how-can-india-secure-its-energy-needs/42317.

46. Jenny W. Hsu, "China's Crude Oil Imports Rise 8.8% in 2015," *Market Watch*, January 26, 2016, http://www.marketwatch.com/story/chinas-crude-oil-imports-rise-88-in-2015-2016-01-26.

47. "India," *U.S. Energy Information Administration*, June 26, 2014, http://www.eia.gov/countries/analysisbriefs/India/india.pdf.

48. See Kang Wu and Charles E. Morrison, "Energy Insecurity Index," in *Asia's Energy Future*, eds. Wu and Fesharaki, 107–113.

49. C. Raja Mohan gives a Realist interpretation of these questions in *Samudra Manthan: Sino-Indian Rivalry in the Indo-Pacific* (Washington, D.C.: Carnegie Endowment for International Peace, 2012), 44.

50. For an elaboration of these concepts, see Danielle Cohen and Jonathan Kirshner, "The Cult of Energy Insecurity and Great Power Rivalry Across the Pacific," in *The Nexus of Economics, Security, and International Relations in East Asia*, eds. Avery Goldstein and Edward D. Mansfield (Stanford, Calif.: Stanford University Press, 2012), 144–176; and Llewelyn Hughes and Seán J. Kreyling, "Understanding Resource Nationalism in the 21st Century," *Journal of Energy Security* (July 2010), http://ensec.org/index.php?option=com_content&view=article&id=253:resource-nationalism-in-the-21st-century&catid=108:energysecuritycontent&Itemid=365.

51. World Nuclear Association, "World Nuclear Power Reactors & Uranium Requirements," last updated and accessed on March 1, 2106, http://www.world-nuclear.org/information-library/facts-and-figures/world-nuclear-power-reactors-and-uranium-requireme.aspx. Japan's production reflects the post-Fukushima crisis and reactor shutdowns in 2014.

52. Kurt M. Campbell, Robert J. Einhorn, and Mitchell B. Reiss, eds., *The Nuclear Tipping Point: Why States Reconsider Their Nuclear Choices* (Washington, D.C.: Brookings Institution, 2004), 344.

53. For further elaboration, see Nikola Mirilovic and Deepa Ollapally, "Realists, Nationalists, and Globalists and the Nature of Contemporary Rising Powers," in *Worldviews of Aspiring Powers*, eds. Nau and Ollapally, 210–234.

2

China

Evolving Attitudes on Nuclear Affairs

Hui Zhang

China is a nuclear weapon state and a rising power entering an era of particularly rapid nuclear energy growth. China's attitudes and approaches toward nuclear issues play an important role in facilitating a safe and secure world. Based on a series of interviews with Chinese officials, experts, and scholars; government documents; and academic publications, this chapter offers insight into the status of Realist, Nationalist, and Globalist schools of thought in China's domestic nuclear debate on several specific issues: nuclear energy, nuclear weapons, arms control, nonproliferation, and nuclear security.[1]

Chinese nuclear energy decision-making is largely the domain of the central government with limited public involvement, although this monopoly was put to the test after the 2011 accident at Japan's nuclear plant in Fukushima. China's rapid nuclear energy development is primarily driven by Realists who see nuclear energy as the way to alleviate the country's severe air pollution while at the same time meeting economic growth targets. Globalists pushing for action on climate change and Nationalists striving for a more self-reliant energy security strategy further move China toward nuclear energy, but these perspectives are generally overshadowed by the Realist argument.

One of the chief findings in this chapter is that despite the reasonable unease some policymakers and analysts have about China's nuclear weapons modernization programs, Beijing's Realist nuclear weapons strategy has consistently maintained a policy that features a "minimum nuclear deterrent" and a No First Use (NFU) pledge to avoid a costly nuclear arms race. China will continue to modernize its nuclear force, but there is no evidence China will abandon its "minimum nuclear deterrence" doctrines. Realists have also adopted an uneasy embrace of arms control, pushing China to see the country's expanding nuclear weapons capability as leverage in the service of its national interests to bring about future negotiations to reduce global nuclear stockpiles. On nuclear nonproliferation and nuclear security, Globalists have successfully advocated that China's commitment to constraining the spread of weapons of mass destruction (WMD) and delivery systems can improve its international image as a responsible rising power. China hopes to speed up its economic integration with the world through actively engaging with international regimes. These divergent perspectives show Chinese elites tend not to follow a uniform school of thought on all nuclear issues. Nonetheless, the broad Realist-Globalist coalition that has emerged in China will likely continue to direct the country's nuclear debate for the foreseeable future.

CHINESE ATTITUDES TOWARD NUCLEAR POWER DEVELOPMENT

For decades, Chinese nuclear energy decision-making has been the exclusive purview of the central government, with little significant public debate. However, these conditions may have changed in the wake of Japan's 2011 Fukushima Daiichi Nuclear Power Plant accident. This section examines the primary motivators and key actors driving China's growing nuclear energy program and how they can be understood through this volume's schools-of-thought framework.

The Development of China's Ambitious Nuclear Power Program

China has emerged as the central focus of the international nuclear industry due to its ambitious nuclear energy plans. This was true before Fukushima and largely remained the case at the end of 2015. As of December 2015 according to the World Nuclear Association, China has thirty nuclear reactor units in operation with a total capacity of 27 GWe (gigawatt electrical) providing about 2 percent of its total electricity generation;[2] a small fleet compared to the nearly 100 operational reactors in the United States providing over 19 percent of U.S. total electricity generation. Many officials and experts in China, however, have urged the leadership to pursue considerably higher targets. For example, an authoritative study in 2011 by the China Academy of Engineering proposed the country install a nuclear power capacity of 200 GWe by 2030 (about 10 percent of its then total electricity generation) and 400–500 GWe by 2050 (about 15 percent of its then total electricity generation).[3]

Though China's nuclear industry is young relative to other states with nuclear energy programs, it leads the world in the pace of new reactor construction. As of December 2015, twenty-two reactors were under construction in China (capable of producing a total of 25 GWe) in addition to between thirty-four and forty-two reactors planned for the future (capable of producing a total of 49 GWe).[4] In March 2006, China's chief administrative authority, the State Council, approved the *Medium-and Long-term Nuclear Power Development Plan (2005–2020)*, which stated a goal to increase total nuclear capacity to 40 GWe by 2020. Later in 2008, China further revised its position from "active development" to "aggressive development"[5] with targets for nuclear power increased to 70–80 GWe by 2020.[6]

Following the Fukushima nuclear accident, however, Beijing conducted safety inspections on all plants in operation and under construction and reassessed the pace of its nuclear industry. In October 2012, the State Council issued a new "Medium- and Long-Term Nuclear Power Development Plan (2011–2020)," which adjusted the pace of development to a less ambitious total nuclear capacity goal of 58 GWe by 2020–2021.[7] Construction of several reactors were delayed, thus it might be difficult to achieve the 2020 goals. Some Chinese officials and experts interviewed by this author suggest China could install only around 50 GWe by 2020. Nevertheless, looking at the electrical capacity of the country's operating reactors and the planned or under construction plants and as long as there are no further delays, China appears to be making significant progress at a quicker pace than the rest of the world.

China's Growing Energy Needs

The State Council's energy policy office acknowledges that "China's per capita energy consumption is still low—only one third of the average of developed countries. But as the economy and society progress and living standards improve, China's energy consumption will continue to rise sharply."[8] China is the world's largest producer and consumer of coal—accounting for half of the world's coal consumption—and in 2011, China became the world's largest generator of power.[9] As seen in table 2.1, the government had planned to install a total electricity generating capacity of 1490 GWe by 2015—a nearly threefold increase since 2005. China's installed electricity generating capacity was estimated to increase to 1700–2000 GWe by 2030 and 2500–2800 GWe by 2050, according to the 2011 China Academy of Engineering study.[10]

Table 2.1. China's Installed and Targeted Electricity Generating Capacity

	2010 GWe Installed (as a percentage of capacity)	*2015 GWe Targets (as a percentage of capacity)*
Total Installed Generation Capacity	970 GWe	1490 GWe
Coal	660 (68%)	960 (64%)
Hydro	220 (23%)	290 (19%)
Gas	26.4 (2.7%)	56 (3.8%)
Nuclear	10.8 (1.1%)	40 (2.7%)
Wind	31 (3.2%)	100 (6.7%)
Solar	0.9 (0.1%)	21 (1.4%)

Source: Information Office of the State Council of the People's Republic of China, *The 12th Five-Year Plan for Energy Development* (Beijing: People's Republic of China, January 1, 2013), http://www.gov.cn/zwgk/2013-01/23/content_2318554.

Note: Some relatively minor thermal energy sources, such as biomass, are excluded here.

Major Drivers for Nuclear Power Expansion

China's ambitious nuclear power program aims to (1) address the country's air pollution crisis, (2) mitigate climate change, and (3) enhance national energy security. Most of the government officials and energy experts interviewed by the author identified air pollution concerns as the primary motivating force and most focused on air pollution's destructive consequences for human health and economic activity. This argument dominates the nuclear energy debate in China and is most closely aligned with a Realist perspective as the nuclear energy option also protects another major Realist priority—the Chinese growth and development model. This approach is in contrast to other solutions offered, such as capping carbon emissions, an approach that in their view could harm the Chinese economy. Self-reliance as an answer for the country's energy security concerns was raised by those in the Nationalist school of thought, while climate change issues were expressed by Globalists.

Addressing Air Pollution

As China's primary energy resource, coal has dominated China's power mix for decades and will continue to play a fundamental role in China's electricity generation in the near future.[11] As shown in table 2.2, coal supplied the majority (66.6 percent) of China's total energy consumption in 2012 alongside oil (18.8 percent). China has attempted to diversify its energy supplies with a strategy to replace some coal power with expanded nuclear energy, other renewable sources, and natural gas. According to China's 2012 white paper on energy policy, "Nuclear power is a high-quality, clean, and efficient modern energy source" and "its development is of great significance for optimizing the nation's energy structure and ensuring national energy security."[12] Despite these efforts, hydropower, natural gas, nuclear, solar, and other renewables account for relatively small shares of China's energy consumption in early 2016.

China's heavy reliance on coal has had a serious impact on the country's ecology and environment. In 2012, two thirds of China's cities could not meet the country's own air-quality standards.[13] Based on a 2012 Asian Development Bank study, over 99 percent of China's 500 largest cities did not meet the World Health Organization's air-quality standards.[14] As a result, the Global Burden of Disease Study estimated outdoor air pollution contributed to 1.2 million premature deaths in 2010 throughout East Asia.[15] The economic costs of environmental pollution are also substantial with the Chinese Academy of Environmental Planning estimating the cost of environmental degradation at $230 billion in 2010 or 3.5 percent of the nation's gross domestic product—a threefold increase since 2004.[16]

In its *12th Five-Year Plan for Energy Development* released in 2013, Beijing planned to increase nonfossil fuel energy consumption to 15 percent of the energy mix by 2020 in an effort to reduce air pollution. As seen in table 2.2, the China Academy of Engineering study recommended an increase of nonfossil fuel energy consumption to about 40 percent of the energy mix by 2050, including a reduction

Table 2.2. Total Primary Energy Consumption from 1995 to 2012 and Projections from 2020 to 2050.

	1995	*2000*	*2005*	*2010*	*2012*	*2020*	*2030*	*2050*
Total Primary Energy Consumption[1]	1,312	1,455	2,356	3,249	3,617	4,000–4,200	4,500–4,600	5,500–5,600
Coal	74.6%	69.2%	70.8%	68.0%	66.6%	56%	45%	33%
Oil	17.5%	22.2%	19.8%	19.0%	18.8%	oil+gas	oil+gas	oil+gas
Gas	1.8%	2.2%	2.6%	4.4%	5.2%	28%	28.6%	27%
Hydro	5.7%	5.9%	5.9%	7.1%	7.6%	7.3%	8.8%	9%
Nuclear	0.4%	0.4%	0.8%	0.7%	0.8%	4%	10%	16%
Other Renewables	0.0%	0.1%	0.1%	0.8%	1.0%	4.8%	8.8%	14.4%

Source: Department of Energy Statistics, National Bureau of Statistics, *China Energy Statistical Yearbook 2013* (Beijing: China Statistics Press, 2013); China Academy of Engineering, *China's Medium- and Long-Term Energy Development Strategy 2030, 2050* (Beijing: Science Press, 2011).

Note: Fuel sources are shown as a percentage of total energy consumption.

[1] Amount in millions of Total Coal Equivalent (TCE)

of the share of coal to 33 percent.[17] Subsequently, Realists have advocated for nuclear power as a way to meet these impressive goals. They contend nuclear power is a mature technology in use for over half a century and can serve as a source for large-scale and stable base-load electricity generation, unlike other renewable energy sources (e.g., wind and solar power) limited by weather factors.[18] Moreover, they emphasize that nuclear power is economically competitive to coal-fired power in some regions, especially in coastal areas where the economy is developing rapidly, as well as remote areas far from solar, wind, and hydropower sites.

Reducing Carbon Emissions

As a result of high coal consumption, China has overtaken the United States as the world's leading energy-related CO_2 emitter each year since 2006. In terms of cumulative emissions between 1850 and the present, China is in second place among the top emitters, just after the United States. China may soon top all countries in cumulative emissions by the mid 2020s.[19] China is unsurprisingly facing domestic and international pressure to act on its emissions. Since the mid-2000s, Chinese leaders have thus paid greater attention to climate change and strove to negotiate a "powerful" and "ambitious" agreement at the 2015 United Nations Climate Change Conference in Paris.[20] In 2007, the National Development and Reform Commission (NDRC) released its *National Climate Change Program* report, which noted China already suffers from the environmental impacts of climate change, such as climbing surface and ocean temperatures, rising sea level, decreasing annual precipitation in some regions, and other worsening extreme events.[21] According to the *Intended Nationally Determined Contribution* China submitted in June 2015 during the lead-up to the Paris climate change summit, Beijing pledged to lower its carbon intensity (the amount of carbon emitted per unit of GDP) by 60–65 percent below 2005 levels by 2030 and increase the share of nonfossil fuel energy sources of the total energy supply to about 20 percent by the same year.[22]

The China Academy of Engineering's *China's Medium- and Long-Term Energy Development Strategy (2030, 2050)* suggested leaders cap carbon emissions around or even before 2030 and then move to reduce the total quantities of greenhouse gases.[23] To realize those goals, many Globalists in the government and Chinese energy sectors suggest accelerating the development of nuclear energy as a reasonable and effective measure to build a low-carbon energy mix and reduce greenhouse gas emissions.[24]

Though many of the Chinese officials and experts interviewed for this chapter believe climate change will have long-term consequences for China, most are skeptical it is an issue of urgent concern, especially when compared to air pollution. These individuals are also anxious about the potential negative impacts greenhouse gas emission caps could have on the Chinese growth and development model. Nevertheless, some Globalists note that by taking leadership on climate change issues—with nuclear power as a central part of the strategy—China can promote itself as a responsible power and improve its international image. This perspective is certainly heard in China, but it is overshadowed in the wider nuclear debate by Realist voices all the same.

Promoting Energy Security

According to the 2012 white paper on energy policy, China is facing "grave challenges to energy security. The country's dependence on foreign energy sources has been increasing in recent years."[25] Since China relies so heavily on imported energy, its energy security is put at risk by an uncertain geopolitical landscape that may constrain China's energy options in the future.[26] In this view, increasing the share of nuclear energy in China's energy portfolio is imperative to diversify its energy supplies, reduce its dependence on coal, and bolster its energy security.[27]

In late 2013, China surpassed the United States as the largest net importer of petroleum and other liquid fuels and the U.S. Energy Information Administration (EIA) expects China's oil consumption to exceed that of the United States by 2034.[28] The 2012 white paper on energy policy noted that "the percentage of imported petroleum in the total petroleum consumption has risen from 32 percent at the beginning of the 21st century to the present 57 percent." The government's current *Five-Year Plan* targets oil imports reaching no more than 61 percent of its demand by the end of 2015. The Chinese government is concerned that "marine transportation of petroleum and cross-border pipeline transmission of oil and gas face ever-greater security risks. Price fluctuations in the international energy market make it more difficult to guarantee domestic energy supply."[29] As discussed in later sections, the Chinese nuclear energy sector remains closely interconnected to the global nuclear industry as Realists in government have favored more pragmatic nuclear cooperation arrangements with an eye toward economic development and air pollution goals.

The Major Players in China's Nuclear Energy Debate

China's nuclear energy decision-making is dominated by government organizations with often complicated and opaque processes. Public opinion ends up playing a limited role, though this has changed in the wake of the Fukushima nuclear accident. This section surveys the major government actors involved in China's nuclear energy debate and how local governments and the general public may seek a louder voice in the decision-making process.

Government Actors

In terms of central government decision-making, the National Energy Commission (NEC), an advisory and coordination body directly under the State Council, has the authority to draft a national energy development strategy and oversee energy-related issues. The National Energy Administration (NEA) under the authority of the National Development and Reform Commission (NDRC) is responsible for implementing NEC policy. Besides the NDRC, the State-owned Assets Supervision and Administration Commission (SASAC) of the State Council has a major role in nuclear power expansion and supervises these Chinese nuclear enterprises: the China National Nuclear Corporation (CNNC), the China General Nuclear Power Group (CGN), and the State Power Investment Corporation (SPI), which is a new company from the merged China Power Investment Corporation (CPI) and the State Nuclear Power Technology Corporation (SNPTC). These are the only three enterprises licensed to own and operate nuclear power plants in China. The SASAC has the responsibility of approving new nuclear plant projects proposed by those operators and requires final approval by the NDRC. In addition, the former SNPTC was mainly responsible for the Generation III Pressurized Water Reactors (PWR) technology imported from overseas. Some Chinese nuclear experts, however, have criticized China's nuclear program for relying too heavily on foreign technology instead of focusing on the development of domestic technology.[30]

Moreover, while the central government has primary responsibility for the nation's nuclear power plan, local governments also have some influence. Historically, local government has welcomed nuclear power projects in its jurisdiction with these officials hopeful the endeavors will bring remarkable benefits to promote regional economy development, including thousands of jobs, investment in their communities, an increased local tax base, and solutions to their electricity shortfalls.[31] As a result, many local governments, along with the proposed operators, lobby relevant nuclear experts and officials at the national level to approve their own proposed projects.

The Role of the Public

The Fukushima nuclear accident temporarily slowed China's ambitious nuclear plans due to increased public concerns about nuclear safety.[32] Before Fukushima, Chinese public participation in nuclear energy decision-making was limited by a lack of transparency on nuclear safety risks and a shortage of relevant nongovernmental organizations at the community level. Based on a survey conducted in 2011, only 20 percent of respondents in the general public had knowledge of certain nuclear power issues (e.g., safety concerns) prior to Fukushima and much of the country at large had never even heard of nuclear power.[33] Research conducted before the Fukushima accident, nevertheless, found local communities near nuclear power plants had mostly embraced nuclear power.[34]

However, the public's interest in the nuclear energy debate post-Fukushima may have changed with many Chinese newspapers, television, and the Internet reporting on the accident. A survey conducted between June and August 2011 in villages near Haiyang of Shandong province, where several of China's nuclear reactor projects were underway, showed that most of the respondents (over 80 percent) had started to pay attention to nuclear power and radiation risks. Also, it found attitudes toward the construction of new nuclear power plants in their vicinity were almost equally divided among supporters (almost 33 percent) and those who opposed such construction (33 percent), with 29 percent still without an opinion.[35]

As general awareness of nuclear safety continues to grow and citizens become more involved, public opinion could have a greater impact on nuclear project decision making. The Ministry of Environmental Protection (MEP) has long required local governments to consult with the public in environmental impact assessment reports, and since 2012 the NDRC had required public consultation on social stability assessments as well.[36] One example of the public's evolving role in nuclear energy decision making is the cancellation of plans to build a six-billon-dollar, large-scale, uranium-processing complex in Jiangmen of Guangdong province in July 2013 after a thousand people protested against the project.[37] Those protestors voiced public health concerns because the project lacked an adequate environmental impact assessment and the ten-day public consultation on social stability was too short. Though industry experts and officials argued the project would pose little risk to health, the public did not trust them. In practice, in recent years, growing public skepticism over official assurances about safety has forced the government to cancel plans to build several major chemical projects in coastal areas.

As China's economy grows, wealthier residents may be more likely to press local officials about safety concerns rather than economic benefits, though some also worry how nuclear plants may devalue real estate prices.[38] Thus, local governments have become more cautious about proposing new nuclear programs or have suddenly expressed reluctance to host nuclear research facilities they may have previously welcomed.[39] If the central government wants to receive sustained public support for its nuclear energy development plans, officials must take measures to increase transparency on nuclear decision-making, improve communication and consultation between officials and the public, foster public education on nuclear issues, and build trust between the government and public.[40]

The opinion of the general public and its expanded participation in nuclear affairs should be one important factor in the government nuclear decision-making process. The influence of this factor, on the other hand, should not be overstated since official nuclear power plans will not be significantly impacted in the near future. The majority within government and the public still widely embrace China's nuclear power development. Moreover, since 2013 the Chinese government has been actively pursuing a "go global" policy with exporting nuclear technology including nuclear reactors and heavy components in the supply chain.[41] The major drivers for such a movement are economic interests and willingness to work with the United States and others to make money. This development further strengthens China's Realist perspective.

CHINA'S NUCLEAR WEAPON POLICY AND MODERNIZATION

Unlike the civilian nuclear sector, where the public is becoming more involved in decision making, the country's Realist approach to its nuclear weapons strategy is conclusively dominated by the Central Military Commission. Nevertheless, the public has widely embraced these official views. Since its first nuclear explosion in

1964, China has maintained it has a "minimum nuclear deterrent" and a No First Use of nuclear weapons pledge aimed at avoiding a costly nuclear arms race through a nuclear strategy based in self-defense.[42] Like its 2006 white paper on defense states, the fundamental goal of China's nuclear strategy is "to deter other countries from using or threatening to use nuclear weapons against China. China remains firmly committed to the policy of No First Use of nuclear weapons at any time and under any circumstances."[43]

In the Chinese calculus, the minimum acceptable nuclear force is one that will survive a first nuclear strike and penetrate a missile defense system to reach its designated targets. The number of the "minimum" nuclear warheads that must reach their target would be relatively constant. However, the total number of warheads required to support an effective nuclear force is changeable, depending on factors such as estimates about the survivability and penetrating abilities of Chinese missiles. As such, China has never openly declared the specific number of weapons needed for its minimum nuclear force. Mao Zedong once stated "In any case, we won't build more atomic bombs and missiles than others." He also said "a few atomic bombs are enough (for China). Six are enough."[44] Six warheads is likely not the specific number in the mind of Chinese leaders today, but a minimum nuclear force with approximately ten warheads reaching a target country may be considered enough to "deter" a nuclear first strike by inflicting unacceptable damages.[45]

Occasional comments from high-level, Chinese, military officials have stimulated debate within and outside the country on China's NFU policy. During a July 2005 meeting with foreign journalists in Hong Kong, Major General Zhu Chenghu of China's National Defense University reportedly expressed the view that China would have no choice but to respond with nuclear weapons if the United States attacked Chinese territory with conventional forces during a conflict over the Taiwan Strait, a view that contradicts China's long-standing nuclear NFU doctrine.[46] Such comments stirred criticism at home and drew strong reactions abroad.

There have also been discussions in China about further defining "nuclear first use." For instance, some experts analyzed a situation in which an adversary uses highly precise conventional forces to target China's nuclear installations, including nuclear power reactors and nuclear weapon sites, and whether China should consider the attack as a "nuclear first strike."[47] Many Chinese security experts, however, argue China would never use nuclear weapons to deter another country's conventional intervention. In fact, China has developed and deployed advanced conventional missiles including versions of the DF-21 that can attack aircraft carriers and penetrate regional missile defense systems. These new conventional forces should make Chinese leaders more confident and less reliant on nuclear weapons to deal with conventional attacks.[48]

Unlike previous white papers on defense, China's 2012 edition did not reiterate China's long-standing NFU policy, which sparked a debate over whether China was changing its nuclear doctrine.[49] However, Colonel Yang Yujun, a spokesman for China's Ministry of Defense, reaffirmed in a briefing in April 2013 that China "has always pursued a No First Use nuclear weapons policy, upholds its nuclear strategy of self-defense, and never takes part in any form of nuclear arms race with any country. The policy has never been changed."[50] While some Western experts and scholars are suspicious of China's NFU pledge—the Pentagon's 2013 annual report on the Chinese military called it "ambiguous"[51]—this author contends China's nuclear force posture has all the features of a meaningful NFU policy. Specifically, it has a much smaller and simpler arsenal with a much lower alert status than what is required for a first-use option. The Second Artillery Corps, the military unit in control of China's strategic missile forces, conducts war planning and training under the assumption that China will absorb a first nuclear blow and use its nuclear forces only to retaliate. All these facts indicate that China's NFU pledge is serious.[52]

It should be noted China's nuclear force posture is determined primarily by its strategy, not financial or technological constraints.[53] China's economic and technological development since the 1980s would allow for a significant expansion of its nuclear force if leaders determined this approach to be in China's strategic interest, but Beijing still maintains a limited nuclear force. Because China's current nuclear policy has proven to be effective and smart, providing savings that can be used on economic development, the Chinese nuclear experts interviewed by this author consider it unthinkable that Beijing would change its policy to pursue extremely expensive weapons parity with the superpowers.

The Nuclear Modernization Debate

While analysts may disagree about the purpose, scope, and strategy behind China's nuclear modernization program, I argue in this section that Beijing has been shown to prioritize quality over quantity. Specifically, China's nuclear modernization has sought to increase the survivability of its nuclear force by replacing older, liquid-fueled missiles with solid-fueled, mobile, ballistic missiles and constructing deep underground tunnels that can act as missile bases. Recently, Western governments and media outlets have questioned Chinese nuclear buildup, in particular, whether Beijing is pursuing nuclear parity with the United States and Russia after the New START treaty was signed in 2010. Such voices were amplified after one study suggested the 3,000-mile-long network of underground tunnels—sometimes called China's "underground Great Wall"—could hide as many as 3,000 nuclear weapons.[54]

In practice, however, China's nuclear force and modernization would be determined mainly by its minimum deterrence and NFU nuclear policies. The vast "underground Great Wall" China has been building since 1985 to protect its smaller nuclear arsenal against nuclear and conventional attacks actually provides additional evidence for Beijing's minimum deterrence posture.[55] Furthermore, Beijing is constrained technically by its inventory of fissile materials. China currently has a military inventory of about 1.8 tons of plutonium and 18 tons of weapon-grade, highly enriched uranium (HEU).[56] The country stopped its HEU production in 1987 and plutonium production by 1990. All of its military HEU and plutonium production facilities have been closed, converted to other uses, or decommissioned.[57] China's current military inventory of fissile materials would not support an arsenal of more than 1,000 warheads.[58] In fact, part of these fissile material stocks would be held as a reserve for future needs. The other five permanent members of the U.N. Security Council—France, Russia, the United Kingdom, and the United States—devote half or less of their fissile material to their nuclear weapons. If this were the case for China, the upper boundary on its arsenal would be around 500 warheads. Published estimates of the size of China's arsenal vary, though credible sources agree it is relatively small. The best estimates suggest China has about 260 total nuclear weapons.[59]

Some Western officials and experts are concerned that as China accelerates the deployment of MIRVs (Multiple Independently Targetable Reentry Vehicles),[60] the size of China's nuclear arsenal will rapidly increase and Beijing would then need to change its "minimum nuclear deterrent" policy and NFU pledge. However, this MIRVs development does not necessarily change China's nuclear strategy. The major driver for China's MIRVs is a reaction to a U.S. missile defense system that could neutralize China's smaller nuclear force.

As noted previously, in the Chinese calculus, the minimum acceptable nuclear force is one that will survive a nuclear first strike and still penetrate a missile defense system to reach its designated targets. The Chinese government insists its nuclear force modernization is therefore necessary to maintain a reliable, second-strike, retaliatory capability. Professor Hu Side, the former president of the Chinese Academy of Engineering Physics—the Chinese equivalent of the U.S. Los Alamos National Laboratory—said, "China's nuclear modernization [is conducted] under the guideline of China's nuclear policy, maintaining the principle of counterattack in self-defense and avoiding [an] arms race." He further emphasized that it is impossible and unnecessary to accomplish China's modernization requirements through a simple increase in the number of nuclear weapons. In this view, modernization efforts will provide assurance in the reliability and safety of its nuclear arsenal, ensure China's investments will be limited to low levels, and be conducted without violating the Comprehensive Nuclear Test Ban Treaty (CTBT).[61] Again, China has demonstrated it prioritizes the overall quality of its nuclear force deterrent over the quantity of any individual element.

While many Western analysts complain that Beijing's nuclear force posture is unnecessarily opaque, Beijing believes the transparency of its nuclear strategy and doctrine is more important than that of its force posture. Furthermore, China contends the opacity of its force posture can serve to enhance the deterrence effect of its small nuclear force by keeping rivals guessing about how many warheads they would need to destroy in a first strike without risking China's second strike force. Though Beijing has not revealed the full details of its plans for modernizing its nuclear force, China's nuclear modernization program will likely continue to be guided by its "minimum nuclear deterrence" policy. Some Chinese arms control proponents mention that if Beijing develops more confidence about the survivability of its small nuclear force, the government might become transparent about its nuclear programs.[62]

Beijing has clarified its nuclear policies by issuing defense white papers since 1998. However, the Chinese public gets information about its nuclear posture mainly through Western publications. While some scholars and security analysts in China challenge the government's official nuclear policies,[63] in particular its unconditional NFU pledge, there are few civil society or nongovernmental groups who engage in critical analysis of China's programs or try to influence its direction.[64] Instead, most Chinese citizens believe nuclear weapons are vital to guarding national security and assuring the independence of sovereign and foreign affairs.

The U.S. Role in China's Nuclear Debate

China's modernization program would likely continue to focus on quality over quantity were it not for U.S. missile defense plans. Some Chinese officials are concerned that even a limited missile defense system could neutralize China's smaller nuclear force. In its 2010 white paper on defense, Beijing argued, "The global missile defense program will be detrimental to international strategic balance and stability, will undermine international and regional security, and will have a negative impact on the process of nuclear disarmament."[65] The document further said that "no state should deploy overseas missile defense systems that have strategic missile defense capabilities or potential, or engage in any such international collaboration."

As previously discussed, China's current arsenal of longer-range intercontinental ballistic missiles (ICBMs) could change significantly if the United States were to deploy a more comprehensive or proven missile defense system. This might include building more total warheads and MIRVs to overcome missile defenses as well as developing decoys and other missile defense countermeasures.[66] However, any expansion of the Chinese nuclear arsenal would still be constrained by its inventory of fissile materials.[67]

Moreover, Washington's nuclear intentions toward Beijing could also influence China's nuclear modernization plans. In particular, China worries the United States might use nuclear weapons against China in a potential conflict over Taiwan. This concern is further exacerbated by the U.S. strategic intention to shift the focus of its military strategy to the Pacific and East Asian region. According to the Pentagon's new strategic defense guidance, "The U.S. military will invest as required to ensure its ability to operate effectively in anti-access and area denial (A2/AD) environments."[68] China and Iran are listed as the major targets.

China will continue to update its nuclear force posture to adapt to other countries' military developments and the international security environment. Since China's economic reforms in 1978, the Realist school has prevailed on the country's nuclear weapons strategy and its nuclear modernization will likely continue to be guided by a NFU and "minimum nuclear deterrence" policy. Despite the end of the Cold War and greater uncertainty in global affairs, China's minimum nuclear deterrent and self-defense-driven nuclear weapons policy survived with a surprising degree of continuity, though that could swing in response to significant changes in U.S. missile defense capabilities or its strategic posture toward China.

CHINESE VIEWS ON NUCLEAR ARMS CONTROL

A coalition of Realists and Globalists in China has emerged to adopt an uneasy embrace of arms control as a means to achieve their national interests. China's official policy has long called for "the complete prohibition and thorough destruction of nuclear weapons," which was reiterated in its 2010 white paper on defense. Beijing maintains that "countries possessing the largest nuclear arsenals bear special and primary responsibility for nuclear disarmament," and thus they "should further drastically reduce their nuclear arsenals in a verifiable, irreversible, and legally-binding manner, so as to create the necessary conditions for the complete elimination of nuclear weapons."[69] Before this can occur, China insists nuclear weapon states should denounce the first use of nuclear weapons, make an unequivocal commitment that under no circumstances will they use or threaten to use nuclear weapons against non-nuclear weapon states or those in nuclear-weapon-free zones, and negotiate an international legal instrument to lock in these pledges.[70] This section outlines competing debates within China on arms control and the emerging of a Realist-Globalist coalition in support of this approach.

Uneasy Embrace of Arms Control in China

Before initiating economic reform in the late 1970s, Beijing showed little interest in nuclear arms control or nonproliferation. Beijing viewed the superpowers' arms control agreements as plots to constrain China's budding nuclear weapons programs. Beijing denounced the Treaty on the Nonproliferation of Nuclear Weapons (NPT) as inherently discriminatory. As a result, China lacked an arms control community and a formal decision-making mechanism for nuclear arms control and nonproliferation issues.[71]

However, since the early 1990s, China has been actively involved in arms control and nonproliferation negotiations and adheres to most global norms and agreements. This Globalist perspective is mainly driven by Beijing's eagerness to integrate into the global economic system, improve its international image, and demonstrate itself to be a responsible stakeholder working in cooperation with international society and, in particular, with the United States. President Barack Obama's vision of a nuclear-weapon-free world—welcomed, at least in public, by Beijing—stimulated debate in the Chinese public on whether or not China should follow suit. Based on the most recent available data, a September 2009 online survey conducted by the *Global Times*, found that about 51 percent of respondents support the call for a nuclear-free world (49 percent disagreed).[72] Others officials and analysts in private are more skeptical of the true intentions behind Obama's vision of a world free of nuclear weapons—articulated during a speech in 2009 at Hradčany Square in Prague—that they argue aims to constrain China's nuclear modernization process.[73] For instance, China may feel it needs to build more warheads in the future to maintain a deterrent in response to a proven U.S. missile defense system, but it may be blocked from doing so by international pressure in the wake of promised declines in Russian and U.S. arsenals.

China's participation in the negotiation of the Comprehensive Nuclear Test Ban Treaty (CTBT) is a prime example of its growing comfort with arms control and the rise of the Globalist school of thought. Early in the talks, many Chinese nuclear experts argued the CTBT would constrain China's nuclear modernization relative to other NPT nuclear weapon states after it conducted only forty-five tests before announcing a testing moratorium in 1996, leaving a limited number of tested warhead designs certified for deployment.[74] However, Beijing ultimately made some compromises and signed the CTBT soon after it opened for signing in 1996.[75] China has not yet ratified the accord, due in part to the CTBT's being rejected by the U.S. Senate in 1999. Beijing's ratification will likely follow Washington's, as indicated by China's recent white paper that "supports the early entry into force" of the treaty.

An important deviation from this trend, however, is that many Chinese officials and arms control experts have begun to see China's own growing nuclear weapon capability as leverage in future arms control negotiations. This is further evidence that China's approach toward arms control is now also closely aligned with the Realist school of thought.

In contrast to the CTBT debate, Beijing's attitude toward the proposed fissile material cut-off treaty (FMCT) indicates a resurgent Realist view anxious about how an FMCT might impact its national security and future nuclear force. While Beijing supports "the early commencement of negotiations" on an FMCT at the Conference on Disarmament (CD),[76] significant improvement in the capabilities of U.S. missile defense will affect China's willingness to participate in FMCT negotiations. If Beijing remains concerned about whether these developments may degrade China's ability to respond with a second strike after an attack, it might decide to build more ICBMs in order to maintain its deterrence and, therefore, might require more plutonium and HEU to fuel those weapons.

On the issue of global disarmament, Beijing does not specify when China itself would participate in this process, but it sees the Chinese arsenal as leverage to jump-start the debate. Many Chinese analysts believe Beijing may wish to wait until the United States and Russia reduce their stockpiles to no more than about 1,000 total warheads each.[77] China may then need to reveal the size of its nuclear force as a way to create the necessary confidence for the United States and Russia to continue their reductions; Beijing could pledge a lower cap of its forces (say around 200 total warheads) with Russia and the United States coming down further. Other Chinese experts and analysts suspect the United States and Russia never actually intend to reach those lower numbers.

Another example of building "leverage" for future arms control is China's actions on the weaponization of space. Starting in the late 1990s, China advocated at the CD for a treaty on the prevention of an arms race in outer space (PAROS) and joined Russia in submitting several proposals in hopes of concluding negotiations as soon as possible. The latest proposal was sent in June 2014.[78] However, Washington has slowed progress toward this negotiation at the United Nations. In January 2007, a Chinese weather satellite was destroyed by a kinetic kill vehicle launched with a multistage solid-fuel missile from China. Beijing perhaps hoped to use this antisatellite test and other reported tests to push Washington to seriously consider the risks of a weapons race in space and to start a PAROS treaty negotiation. Should diplomatic efforts fail, those tests could be used as a hedge and mark the beginning of China's own space weapons program. This approach signals a Realist view that arms control negotiations should be pursued only after gaining the necessary leverage to ensure the final product is in one's national interest.

CHINA'S VIEWS ON NUCLEAR NONPROLIFERATION

China has undergone a dramatic swing in its nonproliferation views over the past several decades, from one of hostility and disinterest toward a stance more aligned with a Globalist perspective. Nonproliferation, as distinct from arms control measures that target the reduction or limitations of existing arsenals, is concerned primarily with preventing the spread of WMD to new states and individuals. From the 1980s to early-1990s, economic interests pushed China to export major quantities of nuclear and missile-related transfers to other countries. China's gradual market reforms since 1978 pushed the country's state-owned defense industry to find new ways to financially support itself, aided by an attitude in Beijing that favored the promotion of overall exports rather than trade controls that could limit economic growth.[79] Moreover, as a newcomer to many of these nonproliferation treaties and regimes, China had limited experience with export control systems.

However, since the early 1990s, China has started to seriously address proliferation concerns by the United States and others.[80] Many Chinese experts identify the primary motivation behind China's newfound interests in nonproliferation as Beijing's desire to have a positive international image and present itself as a responsible rising power.[81] China hoped to speed up its economic integration with the world through actively engaging in international regimes. As a result, China has moved to establish comprehensive export control systems for potential dual-use items that could lead to the proliferation of WMD and related delivery systems. As stated in China's 2010 white paper on defense, "China firmly opposes the proliferation of weapons of mass destruction and their means of delivery, and consistently deals with nonproliferation issues in a highly responsible manner."[82] China has joined all international treaties and international organizations in the field of nuclear nonproliferation. Furthermore, China supports the role played by the United Nations in nonproliferation and has conscientiously implemented any relevant resolutions of the U.N. Security Council.

China maintains three principles governing its nuclear exports: (1) upholding International Atomic Energy Agency (IAEA) safeguards; (2) supporting the peaceful use of nuclear energy; and (3) preventing transfers of these exports to a third country without China's prior consent. In 2009, the Ministry of Commerce promulgated the Measures for Administration of Dual-Use Items and Technologies Subject to Export General License to further strengthen the licensing system for export control. China joined the Nuclear Suppliers Group (NSG) in 2004,[83] signed onto the Zangger Committee in 1997, and ratified the IAEA Additional Protocol in 2002, the first nuclear weapon state to do so. The 2003 white paper on *China's Nonproliferation Policy and Measures* outlined specific requirements for end user and end use certificates, a licensing system, license examination and approval criteria, a catch-all principle, and civil and criminal liabilities for export control violations.[84]

Besides these WMD export control measures, starting in the early 1990s, China has also taken a number of steps to control missile exports. Since 1992, China has committed to adhere to Missile Technology Control Regime (MTCR) guidelines and parameters in the export of missiles and related components. In 2002, China announced its Regulations on Export Control of Missiles and Missile-Related Items and Technologies and the Control List. Since 2004, China has applied to join the MTCR but has not been accepted as a member because the United States is still concerned about Chinese behavior in some of its missile exports.[85]

In short, since China's economic reforms in the late 1970s, Chinese top leaders from Deng Xiaoping to today have favored the Realist school of thought on nuclear weapons policy and arms control. However, Chinese attitudes toward nonproliferation today have evolved toward the Globalist school of thought—especially as China moves forward with its political and economic development. Generally, there have been no significant gaps between the government and public on these issues. It should be noted that even the same expert may have different views on nuclear weapons policy, arms control, and nonproliferation, and could thus belong to a different school of thought for each specific area.

CHINESE VIEWS ON NUCLEAR SECURITY

The central government is primarily responsible for formulating policy and development plans for the nuclear industry on nuclear security matters and public involvement is virtually absent. At the 2014 Nuclear Security Summit, China's President Xi Jinping stressed that increased cooperation on the nuclear security of any one country is beneficial to all nations.[86] Globalists in China have substantially enhanced the country's nuclear security since 9/11 in an effort to improve China's image abroad and its relationship with the United States and not based on any perceived threat from nuclear terrorism[87]

Chinese Assessment of the Risk of Nuclear Terrorism

Chinese accomplishments in nuclear security do not seem to be driven by Chinese perceptions of a nuclear terrorism threat. Indeed, the majority of Chinese nuclear experts are less concerned about nuclear terrorism than their counterparts in the West. Many Chinese experts continue to doubt there is a credible threat to Chinese nuclear materials and facilities because the country's arsenal is thought to be limited in size, tightly monitored, and guarded by well-armed forces.[88] Moreover, most Chinese warheads are stored in remote, highly secret mountainous locations that are very difficult for outsiders to access. Some experts and government officials may endorse Chinese commitments to upgrading nuclear security, but only because they see it as necessary to comply with international requirements. Furthermore, Chinese nuclear experts interviewed for this chapter believe the probability of terrorists' obtaining fissile materials inside China and manufacturing a crude bomb is extremely low. In this view, nuclear terrorism may be a problem for the United States or Russia, but it is not an urgent concern for China.[89] On the other hand, these same individuals feel that a terrorist attack using a radiological dispersal device or "dirty bomb" is a real threat.[90] An abundance of radioactive sources, nuclear waste, and orphaned and spare nuclear materials distributed widely throughout the country—along with a relatively weaker control and management system for these sources—make a dirty bomb a more attractive target.

To the Chinese experts and officials interviewed for this chapter, the risk of sabotage at civilian nuclear facilities, in particular a nuclear power plant, is plausible. In a comprehensive assessment of the risk of sabotage of nuclear facilities, the China Academy of Engineering's 2005 book *Management of Nuclear and Radiological Terrorism Incidents* concluded "one possible route nuclear terrorists would take is to sabotage [a] nuclear facility. Once [it] happen[s], it could seriously affect the environment and public. China has a number of nuclear facilities that have a variety of vulnerabilities if attacked."[91] However, these experts also see the risk of sabotage at a nuclear power plant resulting in a major release of radioactive material as very low because groups hostile to the government lack the means to conduct such attacks and overwhelm China's security system. As the number of China's nuclear power plants expands, this risk may pose a more pressing challenge to China's nuclear security. Furthermore, the nuclear accident at Fukushima has increased the interest of terrorists in targeting China's growing nuclear fleet.[92]

The severe consequences of a nuclear weapon detonation, however, are significant enough that no country, particularly those with nuclear weapons or weapon-usable fissile materials, can ignore the real and urgent threat of nuclear terrorism. China cannot and should not exclude the possibility of outsider and insider threats. One worry is that China faces the challenge of complacency among a large number of senior nuclear experts and nuclear industries who believe China already has strict nuclear security systems that have worked well and "free of accident" over the past fifty years. Too many managers and employees at Chinese nuclear plants do not recognize the importance of advanced and stringent materials protection, control, and accounting (MPC&A) programs and doubt whether it is worth the money and time.[93]

The possibility that nuclear materials could become a victim of insider theft in China cannot be ruled out, especially as China moves closer toward a market-oriented society and continues to fight against corruption.[94] Terrorist attacks by outsiders may also pose a real threat to China's nuclear facilities—in particular the terrorist forces of "East Turkestan," which the Chinese government believes have long been recipients of training, financial assistance, and support from groups like Al Qaeda.[95] Furthermore, a security incident on the scale of Chernobyl would be a global catastrophe and would result in severe consequences that doom prospects for large-scale nuclear power growth everywhere.

Major Drivers for Improvement of China's Nuclear Security

While many Chinese nuclear experts do not believe the threat of nuclear terrorism is urgent in China, some relevant government departments, including the China Atomic Energy Authority (CAEA) and the National Nuclear Safety Administration (NNSA), have taken significant steps to develop and apply approaches to nuclear security and nuclear accounting in the aftermath of 9/11. As some interviewees noted, one major driver for strengthening China's nuclear security is to fulfill international pledges China has made at the Nuclear Security Summits and in other forums.[96] Globalists argue that the participation by China, as a responsible country, in international legal instruments relevant to nuclear security and China's fulfillment of its obligations would improve its international image.

Indeed, China has joined nearly all of the international legal instruments relevant to nuclear security, including the Convention on the Physical Protection of Nuclear Material, the Amendment to the Convention on the Physical Protection of Nuclear Material, the International Convention for the Suppression of Acts of Nuclear Terrorism, and the IAEA Additional Protocol. In 2007, China participated in the Incident and Trafficking Database (ITDB), the IAEA's information system on incidents of illicit trafficking and other unauthorized activities and events involving nuclear and other radioactive material outside of regulatory control.

Another major driver of improvements to China's nuclear security and control system has been the support from international bodies like the IAEA and the cooperation of countries such as the United States. Given the prominence nuclear terrorism has on the U.S. security agenda, Beijing's cooperation with Washington is seen as particularly beneficial to the Sino-U.S. relationship.[97] In recent years, China's cooperation with the U.S. Department of Energy has seen an extensive series of exchanges, including visits to U.S. facilities to observe nuclear security and accounting approaches; in-depth training and discussion workshops on everything from approaches to protect against insider threats to the design of physical protection systems; a joint demonstration of advanced MPC&A technology; work to strengthen security and accounting regulations and inspections in China; and cooperation to build a Center of Excellence (CoE) on Nuclear Security.[98] In addition, China has also cooperated extensively with the IAEA on issues ranging from safety and security training to preparations for the 2008 Olympics, as well as detection, emergency response, regulations, and standards.[99]

In sum, China's moves to improve its nuclear security is mainly driven by three factors: (1) fulfilling its international pledges, (2) improving its image abroad, and (3) international cooperation with the United States and IAEA. These efforts show China's approach to nuclear security is in line with the Globalist school of thought.

CONCLUSION

Within China's domestic nuclear debates, a coalition of Realist-Globalist perspectives has emerged to direct the country's decision making on nuclear energy, nuclear weapons, arms control, nonproliferation, and nuclear security. This broad coalition comes together on a number of policy challenges, but as stated earlier, even the same Chinese official, scholar, or expert may have different views on nuclear weapons policy, arms control, and nonproliferation, and thus could belong to a different school of thought for each specific area.

With relatively ambitious plans for its nuclear energy program, China has become the central focus of the international nuclear industry. Of the three major drivers for China's nuclear energy development—air pollution, climate change, and energy security—the need to alleviate air pollution concerns rises to the top for those interviewed by the author for this chapter. The dominant Realist perspective in China has placed a strong emphasis on protecting the Chinese growth and development model by using alternative energies and measures other than a reliance on carbon emission caps to address air pollution and climate change issues. In this view, nuclear energy represents a solution to promote economic growth while at the same time reducing the amount of energy derived from coal and other fossil fuels.

This is not to say climate change and energy security concerns are nonexistent in China's nuclear debates. On the contrary, most Chinese officials and energy experts see them as important motivations for nuclear energy development. Nationalists argue that a self-reliant energy supply system is more critical than ever in today's uncertain international political and security environment. Globalists who see nuclear energy as a solution to climate change contend China should take responsibility to mitigate global warming, which could in turn have a positive effect for promoting China's international image. The near-universal consensus in China about the need for nuclear energy—whatever the reason—means China will continue nuclear power development in the coming decades. While the Fukushima accident had some impact on the public's opinion about nuclear power and government decisions on some local nuclear projects, national nuclear power plans will not meaningfully change in the near future.

Unlike China's civilian nuclear sector, where public involvement has expanded in recent years, the government conclusively dominates the nuclear weapons strategy debate, though the public generally supports the official position. Since its first nuclear explosion in 1964, China has maintained a policy that features a minimum deterrent and a No First Use pledge, both aimed at avoiding a costly nuclear arms race. Top Chinese leaders have consistently shored up this approach. China will continue to update its nuclear force posture according to other countries' military developments and the international security environment, but there is no evidence China will change its long-standing NFU or "minimum nuclear deterrence" policies.

The Nationalist school of thought prevailed on nuclear weapons questions during the period between the 1950s and late 1970s under Mao Zedong. Since China started economic reform in 1978, the Deng Xiaoping administration and its successors have shifted toward the Realist perspective and prevailed among military and politic communities as well as the public. The Realist school of thought will influence China's nuclear weapons policy for the coming decades.

Though China's official policy on nuclear disarmament has called for "the complete prohibition and thorough destruction of nuclear weapons," since the 1990s, China has addressed capability as leverage on relevant arms control negotiations. This shows China's approach toward arms control is now closest to that of the Realists. Such a perspective will affect Beijing's attitude toward its weapons programs and arms control negotiations.

Meanwhile, China's commitment to the nonproliferation of WMD and delivery systems is driven mainly by a Globalist desire to improve its international image as a responsible rising power. Generally, there have been no significant gaps between the government and public. The Globalist perspective should further grow within China as it moves forward with political and economic development.

Finally, since the 9/11 attacks, China has significantly improved its nuclear security. With public involvement virtually absent, the central government formulates policy and development plans for the nuclear industry. Rather than a perceived threat from nuclear terrorism, China's efforts to strengthen nuclear security are motivated by Globalist ends to improve its image abroad, fulfill international obligations, and collaborate with the United States and the IAEA.

Overall, the Realist-Globalist coalition will likely continue to dominate China's nuclear debates for the foreseeable future and to maintain its remarkable consistency for its nuclear weapons, nuclear energy, arms control, nonproliferation, and nuclear security policies.

NOTES

1. For this chapter, the author conducted interviews in March 2014 and July 2014 with over thirty Chinese officials, experts, and scholars from a number of government departments and academic institutes and universities, including the following: Arms Control and Disarmament Department at the Chinese Foreign Ministry; China Atomic Energy Authority (CAEA), responsible for the development of nuclear security laws, regulations, and standards, and the review and approval of licenses for operators of civilian nuclear facilities; China National Energy Administration (NEA), responsible for nuclear energy development; China's National Nuclear Safety Administration (NNSA), a regulatory body under the Ministry of Environmental Protection; China Academy of Engineering, a consultant group for energy

development; China Institute of Atomic Energy (CIAE), which provides technical support and technology development to CAEA; China Academy of Engineering Physics (CAEP, the Chinese equivalent of the U.S.'s Los Alamos National Laboratory) responsible for nuclear weapon development and policy and for innovation of nuclear security policies and technology—particularly for military facilities and material; China Defense Science and Technology Information Center, a PLA (People's Liberation Army) general armament department; China Institute of Nuclear Information and Economics, Beijing; China's Ministry for Public Security, the agency leading physical protection regulations governing facilities and transport; China Arms Control and Disarmament Association; China Foundation for International and Strategic Studies; the PLA National Defense University; Chinese Academy of Military Science; Peking University; and Tsinghua University.

2. "Number of Nuclear Reactors Operable and under Construction," *World Nuclear Association*, last updated December 23, 2015, http://www.world-nuclear.org/info/Country-Profiles/Countries-A-F/China--Nuclear-Power/.

3. China Academy of Engineering, *China's Medium- and Long-Term Energy Development Strategy 2030, 2050* (Beijing: Science Press, 2011).

4. "Number of Nuclear Reactors Operable and under Construction."

5. Zhang Guobao, "Scientific Development of Electric Power Industry: The Pathway to Win the Challenge," *Qiushi Journal* no.7 (2009).

6. "China Nuclear Power Installed Capacity Expected to Reach 70 GW by 2020, Scientists," *China View*, November 22, 2009, http://news.xinhuanet.com/english/2009-11/02/content_12373871.htm; "China 2020 Nuclear Capacity to Double from Earlier Plan," *Reuters*, March 22, 2010, http://www.reuters.com/article/2010/03/22/us-china-energy-idUSTRE62L13J20100322.

7. "Wen Jiabao Chairs Executive Meeting of the State Council," *Central People's Government of the People's Republic of China*, October 24, 2012, http://www.gov.cn/ldhd/2012-10/24/content_2250357.htm; Information Office of the State Council of the People's Republic of China, *China's Energy Policy 2012* (Beijing: People's Republic of China, October 24, 2012), http://news.xinhuanet.com/english/china/2012-10/24/c_131927649.htm.

8. Ibid.

9. U.S. Energy Information Administration, *China Country Report*, February 4, 2014, http://www.eia.gov/countries/analysisbriefs/China/china.pdf.

10. China Academy of Engineering, *China's Medium- and Long-Term Energy Development Strategy*, ix.

11. It will take several years to see how this develops, but it is worth noting the International Energy Agency in December 2015 saw a decline in the demand for coal in China, arguing "the golden age of coal seems to be over." For more, see Ambrose Evans-Pritchard, "International Energy Agency Sees 'Peak Coal' as Demand for Fossil Fuel Crumbles in China," *The Telegraph*, December 19, 2015, http://www.telegraph.co.uk/finance/economics/12058456/IEA-sees-peak-coal-as-demand-crumbles-in-China.html.

12. Information Office of the State Council of the People's Republic of China, *China's Energy Policy 2012*.

13. Jianyu Jin, "Stricter Central Air Standards Hard for Cities to Introduce," *Global Times*, March 2, 2012, http://www.globaltimes.cn/content/698367.shtml.

14. Qingfeng Zhang and Robert Crooks, *Toward an Environmentally Sustainable Future: Country Environmental Analysis of the People's Republic Of China* (Mandaluyong City, Philip-

pines: Asian Development Bank, 2012), xvii, http://www.adb.org/sites/default/files/publication/29943/toward-environmentally-sustainable-future-prc.pdf.

15. *The Global Burden of Disease Study 2010* was published as part of a special issue of *The Lancet*. For a summary of their findings, see Stephen S. Lim et al., "A Comparative Risk Assessment of Burden of Disease and Injury Attributable to 67 Risk Factors and Risk Factor Clusters in 21 Regions, 1990–2010," *The Lancet* 380, no. 9859 (December 15, 2012): 2224–2260; "Outdoor Air Pollution among Top Global Health Risks in 2010," Health Effects Institute, December 13, 2012, http://www.healtheffects.org/International/GBD-Press-Release.pdf.

16. Edward Wong, "Cost of Environmental Damage in China Growing Rapidly amid Industrialization," *New York Times*, March 29, 2013. http://www.nytimes.com/2013/03/30/world/asia/cost-of-environmental-degradation-in-china-is-growing.html.

17. China Academy of Engineering, *China's Medium- and Long-Term Energy Development Strategy*, xi.

18. Communication with Chinese nuclear experts, July 2013.

19. Robert Stavins, "Chinese and US Climate Interests are Converging," *China Dialogue* (June 5, 2014), https://www.chinadialogue.net/blog/6952-Chinese-and-US-climate-interests-are-converging/en.

20. "We Have a Climate Deal: India and China React to the Paris Agreement," *Rising Powers Initiative—Policy Alert #115*, December 21, 2015, http://www.risingpowersinitiative.org/wp-content/uploads/policyalert115.pdf; "COP21: Rising Powers Seek Solutions for Climate Change," *Rising Powers Initiative—Policy Alert #114*, December 10, 2015, http://www.risingpowersinitiative.org/wp-content/uploads/policyalert_COP21.pdf.

21. National Development and Reform Commission (NDRC), People's Republic of China, *China's National Climate Change Programme* (Beijing: NDRC, June 2007), http://en.ndrc.gov.cn/newsrelease/200706/P020070604561191006823.pdf.

22. Department of Climate Change, National Development, and Reform Commission of China, "Enhanced Actions on Climate Change: China's Intended Nationally Determined Contributions," June 30, 2015, http://www4.unfccc.int/submissions/INDC/Published%20Documents/China/1/China's%20INDC%20-%20on%2030%20June%202015.pdf.

23. China Academy of Engineering, *China's Medium- and Long-Term Energy Development Strategy*, ix.

24. Communication with Chinese nuclear experts, July 2013.

25. Information Office of the State Council of the People's Republic of China, *China's Energy Policy 2012*.

26. Communication with Chinese nuclear experts, Beijing, August 2013.

27. Ibid.

28. "China Country Report," *U.S. Energy Information Administration*, last updated May 14, 2015, https://www.eia.gov/beta/international/analysis_includes/countries_long/China/china.pdf.

29. Information Office of the State Council of the People's Republic of China, *China's Energy Policy 2012*.

30. Interviews with Chinese nuclear experts, July 2013.

31. Ibid.

32. Zhu Yue, "China's Nuclear Expansion Threatened by Public Unease,"*ChinaDialogue.net*, September 23, 2014, https://www.chinadialogue.net/article/show/single/en/7336-China-s-nuclear-expansion-threatened-by-public-unease.

33. Q. Hou and X. L. Liu, "Fukushima Nuclear Accident Highlights the Weakness of China's Publicity of Nuclear Science," *Chinese Nuclear Industry* 6 (2011): 64–66.

34. Li Xiaojuan et al., "Risk Perception of the Public Living in Vicinity of Nuclear Power Plant," *Chinese Journal of Radiological Medicine and Protection* 28 no. 3 (2008): 269–273.

35. Guizhen He et al., "Nuclear Power in China after Fukushima: Understanding Public Knowledge, Attitudes, and Trust," *Journal of Risk Research* 17 no. 4 (2014): 435–451.

36. National Development and Reform Commission, *Notification of the Interim Measures of Social Stability Risk Assessment* (Beijing: People's Republic of China, 2012), http://www.hnaec.org/upload/News/1/201210/20121012135641l8100.pdf.

37. Wen Yuqing and Yang Fan, "Guangdong Anti-Nuke Marchers Win Concession," *Radio Free Asia,* July 12, 2013, http://www.rfa.org/english/news/china/nuclear-07122013154505.html.

38. Interviews with Chinese nuclear experts in Beijing, March 2014.

39. Ibid.

40. Ibid.

41. "Number of Nuclear Reactors Operable and under Construction."

42. Information Office of the State Council of the People's Republic of China, *China's National Defense in 2010,* (Beijing: People's Republic of China, March 2011), http://www.nti.org/media/pdfs/1_1a.pdf?_=1316627912.

43. Information Office of the State Council of the People's Republic of China, *China's National Defense in 2006* (Beijing: People's Republic of China, December, 29 2006), http://www.fas.org/nuke/guide/china/doctrine/wp2006.html.

44. John Lewis and Xue Litai, *China's Strategic Seapower: The Politics of Force Modernization in the Nuclear Age* (Stanford, Calif.: Stanford University Press, 1994)., 232.

45. Hui Zhang, "How US Restraint Can Keep China's Nuclear Arsenal Small," *Bulletin of the Atomic Scientists* 68 no. 4 (July/August, 2012): 73–82; Hui Zhang, "Nuclear Modernization in China," in *Assuring Destruction Forever: Nuclear Weapon Modernization around the World*, ed. Ray Acheson (New York: Reaching Critical Will, March 2012).

46. Joseph Kahn, "Chinese General Sees U.S. as Nuclear Target," *New York Times,* July 16, 2005, http://www.nytimes.com/2005/07/15/world/asia/15iht-china.html.

47. Peng Guangqian and Rong Yu, "Nuclear No-First-Use Revisited," *China Security* 5 no. 1 (Winter 2009): 78–87.

48. Interviews with Chinese nuclear experts, July 2013; Sun Xiangli, "Analysis of China's Nuclear Strategy," *China Security* 1 no. 1 (Autumn 2005): 23–27.

49. James M. Acton, "Is China Changing Its Position on Nuclear Weapons?" *New York Times*, April 18, 2013, http://www.nytimes.com/2013/04/19/opinion/is-china-changing-its-position-on-nuclear-weapons.html.

50. Zhang Jiexian, "Department of Defense: China's Nuclear Strategy Has Not Changed," *People's Daily,* April 26, 2013, http://news.china.com.cn/live/2013-04/26/content_19698408.htm.

51. "Annual Report to Congress: Military and Security Developments Involving the People's Republic of China 2013," *Office of the Department of Defense*, 2013, http://www.defense.gov/Portals/1/Documents/pubs/2013_China_Report_FINAL.pdf.

52. Hui Zhang, "China's Perspective on a Nuclear-Free World," *Washington Quarterly* 32 no. 2 (April 2010): 139–155.

53. Hu Side, "The Road toward China's Nuclear Weapons," *Global Science* 12 (2007).

54. Phillip A. Karber, *Strategic Implications of China's Underground Great Wall* (presentation, Georgetown University, Washington, D.C., September 26, 2011), http://www.fas.org/nuke/guide/china/Karber_UndergroundFacilities-Full_2011_reduced.pdf; William Wan, "Georgetown Students Shed Light on China's Tunnel System for Nuclear Weapons," *Washington Post*, November 29, 2011, http://www.washingtonpost.com/world/national-security/georgetown-students-shed-light-on-chinas-tunnel-system-for-nuclear-weapons/2011/11/16/gIQA6AmKAO_story.html; Bret Stephens, "How Many Nukes Does China Have?," *Wall Street Journal*, October 24, 2011, http://online.wsj.com/article/SB10001424052970204346104576639502894496030.html.

55. Hui Zhang, "The Defensive Nature of China's 'Underground Great Wall,'" *The Bulletin of the Atomic Scientists*, January 16, 2012, http://thebulletin.org/defensive-nature-chinas-underground-great-wall.

56. Hui Zhang, "China's Nuclear Modernization," in *Assuring Destruction Forever: 2015 Edition,"* ed. Ray Acheson (Geneva, Switzerland: Reaching Critical Will, May 2015), http://www.reachingcriticalwill.org/images/documents/Publications/modernization/assuring-destruction-forever-2015.pdf.

57. For more, see Hui Zhang, "China," *Global Fissile Material Report 2010: Balancing the Books: Production and Stocks* (Princeton, N.J.: Princeton University, 2011), 97–106, http://belfercenter.ksg.harvard.edu/files/Hui-Zhang-China-Chapter-Global-Fissile-Materials-Report.pdf; Hui Zhang, "China's HEU and Plutonium Production and Stocks," *Science & Global Security* 19 no. 1 (January–April 2011): 68–89, http://belfercenter.ksg.harvard.edu/files/huizhangSGS2011.pdf.

58. For a compact thermonuclear warhead, assuming the average numbers for U.S. and Russian warheads are about 4 kg of plutonium in the primary and about 20 kg HEU in the secondary, then 1.8 tons of plutonium could produce about 450 warheads, which could also use about 9 tons of HEU in their secondaries. The remaining 9 tons of HEU might produce about 300 more warheads (assuming 10 kg of HEU for the primary and 20 kg for the secondary). Thus, a stockpile of 1.8 tons of plutonium and 18 tons of HEU could support about 750 thermonuclear warheads. Even using all of China's fissile material inventory would not support an arsenal of more than 1,000 warheads.

59. Hans Kristensen and Robert Norris, "Chinese Nuclear Forces, 2015," *The Bulletin of the Atomic Scientists* 71 no. 4, 2015, http://thebulletin.org/2015/july/chinese-nuclear-forces-20158459.

60. According to an April 2015 Pentagon report on Chinese military power, China deploys a MIRVed version of the DF-5 ICBM (liquid-fueled and silo-launched DF-5B). Also, it is reported that China is testing the mobile and solid-fueled DF-41, another MIRV-equipped ICBM. For more, see Office of the Secretary of Defense, *Annual Report to Congress: Military and Security Developments Involving the People's Republic of China 2015* (Washington, D.C.: Department of Defense, April 2015); Bill Gertz, "China Tests New Long-Range Missile with Two Guided Warheads," *Free Beacon*, August 18, 2015, http://freebeacon.com/national-security/china-tests-new-long-range-missile-with-two-guided-warheads/.

61. Hu Side, "The Road toward China's Nuclear Weapons."

62. Communications with Chinese nuclear arms controller, August, 2013.

63. Peng and Rong, "Nuclear No-First-Use Revisited."

64. Interviews with Chinese security analyst, March 2014.

65. Information Office of the State Council of the People's Republic of China, *China's National Defense in 2010*.

66. Hui Zhang, "Action/Reaction: U.S. Space Weaponization and China," *Arms Control Today*, December 2005, http://www.armscontrol.org/act/2005_12/Dec-cvr.asp.

67. For more, see H. Zhang, "China's HEU and Plutonium Production and Stocks."

68. "Sustaining US Global Leadership: Priorities for 21st Century Defense," *Department of Defense*, January 2012, http://www.defense.gov/news/Defense_Strategic_Guidance.pdf.

69. Information Office of the State Council of the People's Republic of China, *China's National Defense in 2010.*

70. Ibid.

71. Interviews with Chinese nuclear experts, July 2013.

72. Guo Qiang, "US' Nuke-Free World Plan Stirs Debate," *Global Times*, September 24, 2009, http://www.globaltimes.cn/content/471171.shtml.

73. For more, see H. Zhang, "China's Perspective on a Nuclear-Free World."

74. Interviews with Chinese nuclear expert, July 2013, Beijing; Sun Xiangli, "Implications of a Comprehensive Test Ban for China's Security Policy," *Center for International Security and Cooperation and Arms Control*, June 1997, http://cisac.fsi.stanford.edu/sites/default/files/sun.pdf.

75. For more, see Sun Xiangli, "Implications of a Comprehensive Test Ban for China's Security Policy."

76. Information Office of the State Council of the People's Republic of China, *China's National Defense in 2010.*

77. Interviews with Chinese nuclear experts, July 2013.

78. Hui Zhang, " Chinese Perspectives on Space Weaponization," in *Russian and Chinese Responses to U.S. Military Plans in Space*, eds. Pavel Podvig and Hui Zhang (Cambridge, Mass.: American Academy of Arts & Sciences, March 2008); "China, Russia Update Draft Treaty to Prevent Arms Race in Outer Space," *Xinhua*, June 11, 2014, http://www.chinadaily.com.cn/world/2014-06/11/content_17579161.htm.

79. Interviews with Chinese arms controller, August 2013. For more, see Jishe Fan, "China Nonproliferation Export Control System: Evolution and Revolution" (presentation, Managing the Atom Seminar Series, Belfer Center Library, Harvard University, Cambridge, Mass., February 26, 2008), http://belfercenter.ksg.harvard.edu/files/Jishe_Fan_Export_Control_MTA_Talk_1.pdf.

80. The United States and others are still concerned about China's proliferation activities—specifically nuclear and missile cooperation with Pakistan. While Beijing often criticizes those accusations as groundless and irresponsible and insists there have been no violations of any international agreements, China needs to strengthen the implementation of related guidelines and regulations of nuclear and missile control. Shirley A. Kan, *China and Proliferation of Weapons of Mass Destruction and Missiles: Policy Issues* (Washington, D.C.: Congressional Research Service, January 5, 2015), https://www.fas.org/sgp/crs/nuke/RL31555.pdf.

81. Interviews with Chinese nuclear experts, August 2013.

82. Information Office of the State Council of the People's Republic of China, *China's National Defense in 2010.*

83. Recently it has been reported China would sell a new round of nuclear reactors to Pakistan (as discussed in chapter 9), which caused some NSG members to complain that the transfer could violate Beijing's commitment as an NSG member. However, Beijing has argued that these reactors should be "grandfathered" into a long-term China-Pakistan nuclear cooperation agreement signed before China joined the NSG. For more, see Timothy Westmyer, "Nuclear Debates in Asia Digest: China's Nuclear Energy Technology Sales to Pakistan," Rising Powers

Initiative—Sigur Center for Asian Studies, April 3, 2013, http://www.risingpowersinitiative.org/nuclear-digest-china-nuclear-pakistan/.

84. Information Office of the State Council of the People's Republic of China, "China's Nonproliferation Policy and Measures," December 3, 2003, http://www.china.org.cn/english/2003/Dec/81312.htm.

85. Many Chinese argue that the real reason for Washington's rejection of Chinese MTCR membership is that Washington is concerned China may gain access to sensitive missile design and high technology information for a space-launch vehicle program that could contribute to ICBMs. For more, see Jonathan Davis, "Export Controls in the People's Republic of China," Report of the Center for International Trade and Security, University of Georgia, Social Science Research Network, 2005, http://papers.ssrn.com/sol3/papers.cfm?abstract_id=1918230.

86. H. E. Xi Jinping, "Statement by H. E. Xi Jinping, President of the People's Republic of China at the Nuclear Security Summit," The Hague, Netherlands, March 25, 2014, http://se.china-embassy.org/eng/wjdt/t1140583.htm.

87. Hui Zhang and Tuosheng Zhang, "Securing China's Nuclear Future," (Cambridge, Mass.: Harvard Kennedy School, March 14, 2014), http://belfercenter.ksg.harvard.edu/files/securingchinasnuclearfutureenglish.pdf.

88. Pan Ziqiang et al., eds., *Management of Nuclear and Radiological Terrorism Incidents* (Beijing: Science Press, 2005).

89. Communications with Chinese nuclear experts, October 2011, July 2013, Beijing.

90. Liu Senling, "Status of Research on Nuclear Security Technology in CIAE" (presentation, Harvard/Peking University Workshop on Nuclear Security, Beijing, China, October 13–14, 2011).

91. Ibid.; Pan Ziquiang et al., eds., *Management of Nuclear*, 91.

92. For more see Jonathan Tirone, "Fukushima Shows Nuclear Terrorism Risks at UN Meeting," *Bloomberg*, July 3, 2013, http://www.bloomberg.com/news/2013-07-01/fukushima-shows-nuclear-terrorism-risks-at-un-meeting.html.

93. Communications with Chinese nuclear experts in Beijing, October 2012.

94. H. Zhang and T. Zhang, *Securing China's Nuclear Future*.

95. The Ministry of Public Security of the People's Republic of China detailed "the first determined lists of 'East Turkestan' terror organizations and terrorists" on December 15, 2003, http://www.mps.gov.cn/n16/n983040/n1988498/1988553.html.

96. Interviews with Chinese nuclear experts, October 2012 and August 2013.

97. For more, see H. Zhang and T. Zhang, *Securing China's Nuclear Future*; Interviews with Chinese nuclear experts, August 2013.

98. "Construction on China-U.S. Nuclear Security Center Begins," *Xinhua*, October 29, 2013, http://news.xinhuanet.com/english/china/2013-10/29/c_132841259.htm.

99. Zhu Zhixuan, "China's Nuclear Security and International Cooperation" (presentation at Harvard/Tsinghua Workshop on Nuclear Policies: Nuclear Energy and Nuclear Security, Beijing, China, March 14–15, 2010); China Foundation for International and Strategy Studies (CFISS), *Combating Nuclear Terrorism—Non-State Actors' Nuclear Proliferation and Nuclear Security* (Beijing: Social Science Academic Press, 2012).

3

India

The Nuclear Debate of a Rising Power

Deepa M. Ollapally and Rajesh Rajagopalan

Two startling historic events—the 1998 South Asian nuclear tests and the 2008 U.S.-India nuclear deal—ignited a domestic nuclear debate in India that had been more or less dormant since the country's first atomic test in 1974. Meanwhile, geopolitical rivalry with India's two main adversaries, China and Pakistan, has intensified. India's nuclear debate is particularly important given the country's unique position in the global nuclear order after the 2008 deal conferred an exception for India to remain outside the Nuclear Nonproliferation Treaty (NPT) and still enjoy access to international nuclear trade. With the ascent to power of the Nationalist-leaning Bharatiya Janata Party (BJP) in May 2014 after a gap of ten years, the direction of India's nuclear weapons strategy and nuclear energy policy deserves renewed attention.

The findings presented in this chapter challenge conventional wisdom and existing international relations arguments regarding Indian nuclear policy. India, like almost all the other countries studied in this book, is only partly shaping its nuclear decisions in reaction to the geopolitical environment. India's leadership refused to build nuclear weapons after China went nuclear in the early 1960s and decided to restart the weapons program reluctantly in the late 1980s after realizing the United States and the international community had failed to stop Pakistan from weaponizing. India's nuclear weapons program since 1998 also demonstrates that while India is building up capabilities to counter both China and Pakistan, it is at a far more sedate pace than many analysts feared. While not entirely ignoring international systemic pressures, India's slow nuclear weapons program suggests that much greater attention needs to be paid to domestic variables.[1] India's approach is being conditioned by these key domestic factors: rising bureaucratic pluralism, growing political pluralism, and entrenched civilian control over nuclear decision-making. In the Indian case, this has been in part the consequence of India's fairly comfortable international position that reduces structural pressures on strategic policy.

Our chapter traces government and strategic elite discourse on India's long-running nuclear program and then looks at how different actors view these three major issues: (1) nuclear energy development, (2) nuclear weapons and nuclear doctrine, and (3) engagement with nuclear nonproliferation regimes. We assess the balance of these debates—what we see as the emergence of a dominant Realist-Globalist coalition today—and project where the center of gravity of India's nuclear debate is likely to reside.

STAKING OUT THE SCHOOLS OF THOUGHT IN INDIA ON NUCLEAR ISSUES

The spectrum of opinion on nuclear issues has been widening in India since the mid-1990s.[2] Until then, policy perspectives could be broadly characterized as being either pro-nuclear (Nationalist) or anti-nuclear. Though there were differences within the pro-nuclear school—for example, some supported nuclear energy without necessarily supporting nuclear weapons—these were not very pronounced. Subsequently, however, the pro-nuclear school has split into what we characterize as the Nationalist, Realist, and Globalist perspectives as seen in table 3.1. The anti-nuclear school has largely remained the same, and we characterize them as the Leftists and Rejectionists to distinguish the basis of their opposition to nuclear technology, since their political objectives are not the same. Still, discourse on nuclear matters has traditionally been largely confined to the strategic and policy elite. Even within this narrow band, as an earlier work on India's nuclear debate points out, with the exception of a brief debate after the Chinese nuclear test in 1964, "nuclear policy did not surface in any serious, sustained manner as a topic for elite debate in the first four decades of India's independent existence."[3] In delineating the range of opinion, we refine some of our earlier work that laid out more general foreign policy schools of thought.[4]

The Nationalists emphasize autonomous and indigenous nuclear development, will not countenance any limits or international controls on the Indian nuclear program, and want a more expansive nuclear weapons and energy program. Hard Nationalists push for a greater deterrent capability and are not convinced by India's No First Use (NFU) policy. Nationalists as a whole opposed the U.S.-India nuclear deal because they argued that it will constrain the Indian nuclear program and be the equivalent of forcing India into the NPT through the backdoor. This group includes civilian strategists but also significant sections of India's atomic establishment and political and bureaucratic elites. They are found in both major political parties, Congress, and BJP, but are more strident when serving as opposition. Hard Nationalists find a more conducive environment within the BJP and allied newspapers like *The Pioneer*. For example, the BJP's 2014 election manifesto promised to "revise and update" India's nuclear doctrine, which many commentators took to mean that the BJP was planning to change India's NFU policy, though the party backed away from

Table 3.1. Schools of Thought in India's Nuclear Debate

	Pro-Nuclear				*Anti-Nuclear*	
	Nationalists	*Hard Nationalists*	*Globalists*	*Realists*	*Leftists*	*Rejectionists*
Nuclear Energy	Support more nuclear energy development	Support more nuclear energy development	Support based on energy markets	Support more nuclear energy development	Mixed, but oppose external constraints	Oppose
Nuclear Weapons	Support more expansive nuclear weapons program	Strongly support more expansive nuclear weapons program • Anti-NFU • More testing	Strongly oppose	Maintain limited nuclear weapons capabilities • Under civilian control • Pro-NFU • No more tests	Oppose	Strongly oppose
Constraints on Nuclear Programs	Oppose constraints or international controls	Strongly oppose constraints or international controls • Anti-CTBT • Anti-FMCT • Anti-NPT	Support constraints through arms control, nonproliferation, disarmament • Pro-CTBT • Pro-FMCT	Support constraints to serve broader strategic interests by tolerating CTBT and FMCT • Pro-disarmament in theory	Support constraints and disarmament but strongly oppose international controls as discriminatory	Support constraints and disarmament but strongly oppose international controls as discriminatory
U.S.-India Nuclear Deal	Opposed	Opposed	Mixed	Supported	Opposed	Opposed

it later.[5] Indeed, despite winning a clear majority in Parliament, there has been no indication of any change in India's nuclear doctrine so far.

The Realists emerged after the 1998 nuclear tests, though elements of their thinking can be traced back to the Comprehensive Nuclear Test Ban Treaty (CTBT) debate in India, when some of them suggested that India should test and join the CTBT. The Realists want India to maintain strong nuclear weapons and energy programs, but are willing to make compromises if this would serve larger strategic interests. While they firmly oppose the NPT unless India is allowed in as a weapons state—and many even support the idea of eventual global nuclear disarmament—they are willing to countenance other arrangements such as the CTBT and a fissile material cut-off treaty (FMCT). They would prefer an accommodation with the global nuclear regime and strongly supported the U.S.-India nuclear deal for this reason. Proponents are found among strategic thinkers at various think tanks, as media commentators and in academia, and not surprisingly among some important former government officials and diplomats.

The Globalists strongly support international arms control, nonproliferation efforts, and disarmament. They oppose India's nuclear arsenal, rejecting arguments that these weapons can be stable in South Asia, and suggest that India was motivated by reasons other than national security in its pursuit of a nuclear bomb. They were split on the U.S.-India nuclear deal, though most of them opposed it. A significant section of Indian business leaders, some respected public intellectuals, and a smattering of academics hold this view. On nuclear energy, they favor decisions to be made based on cost-benefit analysis and rely on international energy markets.

Among the anti-nuclear groups outlined in table 3.1, the Leftists are strongly opposed to nuclear weapons, but also to the NPT and nuclear nonproliferation efforts, and take strong exceptions to Western technology controls applied to the developing world.[6] They were vociferous against the U.S.-India nuclear deal. They tend to be opponents of both the Indian nuclear program and Western efforts to control it. India's left-leaning political parties, especially the Communist Party of India–Marxist (CPI-[M]), have carved out this position, which is shared by some independent public intellectuals, commentators, and newspapers like the left-leaning *Hindu*.

The Rejectionists strongly oppose any nuclear technology for any purpose. They support grassroots opposition to the Indian nuclear program and opposed the U.S.-India nuclear deal. However, unlike other perspectives, the Rejectionists are consistently anti-nuclear. For example, the key difference between the Rejectionists and the Leftists is that Rejectionists are opposed to any comport with nuclear technology, a position not shared by Leftists. The Rejectionists are found primarily in a few anti-nuclear movements such as the Coalition for Nuclear Disarmament and Peace (CNDP), Greenpeace India, and the Movement in India for Nuclear Disarmament (MIND), a small group of public intellectuals, as well as grassroots movements against specific nuclear-related installations such as the People's Movement Against Nuclear Energy (PMNE), which worked mostly to oppose the Kudankulam Nuclear Power Project in southern India, and the Movement Against Uranium Project (MAUP), which mostly works to generate local opposition to uranium mines.

This is not to say that individuals cannot hold multiple viewpoints. For example, a Nationalist with concern for the country's autonomy might still be very much in favor of global nuclear disarmament. A Realist too might be willing to consider this under certain circumstances. There is also comingling of Leftists and Rejectionists. Nevertheless, these categorizations can help one better understand shifts in India's multifaceted domestic nuclear debates.

INDIA'S NUCLEAR PROGRAM: EVOLUTION OF THE NARRATIVES

A number of factors went into the evolution of India's narratives on the nuclear program, the most important of which was undoubtedly the thinking of India's first prime minister, Jawaharlal Nehru, on nuclear weapons. There were also other important factors, including the impact of the bureaucracy, India's drive toward nuclear energy, and most recently, the U.S.-India nuclear deal. Each of these factors is discussed below. Over time, there has been a clear shift in India's nuclear debate from a dominant hybrid Nationalist-Globalist coalition to a more Realist-Globalist position following the liberalization of India's nuclear energy program since 2002.

Nehru and India's Nuclear Nationalism

India's nuclear program was never just about nuclear energy or nuclear weapons. Begun soon after independence in 1947, it was a symbol of India's postcolonial autonomy and a world-class technological achievement, thus a source of national pride.[7] At the same time, the Indian government adopted a position against the U.S.-Soviet nuclear bipolarity of the time and became a vocal proponent of nuclear disarmament. The pursuit of nuclear autonomy coupled with a strong disarmament commitment embodied what can be termed a seemingly incongruous Nationalist-Globalist divide in Nehru's thinking.[8] In 1954, India was the first country to propose a treaty that banned all nuclear weapons testing, a position with broad appeal in the country.[9]

Nehru gave virtually unconditional support to Homi Bhabha, his friend and the architect of the nuclear program. Bhabha, a nuclear physicist trained at Cambridge University, convinced Nehru that the technical capabilities offered through nuclear science would be important to lifting India from a poor third-world country to the ranks of more advanced industrial states. Nehru was particularly receptive to the idea of developing technologies that were seen as critical bridges to Indian self-reliance.[10] Bhabha himself favored an Indian nuclear weapons program more, though Nehru's personal views were strongly against nuclear weapons and instead preferred to build the competences India needed only in case it was eventually forced to go down the weapons path.

Nehru's influence continued after his death in 1964. India stood out for its refusal to sign what was viewed as a discriminatory NPT, while stridently arguing against nuclear weapons. Its opposition to the NPT was framed in terms of a demand for nuclear disarmament, a Nehruvian legacy, giving a Globalist cast to Indian foreign policy. When the country conducted a nuclear test in 1974, it went to great pains to describe it as a "peaceful nuclear explosion" (PNE), a term that carried some credence in the nuclear lexicon of the day despite the seeming disingenuousness.[11] India has continued to champion nuclear disarmament even after becoming an overt nuclear weapon power in 1998, calling for a nuclear weapons convention.[12]

Bureaucratic Primacy and Its Impact on Nuclear Thinking

India's nuclear energy program has developed under the control of a fairly concentrated bureaucracy. Bhabha had recommended to Nehru that the administration of the program be entrusted to a small, high-powered body that answered only to the prime minister. Thus was created the Atomic Energy Commission (AEC) in 1948 and the groundwork for an atomic energy program that has enjoyed top political support across the political parties.[13] The Department of Atomic Energy (DAE), established in 1958 and always headed by the prime minister, provided atomic bureaucratic leaders with complete autonomy in matters of staff, construction, and finance, free of other oversight. Even some supporters of nuclear energy have questioned this setup.[14]

At times even prime ministers found it difficult to deal with highly autonomous atomic energy officials. When Rajiv Gandhi came to power in 1984 with an economic liberalization agenda and tested the idea of changing government policy to import nuclear reactors to meet the country's rising energy demand and compensate for a lagging domestic nuclear energy production, he was met with resistance and could not proceed at the time.[15] The imports from an NPT state (the Soviet Union in this case) would have been financed through soft loans, but they would have also entailed increased safeguards on India's nuclear activities. As then head of AEC, Raja Ramanna described his reaction as:

> I put my foot down . . . . [It] was totally unacceptable to the BARC [Bhabha Atomic Research Centre] establishment. . . . The government maintained that imported reactors were being brought in only to supplement our efforts. With such simplistic arguments everything was thrown to the winds—self-reliance, morale, protection from unilateral safeguards, freedom from dependence on fuel, spare parts and so on. For the first time, I got the impression that the views of the AEC played a secondary role in the decision-making process and that the policies were formulated by a handful of bureaucrats, who thought they knew everything.[16]

While the DAE is the most important actor on nuclear issues, other departments are involved in certain specific issues. The military dimension of India's nuclear program is a spin-off of its civilian side, but control over the former has largely resided in the DAE and the political leadership. The Ministry of External Affairs, and in particular the Disarmament and International Security Affairs (DISA) division, is involved with all international aspects of nuclear energy and arms control issues. The Nuclear Power Corporation of India Limited (NPCIL) is in charge of nuclear power generation and all civilian nuclear power plants.

Since the 1990s, India's nuclear decision-making is taking place under greater bureaucratic and political pluralism. First, coordination between different Indian bureaucratic departments tends to be much more difficult than in other countries because the nuclear science establishment and Indian foreign service tend to operate in a competitive mode. India set up a National Security Council Secretariat (NSCS) in the late 1990s following similar institutional innovations in other states, but this does not appear to have made much difference in terms of coordinating security policy. Second, Indian politics have made coordination more complex. For almost a quarter century since the 1989 general elections, India has had either minority or coalition governments, though the 2014 general elections bucked this trend by giving a simple majority to the BJP. This rising political pluralism has posed challenges because of the lack of a larger political consensus on key nuclear issues such the U.S.-India nuclear deal.

Indian military forces are also involved in decision-making with regard to India's nuclear weapons, but only sporadically. The nuclear weapon cores are under the control of the DAE during peacetime and only transferred to the military if ordered by the political high command in times of war or crisis. The tight civilian oversight of India's nuclear weaponry and the military's secondary, if not negligible, role has been a consistent theme over time and the source of much angst in the Indian military.[17] The scientific establishment has been successful in defending their autonomy from the military brass in part because this arrangement fit the strong proclivity of political leadership to prevent the Indian military from having too much say in even national security decision-making, save for very narrow, professional ones, such as promotions within the services.[18] Overall, the result has been to privilege the thinking of nuclear scientists on nuclear matters, which has tended to be a hybrid of Nationalist (self-reliance) and to a lesser degree Globalist (universal disarmament) positions. With India's nuclear sector opening up more since 2002, the overweening influence of the nuclear bureaucracy has declined, but its opinions continue to matter. They played a significant role, for example, in setting the Indian position during the negotiations over the U.S.-India nuclear deal as well as the continuing debate over Indian nuclear liability law.

The Nuclear Energy Drive

Decisions about how to proceed with India's nuclear energy program were hardly guided by economics, or even later by aspirations for clean energy or international integration as a Globalist might espouse. Nor has the Indian energy program been up for international bargaining like some Realists might suggest. The dominance of the Nationalist worldview until into the 1990s was ironically one of the results of international technology sanctions placed on India after its first atomic test in 1974, something many in India came to view as unfairly targeting them since they lacked a nuclear patron, unlike Pakistan (with China) and Israel (with the United States). This further isolation made indigenization a necessity and elevated the country's nuclear scientists and administrators as critical guardians of India's high technology base and possible strategic assets. Their accomplishments became a source of national pride and prowess for most groups except the Leftists and Rejectionists, and questions of cost overruns, project delays, and lack of transparency were relegated to secondary concerns.

India's isolated status also provided the atomic establishment a rationale for falling short. The shift in Nuclear Suppliers Group (NSG) rules from requiring a country to submit to partial International Atomic Energy Agency (IAEA) safeguards to more intrusive, full-scope, IAEA safeguards for any nuclear commerce in essence meant that India would either have to completely open its nuclear program to the world or forsake nuclear commerce with the dominant NSG cartel. This strengthened the hands of the Indian nuclear establishment because no Indian political leader or group would tolerate these constraints, leaving the nuclear establishment in a much stronger position domestically despite its lackluster performance.

Nuclear power in India had begun with runaway ambition that appears naïve today, judging by what was eventually constructed. Bhabha and his associates forecasted reaching an installed capacity of 1,000 megawatts (MW) in a first generation of natural uranium fuel reactors by 1965; 1,600 MW with fast-breeder reactors using a uranium-233 thorium cycle by 1974; 5,000 MW by 1979; and 12,000 MW by 1984.[19] The 2014–2015 annual report of the Department of Atomic Energy puts India's current installed capacity as of 2015 at approximately 5,680 MW—just about the goal articulated for 1979.[20] According to the International Atomic Energy Agency, nuclear power accounted for only 3.5 percent of India's electricity generation in 2014.[21] The country's objective of reaching 14,600 MW nuclear capacity on line by 2020 and its aim to supply 25 percent of electricity from nuclear power by 2050 seem remote.[22]

Nuclear Energy Liberalization and the U.S.-India Deal

When India's isolation from nuclear commerce came to an end in 2008, it reflected a rising domestic challenge from both Realists and Globalists to the dominance of Nationalists in nuclear energy decision-making. The situation started to shift in 2002 when the United States held out the possibility of a civil nuclear energy agreement between the two countries, ultimately resulting in the U.S.-India nuclear deal or "123 Agreement" in 2008. Few foreign policy issues have evoked such passion, with the debate touching on deeper conflicts—in particular, India's relationship with the United States and India's nuclear autonomy.[23]

The debate over the U.S.-India nuclear deal supportively illustrates different schools of thought in India. Nationalists wanted complete Indian control over the full fuel cycle, including all aspects of enrichment and reprocessing (ENR) technologies. Realists, on the other hand, were willing to consider limitations on Indian ENR technology if such a compromise would bring other benefits, such as access to the technology control regimes. The Globalists were also willing to make such compromises, though this is not as much a strategy to get other benefits as it is to bring India under some international controls. The Leftists, like the Nationalists, opposed the deal primarily because they thought it would bring the two countries closer strategically. Though the Leftists were generally anti-nuclear, in this instance they were willing to join with Nationalists, including Hard Nationalists who were strongly pro–nuclear weapons, in order to try to defeat the nuclear deal. Finally, the Rejectionists were opposed to the deal because it would have led to the expansion of India's nuclear program. The debate resulted in some strange coalitions: Nationalists, Leftists, and even Rejectionists joining together against the nuclear deal, while Realists and Globalists supported it.

The main arguments against the deal were that over time India would lose its well-guarded strategic autonomy—a huge concern for Nationalists—and the United States could not be counted on to live up to promises made in the deal.[24] Of particular concern was the wording of the 123 Agreement that was seemingly ambivalent about India's freedom to reprocess spent fuel from reactors using imported uranium.[25] Another concern focused on India's needing to separate its civilian nuclear facilities for inspection by the IAEA, which, it was argued, will limit India's nuclear weapons capacity. A further complaint centered on the reliability of India's fuel supply given past experience with the United States abruptly stopping shipments to India's Tarapur reactor in the aftermath of the 1974 nuclear test. Rejectionists were the only ones who questioned the very basis of the desirability for nuclear energy.

The Left, part of the national ruling coalition, led the opposition in the Parliament and withdrew support from the government over the issue, though the government managed to survive.[26] The Left made common cause against the deal with hardline Nationalists from the BJP, a party the Left otherwise despised. Although strategic analysts were split, the most vocal were those Nationalists against the deal. The government also faced resistance from its own nuclear bureaucracy, especially BARC, AEC, and NPCIL, who feared a loss of power and autonomy in their operations and encroachment from foreign companies and international regimes down the road.

The Realist-Globalist counter arguments were advanced by Prime Minister Manmohan Singh's office with backing from Congress party leaders, several former diplomats, and strategic analysts, as well as key sections of the economic elite. They maintained the deal heralded the entry of India into the international club of nuclear weapon states without having to sign the NPT and that India would no longer be a nuclear pariah, suffering sanctions that crippled its nuclear program for almost five decades. One of the more popular arguments Prime Minister Singh himself articulated was the necessity of nuclear power for India's energy security.[27]

Though the Singh government managed to survive and successfully complete the nuclear deal with the United States, the issue continues to divide Indian perspectives on nuclear issues. The same Nationalist-Leftist-Rejectionist combination that opposed the deal came together later to oppose the government on a new nuclear liability bill.[28] They succeeded in passing a law holding manufacturers potentially liable for nuclear accidents, which has made both Indian and foreign vendors wary of participating in the Indian nuclear energy program, a point even the former head of the Indian nuclear program admitted.[29] Any attempt to change the liability bill will likely set off another debate, but it is unclear whether the earlier Nationalist-Left group will stay together, now that the BJP is in power. During President Barack Obama's visit to Delhi in January 2015, the governments announced what amounts to an administrative understanding on how they might get around the nuclear bill. However, important ambiguities remain, and it is not at all clear that American companies will do any business without a new legally binding bill.[30]

At the end of 2015, Indian nuclear energy interests received a boost from the successful conclusion of a seven-year-long negotiation between India and Japan for a bilateral civil nuclear agreement, opening up access to a new generation of reactors from Japan. This announcement by Prime Ministers Modi and Shinzō Abe at their summit in India also breathed new life into Realist aspirations for a tighter strategic link between India and Japan as a hedge against China. Remarks by both Modi and Abe made it clear that the civil nuclear agreement was not just about commerce and clean energy, but also a step toward enhancing mutual security cooperation.[31] Negotiations had been a struggle, given Japan's strong reluctance to engage in nuclear trade with any country that had not signed the NPT. According to one senior Indian official, "Having the Japanese on your side makes nuclear arrangements with other countries much easier."[32] It is too early to tell if this diplomatic breakthrough will translate into marked increases in nuclear energy generation for India or not.

THE STATUS OF ATOMIC BOMBS IN INDIA'S NUCLEAR DEBATE

The differences between the various schools of thought are illustrated most prominently when it comes to questions of nuclear weapons. The Nationalists see great danger in nuclear adversaries who will be deterred only by a robust Indian deterrent. As a result, this school proposes an expansion of both the quantity and quality of India's own atomic arsenal. Realists see some danger but want to maintain relatively limited capabilities, because they see nuclear weapons as a political, rather than military, tool and hold no value in matching others' capabilities. Globalists see other nations as reacting to Indian capabilities and seek a negotiated elimination—or at least a reduction of such capabilities—by all sides as well as confidence-building measures (CBMs) to reduce nuclear dangers. The Leftists and Rejectionists are opposed to Indian nuclear weapons and argue that India should eliminate its nuclear weapons and campaign for global nuclear disarmament, though they also oppose India's joining arms control and nonproliferation agreements such as the CTBT or the NPT, which are seen as symbols of Western imperialism.[33]

From debates about the kind of capabilities needed, these differences carry over into deliberations about nuclear strategy as well. Nationalists tend to be in favor of much tougher strategies, such as "massive retaliation," and they reject India's NFU doctrine and other "restraints" as nuclear impotence.[34] They want India to have clear nuclear supremacy over Pakistan and much more qualitatively superior capability vis-à-vis China. Hard Nationalists in particular call for greater control and leeway for the military over nuclear weapons. Realists, on the other hand, argue for a more careful nuclear strategy that emphasizes strong civilian control over nuclear weapons and credible but limited nuclear deterrent forces, and they strongly support an NFU.[35] As they oppose India's nuclear deterrent, the other three perspectives focus on the difficulties with existing nuclear strategies and the flaws of nuclear deterrence.

In the following section, we first consider the evolution of these views toward a Realist perspective, especially over the last two decades, and outline how this is reflected in issues of nuclear weapons capability, strategy, and multilateral diplomacy. The Realist view is now buttressed by the Globalist side to form the dominant combination in nuclear debates.

The Evolution of Indian Perspectives on Nuclear Weapons

As mentioned, the nuclear debate was not a very active one in the first several decades in India, but to the extent there was a debate, it was dominated by the pro-nuclear Nationalist-Globalist perspective (i.e., India should have a capability to build nuclear weapons, but not necessarily go ahead and build them immediately). It should also be noted that the nuclear debate was even thinner in the public realm, and so we focus a good part on the perspectives within the government rather than on the thinking outside.

There was little debate about whether India should build nuclear weapons in the first decade following independence in 1947. Though Prime Minister Nehru was aware that the Indian pursuit of nuclear technology also gave it the capacity to build nuclear weapons—and some of his actions in keeping the Indian program highly secretive has been at times interpreted as ambivalence toward the possibility of India's building nuclear weapons in the future[36]—much of his energy was spent campaigning for nuclear disarmament. Following India's humiliating defeat in the 1962 border war with China (which Indians and Nehru in particular viewed as a deep Chinese betrayal)[37] and China's atomic tests two years later, India became embroiled for the first time in a wide-ranging debate about its nuclear weapons option. Bhabha campaigned publicly for building nuclear weapons, even asserting in a speech to the IAEA that it would be difficult to follow a policy of restraint after the introduction of nuclear weapons in the neighborhood. However, the new Indian Prime Minister, Lal Bahadur Shastri, was so opposed to it that he apparently secretly appealed to the British government to supply him with cost estimates to disprove Bhabha's claim that a weapons program would not be expensive.[38]

Though the issue was debated in Parliament, a series of misfortunes—an India-Pakistan war, the death of both Shastri and Bhabha, and political and economic crises in India—ended the public debate. Nevertheless, it remained an important policy issue within the government, because of both India's active participation in the NPT negotiations as one way of resolving its nuclear dilemma, and the country's continuing search for security guarantees from the existing nuclear powers.[39] The powerful chairman of the AEC in this post-Bhabha period, Vikram Sarabhai, was not convinced India required nuclear weapons. He had important allies who shared this belief: the director of the newly created Institute for Defence Studies and Analyses (IDSA, a semiautonomous government think tank), Major General Som Dutt, and Army Chief General K. M. Cariappa. Thus, there is little evidence that India had any interest in a nuclear weapons program at the time.[40]

Despite heightened security anxieties after the 1962 border war and China's newfangled nuclear arsenal (see chapter 2 for details on Chinese nuclear proliferation), no precipitous action was taken. India's nuclear decision-making process stayed firmly in the hands of the civilian nuclear scientists and political leaders without devolving to the military as one might have expected at this point. This despite the fact that India's stunning defeat at China's hands was almost universally viewed as a result of strategic misreading and tactical miscalculation by the civilian defense minister and other political leadership and the near total lack of military input.[41] During debates in the mid-1960s, elements of a Realist argument made an appearance for the first time, led by K. Subrahmanyam, who saw flaws in the moralist Globalist stance on nuclear weapons: "The main difficulty is that all the criticisms advanced against India developing nuclear weapons are essentially sentimental and are, therefore, not always rational."[42] The military was still out of the picture in this debate. Among the political parties, only the Nationalist-leaning small Jan Sangh (precursor to today's BJP) urged the country to pursue nuclear weapons.

The most that the political leadership would countenance was a modest move toward research on subterranean explosives.[43] This subsequently led to explorations of a Peaceful Nuclear Explosion (PNE). A cost-benefit study commissioned by the AEC on PNE applications—such as massive construction, energy, mining, and canal projects—made public in 1971 concluded that PNEs made economic sense for India, while the development of nuclear weapons would not.[44] The 1974 test came a full decade after China's nuclearization, during which time China had conducted fifteen nuclear tests in the neighborhood.[45]

Following the 1974 test, India's political decision-makers slipped into a gray area in terms of nuclear posture, one that required neither explicit doctrine nor deployment of weapons. This supported the government's decision not to proceed with a weapons program after the test. The so-called "recessed deterrence" was particularly well suited to India's rather unique position of resisting the NPT but closely adhering to its spirit.[46] India's leadership believed that going nuclear would have been a betrayal of a cause and would hurt India's moral stature internationally. The dominant elite's normative distaste acting as one important brake on nuclear weaponization is fairly well recognized.[47] At the same time, they also liked the idea of "keeping the option open" while being able to express their continuing grievance against the unequal and unjust privileges for the five NPT Nuclear Weapon States.[48]

The paramount aim for Nationalists continued to be nuclear autonomy, with little interest in explicit security calculations. Besides, peculiar to India, the country's would-be bomb makers were mostly academically oriented South Indian Brahmin scientists with little interest in tainting their world-class technical achievements with military practicalities.[49] Realist critiques focusing on security and strategy began coming in various guises in the mid-1990s. One argument held that the credibility of India's "recessed" nuclear capability was being eroded due to the lack of follow-up testing. Others pointed out that India's testing restraint was going unrewarded without a quid pro quo benefit gained. More importantly, they contended that the very idea of global disarmament was more or less extinguished by the nuclear club after the unconditional and indefinite extension of the NPT in 1995. In this view, nothing was left to hold the club's "feet to the fire" to meet their part of the grand nuclear bargain, principally good faith efforts to negotiate an end to nuclear weapons globally. On top of all this, the outcome at the Conference on Disarmament (CD) on the CTBT ultimately placed India in the position of having to effectively give up the "nuclear option" if it signed the treaty. Ironically, time had run out for its "option" thanks to India's own reluctance to test.[50]

To Indian Nationalists, the lack of recognition by the international community of India's restraint, along with the hypocrisy of the nuclear powers after the 1995 NPT extension, was particularly grating.[51] According to Nationalists, India was practicing exemplary behavior on nonproliferation, yet it continued to be the target of unrelenting sanctions. The preamble of the Draft Indian Nuclear Doctrine (written after the 1998 test) is rife with consternation and criticism against the nuclear weapons powers for "virtual abandonment of nuclear disarmament" and the need for India to guard its autonomy "in a world where nuclear weapons for a select few are sought to be legitimized for an indefinite future."[52]

Despite India's overt nuclearization, the Draft Indian Nuclear Doctrine provided full civilian control over nuclear decision-making in a conflict. As Brajesh Mishra, then National Security Advisor, emphasized at the public release of the doctrine in 1999, "A cardinal principle regarding the use of nuclear weapons is that of civilian control" and "only the elected civilian leader of the country is empowered to authorize the use of nuclear weapons."[53]

Aftershocks from the 1998 South Asia Nuclear Tests

The split within the Nationalist-Globalist consensus became pronounced once India conducted the 1998 nuclear tests and declared itself to be a nuclear power. This was precipitated by the need to decide key questions that India had not fully addressed until this point, including the following: what kind of a nuclear force it would build; what strategy it would adopt to use its nuclear weapons; what kind of civil-military relations it would need to manage command and control issues; and what relationship it would seek with the existing global nuclear regime. Significant differences developed on all these questions, especially between the Nationalists and the Realists. Within the Nationalist group, too, divisions developed over these specific questions, with Hard Nationalists advocating more extreme positions than standard Nationalists. Hard Nationalists were a much smaller group and outside the corridors of power.

One key dispute was over the size of the nuclear arsenal. Hard Nationalists wanted a large nuclear force that would match at least the Chinese nuclear arsenal, which meant several hundred nuclear warheads and long-range missiles that could target virtually any other power. As one leading proponent of this position put it, "India is still some distance from acquiring the technical capacity to terminate a Chinese nuclear threat and needs to plug this vulnerability on a war footing. A deterrent against Pakistan was never the central mission of Indian nuclear strategy."[54] Other Hard Nationalists wanted India to develop a thermonuclear arsenal.[55] The Hard Nationalists also objected to India's unilateral moratorium on nuclear tests, because they argued that additional nuclear tests were required before India could be confident of its existing arsenal or could build more advanced warheads. Nationalists even within the atomic scientific community went along with the moratorium, expressing confidence in their test data collection and computer modeling capabilities.

The Realists, in contrast to the Hard Nationalists, argued that India needed only a minimum number of nuclear weapons, enough to provide for a secure second strike capability.[56] They exhibited little concern with the need for further testing and supported the government's testing moratorium; many even supported signing the CTBT in 1999, though this effort ultimately failed.[57] For Realists unsympathetic to the concerns of Hard Nationalists, nuclear weapons were not as much militarily useful instruments as they were useful political ones.

These differences were not limited to just the size of the arsenal, but also the strategy for employing nuclear weapons. The Hard Nationalists were always dismissive of the Indian posture toward the No First Use (NFU) doctrine, seeing it more as a peacetime posture that would not survive if war breaks out.[58] They see the NFU pledge as a continuation of India's irritating ambivalence toward nuclear weapons, which should have ended with India's overt weaponization. On the other hand, Realists strongly supported the Indian NFU pledge as a stable option that reduces the pressures on India in a crisis.[59] For Realists, India's nuclear strategy is a pragmatic recognition of India's conventional superiority and the difficult choices command and control faces with these dangerous weapons.

Realist views on the size, posture, and doctrine for nuclear weapons just described have held sway since 1998 despite the spectacular rise of China next door, with which India has an ill-defined, unresolved, and at times, tense, border dispute.[60] However, since 1998, China has become at least as equally important as Pakistan for Indian Realists, most notably in terms of delivery vehicles. Without acceding to the Hard Nationalist agenda on nuclear weapons, Realists have argued for gaining "minimum deterrence" against China's superior capabilities by stepping up the development of delivery systems capable of reaching China.[61] Increasing the range of Agni missiles thus has been a priority, but even here, internal bureaucratic pressures have also played a part.[62] Meanwhile, the Indian Navy's first indigenous ballistic missile nuclear submarine, the *Arihant* class, is still a test platform and expected to ramp up delivery capability only slowly.[63] Realists describe the overall India-China nuclear balance as one of stable nuclear deterrence and see little threat to this strategic stability.[64] The Realist dominance in China as described in chapter 2 and continuing commitment to minimum deterrence and NFU augur well for this bilateral nuclear balance.

Nationalists and many Realists, however, bristle at China's assistance to Pakistan in nuclear and missile technology affairs, but hardly anyone sees nuclear arms as the antidote to that nettlesome relationship. On the other hand, Pakistan's own moves on its nuclear posture in recent years to acquire tactical or "war-fighting" nuclear weapons, is posing a strategic dilemma for India.[65] India's conventional superiority is blunted while its NFU doctrine acts as a self-deterrent, a situation that Hard Nationalists abhor. Still, most Realists would continue to place their confidence in nuclear weapons as ultimately political instruments, calling for carefully managing the relationship with Pakistan, and do not seem to be in any hurry to change this view.[66] Simply put, on tactical versus strategic weapons, many Realists appear to hold the view that "a nuclear weapon is a nuclear weapon," and moreover, that Indian nuclear doctrine should not be allowed to be driven by Pakistan. Instead, they privately urge the Americans and international community to condemn Pakistan's move toward this destabilizing nuclear posture.

These debates are largely between the Hard Nationalists and the Realists, because the Globalists, Leftists, and Rejectionists—while disagreeing on other issues—all tend to be critical of these discussions and see them as unnecessary militaristic posturing.

NUCLEAR NONPROLIFERATION AND ARMS CONTROL

As with nuclear weapons issues and to a more limited extent on nuclear energy issues, there was a broad Indian consensus that tilted toward the Nationalist-Globalist position on issues having to do with nuclear disarmament and arms control well into the early 1990s. While this consensus included unrelenting opposition to the NPT and all its associated elements, it certainly did not translate into supporting the proliferation of nuclear weapons around the globe. India has consistently opposed the spread of nuclear *weapons* technology even as it backed the transfer of civilian nuclear technology as the legitimate right of all countries. In line with this slightly complicated position, it also insists that all countries that have signed the NPT have an obligation to live up to their commitments. These are all positions that have enjoyed strong support across the spectrum of Indian perspectives on nuclear issues.

Nevertheless, some differences on nonproliferation and arms control issues have begun to crop up, though these variances are not as great as the ones on nuclear weapons policy. Immediately after the 1998 nuclear tests, the Indian government sought to generate public support to change Indian policy on CTBT and sign the treaty. Since India had already tested and declared a unilateral test moratorium, Realists and Globalists argued there was little to lose by signing the CTBT and much to gain as India improved its relations with key global interlocutors. While Realists like K. Subrahmanyam and C. Raja Mohan and others supported the move as a pragmatic policy that would bring significant benefits, the Vajpayee government finally decided not to go down this path.[67] The lack of any major multilateral initiatives since the CTBT negotiations accounts for lack of further discordance between the schools of thought on these issues.[68] The Conference on Disarmament (CD), which is primarily responsible for negotiating arms control at the United Nations, has been deadlocked for close to two decades, and thus no major multilateral arms control issues have developed. On issues where India has taken a stand—against nuclear terrorism, on nuclear disarmament, and on other such issues—there have not been significant differences within the Indian strategic community.

The only issue on which some differences are visible is India's efforts to develop a closer relationship with the global nuclear regime in the aftermath of the U.S.-India nuclear deal and the NSG exemption. The Hard Nationalists have been somewhat suspicious of these efforts, because they worry that it might be a subterfuge to trap India into the NPT. On the other hand, even the Hard Nationalists support India's entry into nonproliferation technology control regimes such as the NSG and the Missile Technology Control Regime.

Differences could potentially emerge, however, if the CD deadlock ends and serious negotiations were to begin on a FMCT. Hard Nationalists can be expected to oppose India's signing a FMCT because that would limit the Indian fissile material capacity; Realists and Globalists could be expected to support such a deal, because they want India to have a small nuclear arsenal that does not require a large fissile material stockpile. On the other hand, it is not clear what India's official position on a FMCT would be either or whether India has even determined what its interests are. All that can be said with certainty is that India can be expected to join the treaty negotiations whenever they begin.

THE PUBLIC'S CLOUT IN INDIA'S NUCLEAR DEBATES

Opposition from civil society and grassroots organizations to India's nuclear weapons program has not been significant. Polls before the 1998 nuclear tests showed substantial support for the government's Nationalist-Globalist "nuclear option" orientation. A widely noted 1995 survey found that 57 percent of respondents wanted India to keep its nuclear options open, while only 33 percent favored an overt nuclear weapons capability and 8 percent supported the unilateral renunciation of the nuclear option.[69] In a poll conducted in six metropolitan areas immediately after the tests, 91 percent expressed support.[70] Over the subsequent months, however, the level of support declined.[71] Still, the Indian public gives a high level of importance to nuclear weapons (79 percent) in achieving their country's aims in the world.[72] It is essential to note that nearly an equal percentage holds that India's image in the world is very important to achieve national goals.[73] The positive public reaction to the tests did not keep the BJP from being voted out in 2004, with polls showing economic concerns at the forefront.

In contrast, on nuclear energy, local opposition has been rising alongside Rejectionist opinion becoming more vocal.[74] Opposition to nuclear power in India stems largely from threats to the livelihood of local populations such as fishing communities and farmers and less from safety and environmental concerns, though these issues have increased salience after the 2011 nuclear accident in Japan at Fukushima. Since the U.S.-India nuclear deal, efforts to establish four new sites for imported reactors met with local resistance.[75] Uranium mining in Jharkhand, Andhra Pradesh, and Meghalaya has also faced protests. A few national and international NGOs, along with Leftist political parties, have supported this opposition, especially by pursuing public-interest litigation in Indian courts. In the aftermath of the Fukushima disaster, public opinion studies have found that while a majority of Indians continue to support nuclear energy, the level of support dropped from earlier polls.[76] Government officials in turn seem to be showing somewhat greater receptiveness to safety and regulatory questions, especially in light of Japan's unexpected regulatory capture and the disastrous result (see chapter 4 for more on this point).

In 2011, the government introduced the Nuclear Safety Regulatory Authority (NSRA) bill in the Lok Sabha (Lower House), seeking to strengthen the regulatory framework currently implemented by the Atomic Energy Regulatory Board. The Modi government is expected to re-introduce the bill.[77] A major difference from existing practice is that the NSRA would be established by the Parliament and report to the Parliament, rather than to the Atomic Energy Commission. Arguments in favor of change are being made along the lines that while a Nationalist outlook might be rationalized for the early sanctions era, a much more Globalist view of the regulatory system is called for, especially as India seeks to expand its nuclear power base.[78]

India's failure to meet its own ambitious nuclear energy goals is ironically the result of the collective action of two traditionally divergent groups: Rejectionists who are resolutely anti-nuclear and pro-nuclear Nationalists who oppose reforming nuclear liability laws that constrict nuclear commerce. Rejectionist opposition is only likely to delay the program; the 2013 Supreme Court ruling in favor of the government on the Kudankulam nuclear plant ruled, "India cannot afford to be a nuclear isolated nation" in view of its huge electricity demand.[79] However, Nationalist opposition to amending the nuclear liability bill, and to a lesser extent the resistance of local communities, will in all likelihood delay the Indian nuclear power program as well as keep India from reaching the ambitious nuclear power goals that Nationalists and Realists support.[80]

CONCLUSION

We expect India's nuclear policy on both the energy and the weapons front to continue on the same trajectory as the decade 2005–2015. We identify several important trends in the nuclear arena. There has been a shift in the center of gravity on nuclear issues from a prolonged Nationalist-Globalist worldview, to a Nationalist-Realist orientation during the 1990s, and now to a more Realist-Globalist outlook over the past 2005–2015 decade. We see considerable continuity on nuclear weapons doctrine and strategy and only moderate change in nuclear energy policies. Thus when it comes to prospects for India's contribution to regional stability in the nuclear arena, writ large, we tend to be optimistic.

To the extent that there might be some changes, it is likely to come on the nuclear liability issue, on which it is possible there will be further tinkering to make it less onerous on suppliers, as indicated in January 2015.[81] Modifying the basic law itself is difficult for two reasons: (1) the current ruling party was the one that insisted on the hardest of the provisions in the nuclear liability bill while it was in the opposition;[82] and (2) even if Indian Prime Minister Narendra Modi shifted toward a Realist stand, the BJP as of 2016 does not have sufficient strength in the upper house of the Indian Parliament to carry it out.

We do not expect either the discourse or policy to change on nuclear weapons issues either. Even with Pakistan continuing to hew to a Nationalist outlook as chapter 9 concludes, it is hard to see any Pakistani action leading to major nuclear reassessments by India.[83] A substantial and sudden deterioration in relations with China could push the debate closer to a Hard Nationalist view. Despite the entreaties of the Nationalists, India's nuclear weapons development policy has been pragmatic and cautious, with slow but steady evolution in Indian capabilities.[84] Even with the BJP's coming to power, this is unlikely to change as there is a general policy consensus that leans toward the Realist side of the spectrum more than toward the Hard Nationalist side. In addition, the outlines of India's current path were set by the earlier BJP government (1998–2004) led by Atal Vajpayee, which gives it considerable political credibility. For example, despite hinting in its election manifesto about reconsidering India's nuclear doctrine, especially its NFU pledge, the BJP quickly backed away.[85] Civilian leadership will continue to dictate nuclear weapons policy, which places inherent limits on force posture and assures positions closer to those of Realists. The debate over the U.S.-India nuclear deal demonstrated that Nationalists are not completely spent, but the conclusion of the debate also showed the rise of the Realist strand.

NOTES

1. For works that discuss the role of domestic factors in India's nuclear development from different angles see, for example, George Perkovich, *India's Nuclear Bomb: The Impact on*

Global Proliferation (Berkeley: University of California Press, 1999); Vipin Narang, *Nuclear Strategy in the Modern Era: Regional Powers and International Conflict* (Princeton, N.J.: Princeton University Press, 2014); and Itty Abraham, *The Making of the Indian Atom Bomb: Science, Secrecy and the Postcolonial State* (New York: Zed Books, 1998).

2. During 2013 and 2014 in Delhi and Bangalore, the authors conducted over two dozen interviews with a cross-section of Indian strategic elites, policymakers, and former officials to assess viewpoints on a variety of nuclear issues.

3. Priyanjali Malik, *India's Nuclear Debate: Exceptionalism and the Bomb* (New Delhi: Routledge, 2010), 26–27.

4. Deepa M. Ollapally and Rajesh Rajagopalan, "India: Foreign Policy Perspectives of an Ambiguous Power," in *Worldviews of Aspiring Powers: Domestic Foreign Policy Debates in China, India, Iran, Japan, and Russia*, eds. Henry R. Nau and Deepa M. Ollapally (New York: Oxford University Press, 2012), 81–100.

5. Zachary Keck, "Is India about to Abandon Its No First Use Nuclear Doctrine?" *The Diplomat*, April 9, 2014, http://thediplomat.com/2014/04/is-india-about-to-abandon-its-no-first-use-nuclear-doctrine.

6. For a good statement of this somewhat paradoxical stance, see N. Ram, *Riding the Nuclear Tiger* (New Delhi: LeftWord Books, 1999).

7. Priya Chacko, "The Search for a Scientific Temper: Nuclear Technology and the Ambivalence of India's Postcolonial Modernity," *Review of International Studies* 37 no. 1 (January 2011): 185–208; Abraham, *The Making of the Indian Atom Bomb*.

8. Nehru's thinking has been variously described as being conflicted between Realist and Idealist. Without putting too fine a point on it, we think that Globalist is less normative, encompasses more attributes, and is more useful for comparing across cases. See Pratap Bhanu Mehta, "Still under Nehru's Shadow?: The Absence of Foreign Policy Frameworks in India," *India Review* 8 no. 3 (2009): 209–233.

9. United Nations, "Treaty Banning Nuclear Tests in the Atmosphere, in Outer Space and under Water," United Nations, October 10, 1963, http://www.un.org/disarmament/WMD/Nuclear/pdf/Partial_Ban_Treaty.pdf. In a discussion on disarmament in Parliament, one member of Parliament noted this level of consensus in the country. "As far as the question of suspension of tests of atomic and hydrogen weapons and their ultimate ban goes, there is the completest unanimity in the country . . . . [T]he unanimity covers not only the common people but also the Government." Bhupesh Gupta, "Rajya Sabha Debates," *Rajya Sabha*, May 24, 1957.

10. As Nehru noted in a speech, "The gap between the advanced and backward countries has widened more and more. It is only by adopting vigorous measures and by putting forward our utmost effort into the development of science that we can bridge that gap." Quoted in C. V. Sundaram, L. V. Krishnan, and T. S. Iyengar, *Atomic Energy in India: 50 Years* (New Delhi: Department of Atomic Energy, 1998), 255. The book is well-balanced despite the lead writer being an insider.

11. PNEs were being seriously discussed in IAEA and other scientific conferences.

12. Balasaheb Vikhe Patil, "Statement by Mr. Balasaheb Vikhe Patil, Member of Parliament and Member of the Indian Delegation at the 61st Session of the First Committee of the United Nations General Assembly on October 6, 2006," Permanent Mission of India to the UN, October 6, 2006, https://www.pminewyork.org/adminpart/uploadpdf/22238ind1264.pdf.

13. Deepa M. Ollapally, "India's Strategic Doctrine and Practice: The Impact of Nuclear Testing," in *India's Nuclear Security*, eds. Raju G. C. Thomas and Amit Gupta (Boulder, Colo.: Lynne Rienner Publishers, 2000), 71–72.

14. In a comparison with the United States, it was pointed out that the Atomic Energy Commission of India "combine [*sic*] in itself the function of the Atomic Energy Commission, the NASA, as well as such industrial giants such as Westinghouse, General Dynamics, Boeing etc. in the United States." K. Subrahmanyam, "The Path to Nuclear Capability," *IDSA Journal* 3 no. 1 (July 1970): 99–101.

15. Raja Ramanna, a central architect of the 1974 test and one of the only insiders to write a book on his work, emphasizes that "although we were prepared to receive assistance and advice from all friendly countries, foremost in our minds was the dictum that we were not to take orders from anybody in the choice of our programmes nor undermine our capability of tackling problems in-house." Raja Ramanna, *Years of Pilgrimage: An Autobiography* (New Delhi: Viking, 1991), 61.

16. Ramanna, *Years of Pilgrimage*, 109. The bureaucrats Ramanna was referring to were civil servants outside the nuclear bureaucracy. One year later, Gandhi was able to appoint Ramanna's successor, who was less resistant.

17. For a passionate military viewpoint, see Verghese Koithara, *Managing India's Nuclear Forces* (Washington, D.C.: Brookings Institution Press, 2012). See also the *Kargil Review Committee Report* that says that the Indian military's lack of involvement in nuclear decision-making puts the Indian military at a disadvantage vis-à-vis the Pakistan military, which is the custodian of Pakistan's nuclear weaponry. Kargil Review Committee, *From Surprise to Reckoning: The Kargil Review Committee Report* (New Delhi: Sage Publications, 2000), 241–242.

18. On the nature and implications of civil-military relations in India, see Anit Mukherjee, "Civil-Military Relations and Military Effectiveness in India," in *India's Military Modernization: Challenges and Prospects*, eds. Rajesh Basrur, Ajaya Kumar Das, and Manjeet Singh Pardesi (New Delhi: Oxford University Press, 2014).

19. Sundaram, Krishnan, and Iyengar, *Atomic Energy in India*, 16.

20. Department of Atomic Energy, Government of India, *Annual Report 2014–15* (Mumbai: Department of Atomic Energy, 2015), 43.

21. World Nuclear Association, "World Nuclear Power Reactors & Uranium Requirements," last updated and accessed on March 1, 2106, http://www.world-nuclear.org/information-library/facts-and-figures/world-nuclear-power-reactors-and-uranium-requireme.aspx.

22. "Nuclear Power in India," World Nuclear Association, last modified August 8, 2015, http://www.world-nuclear.org/info/Country-Profiles/Countries-G-N/India/.

23. For a detailed description of the internal dynamics in India on the nuclear deal, see Dinshaw Mistry, *The U.S.-India Nuclear Agreement: Diplomacy and Domestic Politics* (Delhi: Cambridge University Press, 2014), 151–75.

24. Brahma Chellaney, "One Man's Nuclear Brinkmanship," *Stagecraft and Statecraft*, July 21, 2008, http://chellaney.net/2008/07/20/one-mans-nuclear-brinkmanship.

25. The 123 Agreement on bilateral nuclear cooperation gives New Delhi prior consent to reprocess. However, it stipulates that this right would come into effect only when India establishes a new national facility dedicated to reprocessing safeguarded nuclear material under IAEA safeguards and reaches an agreement with the United States on "arrangements and procedures under which such reprocessing or other alteration in form or content will take place in this new facility." U.S. Department of State, "Agreement for Cooperation between the Government of the United States of America and the Government of India

Concerning Peaceful Uses of Nuclear Energy (123 Agreement)," Washington, D.C., August 3, 2007, http://2001-2009.state.gov/r/pa/prs/ps/2007/aug/90050.htm. For more, see Siddharth Varadarajan, "Reprocessing Request Is the First Test of Nuclear Deal under Obama," *The Hindu*, March 6, 2009.

26. A clear statement of the Left's opposition to the nuclear deal was made by its parliamentary leader. See Prakash Karat, "Why the CPI(M) and the Left Oppose the Nuclear Deal," *The Hindu*, August 20, 2007, http://www.thehindu.com/todays-paper/tp-opinion/why-the-cpim-and-the-left-oppose-the-nuclear-deal/article1895441.ece.

27. K. Subrahmanyam, "Why the Nuke Deal is Crucial," *Times of India*, April 7, 2008, http://timesofindia.indiatimes.com/edit-page/LEADER-ARTICLE-Why-The-Nuke-Deal-Is-Crucial/articleshow/2931066.cms. Furthermore, Singh said that energy security "was a significant theme of my talks [with the United States]." Manmohan Singh, "Lok Sabha Questions: Regarding Recent Official Visit of Prime Minister to the United States of America," Lok Sabha, July 29, 2005, https://www.indianembassy.org/archives_details.php?nid=566.

28. Sandeep Dikshit, "BJP, Left Parties All Set to Oppose Nuclear Liability Bill," *The Hindu*, March 15, 2010, http://www.thehindu.com/todays-paper/tp-national/tp-newdelhi/bjp-left-parties-all-set-to-oppose-nuclear-liability-bill/article728898.ece.

29. "Nuclear Liability Law Needs Clarity: Anil Kakodkar," *Times of India*, November 6, 2014, http://timesofindia.indiatimes.com/india/Nuclear-liability-law-needs-clarity-Anil-Kakodkar/articleshow/45053409.cms.

30. Associated Press, "India Nuke Deal Still Thorny for U.S. Despite 'Breakthrough,'" *Daily Mail*, January 27, 2015, http://www.dailymail.co.uk/wires/ap/article-2928279/India-nuke-deals-thorny-US-despite-breakthrough.html.

31. "The Indo-Japan Summit," *Indian Express*, December 12, 2015.

32. Quoted in Pramit Pal Chaudhuri "Done Deal: Why Nuclear Agreement with Japan Is Good for India," *Hindustan Times*, December 13, 2015.

33. Ram, *Riding the Nuclear Tiger*. For a collection of Leftist and Rejectionist perspectives, see M. V. Ramana and C. Rammanohar Reddy, eds., *Prisoners of the Nuclear Dream* (Hyderabad, India: Orient Longman, 2003); and Itty Abraham, *South Asian Cultures of the Bomb: Atomic Publics and the State in India and Pakistan* (Hyderabad, India: Orient Blackswan, 2010).

34. Perkovich, *India's Nuclear Bomb*, 458.

35. K. Subrahmanyam, "Triple Deterrent: India Must Stick to No-First-Use," *Times of India*, January 26, 2003, http://www.mea.gov.in/articles-in-indian-media.htm?dtl/14633/Triple+Deterrent.

36. Perkovich, *India's Atomic Bomb*, 18–21; Abraham, *Making of the Indian Atomic Bomb*, 48–54.

37. For a discussion on the depth of the feeling of betrayal, see Priya Chacko, *India's Foreign Policy: The Politics of Postcolonial Identity from 1947 to 2004* (New York: Routledge, 2012), 100–104.

38. Susanna Schrafstetter, "Preventing the Smiling Buddha: British-Indian Nuclear Relations and the Commonwealth Nuclear Force, 1964–68," *Journal of Strategic Studies* 25 no. 3 (2002): 87–108.

39. A. G. Noorani, "India's Quest for a Nuclear Guarantee," *Asian Survey* 7 no. 7 (July 1967): 490–502; Andrew B. Kennedy, "India's Nuclear Odyssey: Implicit Umbrellas, Diplomatic Disappointments, and the Bomb," *International Security* 36 no. 2 (Fall 2011): 120–153; Rajesh Rajagopalan, "Multilateralism in India's Nuclear Policy: A Questionable Default

Option," in *Oxford Handbook on Indian Foreign Policy*, eds. David M. Malone, C. Raja Mohan, and Srinath Raghavan (New Delhi: Oxford University Press, 2015).

40. Even strong nonproliferation proponents like Joseph Cirincione, president of the Ploughshares Fund and former director for nonproliferation at the Carnegie Endowment for International Peace, hold this view. See "Nuclear Weapons," Federation of American Scientists, November 8, 2002, http://fas.org/nuke/guide/india/nuke/; Joseph Cirincione, John B. Wolfsthal, and Miriam Rajkumar, *Deadly Arsenals: Tracking Weapons of Mass Destruction* (Washington, D.C.: Carnegie Endowment for International Peace, June 2002), 191–206.

41. There is near-popular consensus on the failure of the political class. For a detailed account of the failing of India's response to the Chinese invasion, see Rudra Chaudhuri, *Forged in Crisis: India and the United States Since 1947* (London: Hurst and Co., 2014), 88–101.

42. See K. Subrahmanyam, "Options for India," *IDSA Journal* 3 (July 1970): 105. India's best-known Realist was not in favor of any crash program.

43. Perkovich, *India's Nuclear Bomb*, 82–85.

44. Sundaram, Krishnan, and Iyengar, *Atomic Energy in India*, 179.

45. "The Nuclear Testing Tally," Arms Control Association, last update January 2016, https://www.armscontrol.org/factsheets/nucleartesttally.

46. "Recessed deterrence" was a term coined around India's ambiguous posture. See Devin Hagerty, "Nuclear Deterrence in South Asia: The 1990 Indo-Pakistani Crisis," *International Security* 20 no. 3 (Winter 1995–1996): 79–114; Jasjit Singh, "Prospects for Nuclear Proliferation," in *Nuclear Deterrence: Problems and Perspectives in the 1990s*, ed. Serge Sur (New York: United Nations Institute for Disarmament Research, 1993), 66.

47. For example, Jacques E. C. Hymans and George Perkovich are two noted outsiders who give credence to this view. For more, see Jacques E. C. Hymans, *The Psychology of Nuclear Proliferation: Identity, Emotions, and Foreign Policy* (Cambridge: Cambridge University Press, 2006), 195–203; Perkovich, *India's Nuclear Bomb*, 448–449.

48. Manjari Chatterjee Miller argues that a "post-imperial ideology" drove India's nuclear policy and its 1998 nuclear tests. For more, see Manjari Chatterjee Miller, *Wronged by Empire: Post-Imperial Ideology and Foreign Policy in India and China* (Stanford, Calif.: Stanford University Press, 2013), 82–105.

49. George Perkovich is among several observers who give importance to the background of India's nuclear scientists. For more, see Perkovich, *India's Nuclear Bomb*, 189.

50. India reportedly had at least one aborted test prior to Pokhran II under the P. V. Narasimha Rao government in 1995. Documents from the National Security Archive suggest that it was India's own reluctance that kept it from testing in 1995, rather than U.S. pressure as commonly assumed. For more, see William Burr, "U.S. Detected Indian Nuclear Test Preparations in 1995, but Photo Evidence was 'Clear as Mud,'" *National Security Archive*, February 22, 2013, http://www2.gwu.edu/~nsarchiv/nukevault/ebb412/.

51. Miller, *Wronged by Empire*, 82–105.

52. "Draft Report of National Security Advisory Board on Indian Nuclear Doctrine," Ministry of External Affairs, August 17, 1999, http://mea.gov.in/in-focus-article.htm?18916/Draft+Report+of+National+Security+Advisory+Board+on+Indian+Nuclear+Doctrine.

53. Brajesh Mishra, "Opening Remarks by National Security Adviser Mr. Brajesh Mishra at the Release of the Draft Indian Nuclear Doctrine," Ministry of External Affairs, August 17, 1999, http://mea.gov.in/in-focus-article.htm?18916/Draft+Report+of+National+Security+Advisory+Board+on+Indian+Nuclear+Doctrine.

54. Brahma Chellaney, "Bangs Turn into Whimper," *Rediff on the Net*, May 11, 1999, http://www.rediff.com/news/1999/may/11bomb1.htm.

55. Bharat Karnad, "A Thermonuclear Deterrent," in *India's Nuclear Deterrent: Pokhran II and Beyond*, ed. Amitabh Mattoo (New Delhi: Har-Anand Publications, 1999).

56. K. Subrahmanyam, "Capping Arsenal or Sour Grapes?" *Rediff on the Net*, March 25, 2006, http://www.rediff.com/news/2006/mar/24ks.htm.

57. C. Raja Mohan, "Towards a CTBT Consensus," *The Hindu*, November 9, 1999.

58. Brahma Chellaney, "Nuclear-Deterrent Posture," in *Securing India's Future in the New Millennium*, ed. Brahma Chellaney (Hyderabad, India: Orient Longman, 1999), 141–222.

59. Manpreet Sethi, *Nuclear Strategy: India's March towards Credible Deterrence* (New Delhi: Knowledge World, 2009), 135–141.

60. The most recent border standoff came just a day before the much-vaunted visit of Chinese President Xi Jinping to India in September 2014. While it cast a shadow on the visit, the two countries managed to quickly and diplomatically contain the incident.

61. In the estimation of Rajesh Rajagopalan, the number of warheads (or fissile material equivalent) is likely to grow at the same three to five per year as it has since the late 1980s.

62. Vipin Narang, "Five Myths about India's Nuclear Posture," *The Washington Quarterly* 36 no. 3 (Summer 2013): 143–157. India's Defense Research and Development Organization has been operating under internal pressure for performance reasons as well.

63. Amit R. Saksena, "Can India Accommodate the INS Arihant?" *The Diplomat*, January 26, 2015, http://thediplomat.com/2015/01/can-india-accommodate-the-ins-arihant.

64. See for example, Gurmeet Kanwal and Monika Chansoria, "Strategic Stability in South Asia" (presentation, Carnegie Endowment for International Peace, Washington D.C., July 23, 2015). Vipin Narang's study supports this view, concluding that Chinese and Indian nuclear postures promote stable deterrence. Narang, *Nuclear Strategy in the Modern Era*.

65. Pakistan's shift toward tactical nuclear weapons became evident after it tested the Nasr missile in 2011.

66. One example of an Indo-Pakistan, nuclear confidence-building measure that has stood the test of time is the 1988 agreement to not attack each other's nuclear facilities and the exchanging of facilities lists each year. The agreement was reached when Benazir Bhutto and Rajiv Gandhi were in power, during a brief period of hope for better ties under the "second generation" leaderships respectively.

67. For a good overview of this debate, see Gaurav Kampani, "CTBT Endgame in South Asia?" James Martin Center for Nonproliferation Studies, January 2000, http://www.bu.edu/globalbeat/southasia/Kampani0100.html.

68. Author interviews indicate a greater receptivity to India's participation in the CTBT and an FMCT from earlier Nationalist skeptics.

69. David Cortright and Amitabh Mattoo, eds., *India and the Bomb: Public Opinion and Nuclear Options* (South Bend, Ind.: University of Notre Dame Press, 1996).

70. The Indian Market Research Bureau conducted this poll. "Overwhelming Majority Supports N-Tests," *Times of India*, May 13, 1998, http://www.hindunet.org/hvk/articles/0598/0039.html.

71. Malik, *India's Nuclear Debate*, 2.

72. Rory Medclaf, "India Poll 2013," *Lowy Institute for International Policy*, May 29, 2013, http://www.lowyinstitute.org/publications/india-poll-2013.

73. Ibid.

74. Some outspoken prominent personalities include the author Arundhati Roy and historian Ramachandra Guha.

75. These are Jaitapur in Maharashtra, Kovvada in Andhra Pradesh, Mithi Virdi in Gujarat, and Haripur in West Bengal. See M. V. Ramana, *The Power of Promise: Examining Nuclear Energy in India* (New Delhi: Viking, 2012).

76. A BBC/GlobeScan opinion poll in 2011, after the nuclear accident at Fukushima, found that only 23 percent favored specifically building new nuclear infrastructure, representing a 10 percent drop since 2005. In the 2005 IAEA/GlobeScan poll, a plurality had favored more nuclear power to specifically combat climate change. Still, in 2011, according to an Ipsos Global @dvisor survey, 61 percent supported the use of nuclear power. For more, see "Opposition to Nuclear Energy Grows: Global Poll," *GlobeScan*, November 25, 2011, http://www.globescan.com/news-and-analysis/press-releases/press-releases-2011/94-press-releases-2011/127-opposition-to-nuclear-energy-grows-global-poll.html; "Global Citizen Reaction to the Fukushima Nuclear Plant Disaster," *IPSOS*, June 2011, www.ipsos-mori.com/Assets/Docs/Polls/ipsos-global-advisor-nuclear-power-june-2011.pdf.

77. Expectations that the government would introduce NSRA Bill 2015 did not materialize. See M. P. Ram Mohan and Els Reynaers Kini, "India's Nuclear Regulators Have Been Audited," *Hindu Business Line,* January 3, 2016.

78. Then–Environment Minister Jairam Ramesh, a proponent of nuclear expansion, suggested revisiting the idea of large-scale "nuclear parks" post-Fukushima. For more, see Ben Doherty, "Indian Anti-Nuclear Protesters Will Not Be Deterred," *Sidney Morning Herald,* April 23, 2011, http://www.smh.com.au/world/indian-antinuclear-protesters-will-not-be-deterred-20110422-1drcv.html. For a 2014 report on India's nuclear security, including the new bill, see Rajeswari Pillai Rajagopalan, *Nuclear Security in India,* (New Delhi: Observer Research Foundation, 2015), http://www.orfonline.org/cms/export/orfonline/modules/issuebrief/attachments/NUCLEAR_SECURITY_IN_INDIA-ORF_Report_1420632409098.pdf. For a 2015 perspective based on an IAEA peer review, see Aniruddh Mohan, "Nuclear Safety and Regulation in India: The Way Forward," *Observer Research Foundation Issue Brief* no. 96 (June 2015), http://www.orfonline.org/research/nuclear-safety-and-regulation-in-india-the-way-forward/.

79. Utkarsh Anand, "Supreme Court Dismisses Plea to Stall Commissioning of Kudankulam Nuclear Plant," *Indian Express,* May 8, 2014, http://indianexpress.com/article/india/india-others/supreme-court-dismisses-plea-to-stall-commissioning-of-kudankulam-nuclear-plant. At the same time, the Court directed the government to comply with a set of safety measures before the plant could be commissioned.

80. This was a recurring assessment offered by several commentators and former officials during author interviews. The installed capacity in 2012 accounted for a mere 1.1 percent. For more, see OECD/IEA, *IEA Energy Statistics, India: Share of Total Primary Energy Supply in 2012*, accessed online October 1, 2015, http://data.iea.org/statistics; "Nuclear Power in India," World Nuclear Association, last modified August 8, 2015.

81. Indrani Bagchi, "N-Liability Law Boomerangs, Spooks Domestic Suppliers," *Times of India,* September 22, 2014, http://timesofindia.indiatimes.com/india/N-liability-law-boomerangs-spooks-domestic-suppliers/articleshow/43118160.cms.

82. "Lok Sabha Passes Civil Liability for Nuclear Liability Bill," *Economic Times*, August 26, 2010, http://articles.economictimes.indiatimes.com/2010-08-26/news/27579949_1_nuclear-liability-bill-nuclear-damage-nuclear-incident.

83. Pakistan's reported moves toward theater nuclear weapons and war-fighting capabilities has been met with a fair degree of equanimity on the Indian side, though there is talk of going further on missile defense systems.

84. This is consistent with Vipin Narang's findings on India's force posture under strong civilian control. For more, see Narang, *Nuclear Strategy in the Modern Era.*

85. "India's Modi: No Change to 'No First Use' Nuclear Weapons Doctrine," *Reuters,* April 16, 2014, http://www.reuters.com/article/2014/04/16/us-india-election-nuclear-idUSBREA3F11220140416.

4

Japan

Tremors in the Nuclear Consensus after Fukushima

Mike Mochizuki

During the early twenty-first century, Japan experienced developments that shook the long-standing policy consensus regarding nuclear energy and nuclear weapons. The most dramatic shock was, of course, the Fukushima nuclear plant disaster in the wake of the March 11, 2011, Great East Japan Earthquake and tsunami. In addition, the North Korean nuclear test in 2006, followed by the failure of international efforts to prevent Pyongyang from pushing ahead with its nuclear weapons program, provoked some in the Japanese policy community to question Tokyo's three non-nuclear principles. Moreover, President Barack Obama's declaration in Prague of his support for nuclear disarmament, at a time when China was modernizing and expanding its nuclear arsenal as well as becoming militarily more assertive, raised concerns in Japan about the future credibility of U.S. extended nuclear deterrence, which had provided a compelling strategic condition for Japan to refrain from seeking a nuclear arsenal of its own.

How have these developments affected the debate within Japan about nuclear issues, and how has this changing debate influenced the trajectory of Japanese policy toward nuclear energy, nuclear nonproliferation, and nuclear weapons and deterrence? In addressing these questions, this chapter will have the following structure. First, it will discuss the nuclear debate and the resulting policy consensus that had existed prior to the March 2011 Fukushima disaster. Next, the chapter will examine how the Japanese nuclear debate changed after Fukushima. Finally, it will analyze how this changing debate has affected the politics and trajectory of nuclear policy in Japan by focusing on key choices the country now faces.

NUCLEAR POLICY CONSENSUS BEFORE FUKUSHIMA

Before Fukushima, Japan's nuclear policy stood on a political consensus involving a bargain between the vigorous pursuit of atomic energy on the one hand and the rejection of nuclear weapons on the other. This consensus was reflected in the "Four Pillars of Nuclear Policy" that Prime Minister Satō Eisaku enunciated in January 1968: (1) the three non-nuclear principles (not to possess nuclear weapons, not to produce nuclear weapons, and not to allow nuclear weapons into Japan); (2) hope for the abolition and destruction of nuclear weapons; (3) reliance on American nuclear deterrence in response to international nuclear threats while maintaining Japan's own defense capabilities to respond to conventional attacks; and (4) vigorous promotion of the peaceful use of nuclear energy.[1] This policy consensus represented an amalgamation of pacifism, the so-called "nuclear allergy," political and security realism, and techno-nationalism; but its emergence was by no means inevitable. Japan, as the only victim of atomic bombings, could have eschewed nuclear energy as well as nuclear weapons. Or given its nascent nuclear weapons research program conducted by the military during World War II, Japan could have eventually opted for nuclear weapons for the sake of national security or even international status. Explaining the path that Japan did pursue requires an examination of domestic debates and political processes.

Pursuit of Atomic Energy

During the Eisenhower Administration, the United States had a keen interest in helping Japan harness atomic power for electricity. As it pursued its "Atoms for Peace" program, Washington recognized the international symbolic significance of assisting the one country that had been the victim of U.S. atomic bombings.[2] But Japanese receptivity to nuclear energy was by no means assured. Although the Japanese government initiated a nuclear energy research program in 1954 and the National Diet passed the Atomic Energy Basic Law in 1955, public resistance to things nuclear remained widespread—especially after the crew of the Lucky Dragon fishing crew was exposed in March 1954 to radioactive fallout from the U.S. hydrogen bomb test near the Marshall Islands. A survey of the Japanese public taken in 1956 revealed that 70 percent felt that nuclear technology was harmful.[3]

The Japanese government, however, worked methodically and assiduously to convince the public to embrace atomic energy.[4] Spearheading this campaign was Shōriki Matsutarō, the media mogul who owned the Yomiuri newspaper and founded the popular Yomiuri Giants baseball team. After World War II, Shōriki was imprisoned by U.S. occupation forces as a suspected Class-A war criminal, but was later released. In 1955, after getting elected to the House of Representatives, he became a member of the Japanese cabinet with the responsibility of overseeing the development of nuclear energy. In January 1956, Shōriki was named the first chairman of Japan's Atomic Energy Commission, and several months later he became the first state minister in charge of the newly created Science and Technology Agency, which initially had primary jurisdiction over Japan's nuclear energy program.

Shōriki, with assistance from Nakasone Yasuhiro (later to become Japan's prime minister in 1982–1987), launched a campaign to mobilize public support for atomic energy. Nakasone at the time chaired the Atomic Energy Committee of the National Diet's lower house. A key part of this effort was a traveling exhibition, "The Peaceful Use of Nuclear Energy," sponsored by the *Yomiuri Shimbun* and supported by the United States Information Service (USIS). This public relations program paid off. By 1958, the percentage of those who saw nuclear energy as harmful dropped to 30 percent, compared to 70 percent two years earlier. Moreover, as a result of Shōriki's public effort to delink atomic energy and nuclear weapons, some influential leaders of the anti-nuclear weapons movement ended up embracing the use of atomic energy to improve the livelihood of Japanese citizens.[5]

With public opposition diminished, the Japanese government and industry collaborated with U.S. suppliers to implement an ambitious nuclear reactor construction plan. Emphasizing the importance of energy security for a resource-poor country like Japan, political and bureaucratic elites saw nuclear power as critical for reducing dependence on overseas energy sources. An ideology of techno-nationalism reinforced this motivation, and as early as the mid-1950s, Japan adopted a policy objective of eventually closing the nuclear fuel cycle through reprocessing. Although Japan's "resource problem" provided a Realist rationale for nuclear power, the need to import energy sources from abroad in itself did not necessitate the big push for nuclear energy. A number of countries that are also energy-resource poor and dependent on foreign supplies have refrained from pursuing the same path.[6]

In 1966, the first commercial nuclear power plant went on line. By1973 when the first oil crisis took place, nuclear power plants were generating 2.6 percent of Japan's electricity; and there were twenty nuclear plants completed or under construction. The twin oil crises of the 1970s intensified Japan's nuclear energy policy as a way to reduce reliance on oil, much of which came from the volatile Middle East region. In the twenty-year period from 1974 through 1993, Japan commenced construction on thirty-three additional nuclear power plants. By 2006, when all of these plants were in operation and a few more were completed, nuclear power accounted for 30.6 percent of the nation's electricity production, even though the total amount of electricity produced had increased by about 260 percent over the 1973–2006 period. Nine of the ten regional utility companies operated nuclear plants, with Okinawa Electric Power the one exception.

As mandated by Japan's Atomic Energy Commission in 1967, long-term nuclear energy plans included fast-breeder reactors and a complete nuclear fuel cycle that involved enrichment and reprocessing in Japan. Although plutonium reprocessing was more expensive than once-through burning of uranium, Japan's nuclear policy community viewed a closed fuel cycle as an effective way to enhance energy security by eventually reducing reliance on foreign suppliers of uranium—a finite resource that could become depleted. The Jōyō experimental fast-breeder reactor located in

Ibaraki Prefecture attained criticality in 1977, and the Monjū prototype fast-breeder reactor achieved criticality in 1994. The small Tōkai reprocessing plant in Ibaraki Prefecture began to reprocess spent fuel in 1981; and construction for a larger commercial reprocessing plant commenced in 1993 at Rokkashō in Aomori Prefecture.[7]

The nuclear energy policy consensus was reflected institutionally in a collusive network that came to known as the "nuclear village." This network consisted of career government officials (especially in the Ministry of International Trade and Industry and the Science and Technology Agency), pro-nuclear energy politicians, utility companies, nuclear vendors, and academics and media organizations supportive of atomic power.[8] At the national level, the main opponents of nuclear energy were the Socialist and Communist parties, labor unions, and umbrella organizations of citizen anti-nuclear groups. After the oil crises of the 1970s and the government's relentless arguments about Japan's energy vulnerabilities, however, the public at large tended to acquiesce to the government's vigorous pursuit of nuclear power. With the national anti-atomic power movement deflated, the main pockets of resistance were in the local communities that were being asked to host nuclear facilities. To overcome this local opposition, the government used what Daniel Aldrich has called "soft strategies and incentives." They included public information programs to convince residents that nuclear plants were safe, public hearings that were skewed in favor of plant construction, the promise of new job opportunities, generous subsidies to local and regional governments, and donations from electric utility companies for local education and elderly care programs.[9]

Nuclear Weapons Denial and Hedge

Japan's policy not to become a nuclear weapons state has been enshrined in the three non-nuclear principles, but this denial of nuclear weapons, like the pursuit of atomic power, was not inevitable. The three principles emerged in incremental steps in the context of a contentious debate.

During the 1955 Diet deliberations regarding the Atomic Energy Basic Law, Nakasone accommodated the views of the anti-nuclear movement by forging a government–opposition party agreement that nuclear fuel would not be used for weapons to kill or injure human beings. Confronted with opposition party criticisms of U.S. plans to deploy the nuclear-capable Honest John missile on Japanese territory, in February 1957 Prime Minister Kishi Nobusuke was compelled to state in the Diet that he would not accept the stationing of U.S. nuclear forces in Japan. These statements notwithstanding, the government led by the conservative Liberal Democratic Party (LDP) did not unequivocally and irreversibly abandon the nuclear weapons option.

Prime Minister Kishi Nobusuke (the maternal grandfather of Prime Minister Abe Shinzō) declared in May 1957 that Japan could possess nuclear weapons under the Constitution if such weapons were minimally necessary for self-defense. This interpretation, which has been reaffirmed repeatedly by top officials, leaves the constitutional door open for nuclear weapons.[10] And contrary to official statements regarding the denial of atomic weapons, prominent conservative leaders like Yoshida Shigeru, Ikeda Hayato, and Satō Eisaku as well as Kishi privately entertained the nuclear option not only to make Japan more secure, but also to enhance Japan's status and influence in the world. After China's first nuclear test in October 1964, Prime Minister Satō Eisaku startled President Lyndon Johnson during their January 1965 meeting by saying that if Chinese Communists possessed nuclear weapons, Japanese should also have them, and that although the Japanese public currently opposed this, they could be "educated" to allow such weapons.[11]

The different strands of statements regarding Japan's nuclear restraint were woven together during the December 1967 Diet deliberations about the terms for the return of the Ogasawara Islands to Japan and the implications for the prospective reversion of Okinawa. In response to persistent questions from Narita Tomomi, chairman of the Japan Socialist Party, Prime Minister Satō articulated the three non-nuclear principles; and the Diet eventually adopted a resolution in support of these principles in 1971. But as later revealed in a declassified U.S. State Department telegram, Satō's support of the non-nuclear principles was based more on political expediency than a firm belief. During a January1969 farewell meeting with outgoing U.S. Ambassador Alexis Johnson, Satō vented that the three principles were "nonsense."[12]

Satō's ambivalent attitude toward the non-nuclear principles was evident in his move in February 1968 to embed them in his four pillars of nuclear policy (mentioned at the beginning of this chapter). As his foreign policy adviser Professor Wakaizumi Kei noted, by offering the four pillars as a set, Satō suggested that the other three pillars (especially U.S. nuclear deterrence) might serve as a condition for maintaining the non-nuclear principles.[13] Soon after Satō announced the four pillars, he established a nongovernmental group of scholars to examine Japan's nuclear options. According to the 1968 and 1970 secret reports that came out of this process, U.S. extended deterrence over Japan and South Korea would continue to be effective even after China acquired the capability to strike the United States with intercontinental ballistic missiles. The study also concluded that even if Japan were to develop a small nuclear arsenal, the country, for reasons of geography and population density, would suffer overwhelming losses from a nuclear attack. Moreover, instead of enhancing national security, nuclearization could alarm the United States and lead to Japan's diplomatic isolation.[14] Although the original motivation for the three non-nuclear principles may have been to placate the leftist opposition parties and the anti-nuclear peace movement, the 1968 and 1970 secret studies provided a Realist rationale for Japan's rejection of nuclear weapons.

Security realism and political pragmatism shaped the way Japan addressed the Nuclear Nonproliferation Treaty (NPT). Although the NPT was open for signatures in July 1968, Japan did not sign the treaty until February 1970 and did not ratify the agreement until June 1976. Conservative critics of the NPT argued that the treaty discriminated between the have and have-not countries and might foreclose Japan's option to acquire nuclear weapons if a deteriorating security environment necessitated such a response. Critics from the leftist parties insisted that the NPT was not only discriminatory, but also inadequate for promoting the global abolition of nuclear weapons. Nuclear industry representatives, however, voiced their concerns that opting out of the NPT would jeopardize Japan's access to nuclear fuel.[15]

In the end, Realist and pragmatic calculations convinced the ruling LDP to ratify the treaty. The party concluded that even under the NPT, Japan could retain a nuclear hedge by developing and maintaining an advanced nuclear technological and industrial base that could be harnessed for nuclear weapons development if Japan's security environment deteriorated. Moreover, by preserving a nuclear hedge while avoiding diplomatic frictions with the United States over the NPT, Tokyo would give Washington a strong incentive to sustain extended deterrence to prevent a nuclear Japan. Finally, ratifying the NPT would help the LDP mollify the anti-nuclear movement and the leftist parties by reinforcing, through an international commitment, the policy to forgo nuclear weapons.

Two issues in the nuclear domain, however, complicated relations with the United States. One was the transit in Japanese waters of U.S. naval vessels that were likely to be carrying nuclear weapons on board. The controversy concerned the interpretation of the third non-nuclear principle of not allowing nuclear weapons into Japan. Anti-nuclear grassroots movements and leftist parties insisted that the third principle prohibited the entry of U.S. nuclear-carrying warships into Japanese ports and harbors. While Washington adhered to its "neither confirm nor deny" policy regarding the presence of nuclear weapons on U.S. naval vessels, the Japanese government tried to finesse the issue by claiming that since the U.S. had not notified Japan about nuclear weapons being aboard its naval ships under the "prior consultations" agreement of the bilateral treaty system, the transit and entry of U.S. naval ships in Japanese harbors and ports did not violate the three non-nuclear principles. But as later revealed in declassified documents, the Japanese government appeared to acknowledge to the United States that the non-nuclear principles did not prohibit the transit of U.S. ships that hypothetically carried nuclear weapons. Furthermore, as part of the U.S.-Japan diplomatic bargain regarding the terms of Okinawa reversion, in a secret memorandum, the Japanese government agreed to permit the reintroduction U.S. nuclear weapons to Okinawa during an emergency.[16] These secret understandings between Tokyo and Washington show how willing the Japanese government was to bend the three non-nuclear principles in order to maintain America's security commitment to Japan.

The other problematic issue related to Japan's plans for reprocessing and a closed nuclear fuel cycle. In the wake of the Indian nuclear test in 1974, the Carter Administration shifted U.S. policy against reprocessing because of the nuclear proliferation risks. This change alarmed Japan, which saw reprocessing and fast-breeder reactors as a possible long-term path to energy self-sufficiency. Having invested heavily in the Tokaimura reprocessing plant, which was about to make a test run, Japanese leaders were angry that Washington appeared to be treating Japan like a second-class country, while allowing European countries like the United Kingdom, France and West

Germany to continue their reprocessing programs. Moreover, U.S. policy imperiled one aspect of Japan's nuclear hedge: potential use of the Tōkai reprocessing plant to produce weapons-grade plutonium.[17] In the end, a compromise reached in August 1977 prevented a rupture in the U.S.-Japan alliance. This agreement included U.S. permission to Japan to operate Tokaimura on a two-year experimental basis in exchange for Japan's delay in constructing a plutonium conversion plant, as well as Tokyo's support for Washington's broader nonproliferation objectives and policies. Subsequently, the Carter Administration extended the above interim agreement three times, allowing Japanese reprocessing to continue. The Reagan Administration veered away from Carter's universalistic approach to nonproliferation and adopted a more flexible policy. In 1981 the U.S. lifted its restrictions on Tokaimura; and in 1988, a U.S.-Japan nuclear cooperation agreement permitted Japan to reprocess its U.S.-origin spent fuel for thirty years without the need for Washington's explicit approval.[18] Tokyo had succeeded in getting Washington to treat Japan, like the major nuclear energy states in Western Europe, as an exception to the general policy of discouraging reprocessing and a closed fuel cycle.

Fraying of the Nuclear Policy Consensus

The Japanese consensus to pursue atomic power and forgo nuclear weapons started to fray during the 1980s. After the 1979 Three Mile Island accident in the United States, Japanese public support for new reactor construction dropped. The Soviet Chernobyl disaster in 1986 reenergized the grassroots anti-nuclear movement in Japan. Then a succession of accidents in Japanese nuclear facilities further weakened public confidence in atomic power.[19] They included the 1978 and 1989 criticality accidents at the Fukushima Daiichi plant, the 1991 radiation leak at Mihama power plant, the 1995 sodium leak and fire at the Monjū reactor, the 1997 fire and explosion at the Tokaimura reprocessing plant, the 1999 malfunction of the fuel loading system at a Fukui Prefecture plant, and the 1999 criticality accident—that resulted in two deaths—at the Tōkai fuel fabrication facility. Reports of cover-ups and falsification of inspection records compounded public mistrust of Japan's atomic energy programs.

According to public surveys conducted by the Cabinet Office, support for an increase in atomic power fell from 57 percent in 1987 to 43 percent in 1999; and those favoring a decrease or abandonment of atomic power increased from 5 percent to 21 percent over the same period. The Japanese public also became more concerned about the safety of nuclear plants. In 1987, even after the Chernobyl disaster, 51 percent felt that Japanese nuclear plants were generally safe compared to 33 percent who believed them to be unsafe. But after the series of Japanese accidents during the 1990s, in the 1999 government survey 69 percent expressed anxiety about nuclear power compared to 25 percent who did not feel anxiety.[20] In this context, the Japanese government had to abandon its ambitious plan announced in 1986 for 120 nuclear plants by 2030.[21]

Regarding the policy consensus on nuclear weapons, challenges came from two directions. During the 1980s, U.S.-Soviet strategic nuclear parity and the Soviet deployment of SS-20 intermediate-range ballistic missiles (IRBMs) provoked some Japanese commentators to advocate Japan's acquiring its own nuclear weapons because of the weakening of U.S. extended deterrence. Then the end of the Cold War caused these voices to get louder. According to nuclearization advocates, the collapse of the Soviet Union would inevitably weaken the American will to defend Japan. Moreover, North Korea's clandestine nuclear program during the 1990s and then its nuclear and missile tests in 2006 further fueled pro-nuclear sentiment. In the 1950s and 1960s, conservative political leaders had quietly considered atomic weapons; but now it was no longer taboo for political leaders to discuss this option publicly.[22] In 2002, then Deputy Chief Cabinet Secretary Abe Shinzō harked back to his grandfather's 1957 constitutional interpretation by saying that Japan could acquire small nuclear weapons under the Constitution. Soon thereafter, then Chief Cabinet Secretary Fukuda Yasuo suggested to the press that Japan could move away from its three non-nuclear principles if international circumstances changed.[23] Both Fukuda and Abe would later become prime ministers.

The other challenge to the policy consensus came from nuclear disarmament advocates. They highlighted the contradiction between Japan's non-nuclear principles and support for nuclear disarmament on the one hand, and its reliance on U.S. nuclear extended deterrence on the other. According to a national survey conducted by the Japan Association for Public Opinion Research, the percentage of those who did not think that it was necessary to rely on U.S. nuclear weapons to protect Japan's security increased from 45.7 in 1985 to 54.7 in 2001.[24] In the context of the end of the Cold War, this view was not limited to anti-nuclear activists. For example, Kaneko Kumao, a career Japanese diplomat who served as director of the Foreign Ministry's Nuclear Energy Division, wrote in 1996 that Japan's defense did not need to be guaranteed with U.S. nuclear weapons. Insisting that conventional arms were sufficient for Japan's defense, he advocated Japan's taking the lead in establishing a "Nuclear Weapons-Free Zone in Northeast Asia."[25]

Despite the above challenges, the nuclear policy consensus remained largely intact. Successive governments reaffirmed the three non-nuclear principles, and Japan supported a permanent extension of the NPT, quickly signed in December 1998 the IAEA Additional Protocol to strengthen nonproliferation, and deepened its institutional commitment to the NPT regime.[26] Although opinion polls suggested that Japan had become more comfortable about openly debating nuclear options, public support for the non-nuclear principles remained robust and in fact tended to increase after the end of the Cold War. According to a November 2006 survey conducted by the conservative *Yomiuri Shimbun*, even though 74 percent were not optimistic that the six-party talks would resolve the North Korean nuclear problem, 80 percent supported maintaining the three non-nuclear principles.[27] At the same time, the Japanese security policy community continued to see the defense alliance with the United States and U.S. nuclear extended deterrence as essential to Japan's security, and Tokyo therefore took steps to enhance U.S.-Japan defense cooperation.

After the nuclear plant accidents in the 1990s, the so-called "nuclear village" ramped up its public relations campaign regarding the necessity and safety of atomic power. This effort appeared to revive public support for expanding nuclear energy. A December 2005 government survey showed that the majority of Japanese (66 percent) continued to be anxious about atomic power. Nevertheless, the percentage of those who favored an expansion (either assertively or cautiously) of atomic power recovered to 55 percent, compared to 43 percent in the 1999 survey.[28] And according to the October 2009 survey, support for an assertive or cautious expansion of atomic power rose to 59.6 percent—the highest level since these government surveys began in 1978.[29]

With public opinion shifting in a favorable direction, in June 2010 the Japanese cabinet led by the Democratic Party of Japan (DPJ) adopted a national energy plan that would increase the percentage of atomic power for electricity generation from 26 percent in 2007 to 50 percent by 2030. A big motivation for the DPJ to embrace atomic power was a desire to reduce dramatically Japan's greenhouse emissions, thereby helping to address the climate change problem. According to the 2010 Strategic Energy Plan, zero emission power sources (a combination of renewable and nuclear energy) for electricity would expand from 34 percent in 2007 to 70 percent in 2030. To meet this target, fourteen new nuclear plants would be built by 2030, and Japan would achieve the full nuclear fuel cycle, including the development of plutonium-thermal power and fast-breeder reactors.[30]

RECHARGED NUCLEAR DEBATE

By the eve of the March 2011 Fukushima nuclear disaster, Japan had become a vigorous proponent of a global nuclear renaissance to deal with climate change. It also tended to see President Obama's call for a world free of nuclear weapons as consistent with the three non-nuclear principles. In short, not only was the nuclear policy consensus reaffirmed domestically, but also this consensus now appeared to push Japan to the forefront of a global trend.

This heady moment for Japan, however, came to a crashing halt on March 11, 2011, with the Great East Japan Earthquake. While the Fukushima nuclear plant explosion and meltdown dissipated public support for an expansion of atomic power, acute concerns about Chinese assertiveness and military modernization and North Korean moves to develop a nuclear arsenal dampened the dream of global nuclear disarmament. The shock of Fukushima and anxieties about China and North Korea tore apart the nuclear policy consensus and triggered a new round of intense debate about nuclear issues. Although this new debate has numerous strands, it has tended to cluster around three general perspectives: (1) anti-nuclear Globalism; (2) pro-nuclear Nationalism; and (3) nuclear Realism. These schools of thought gravitate

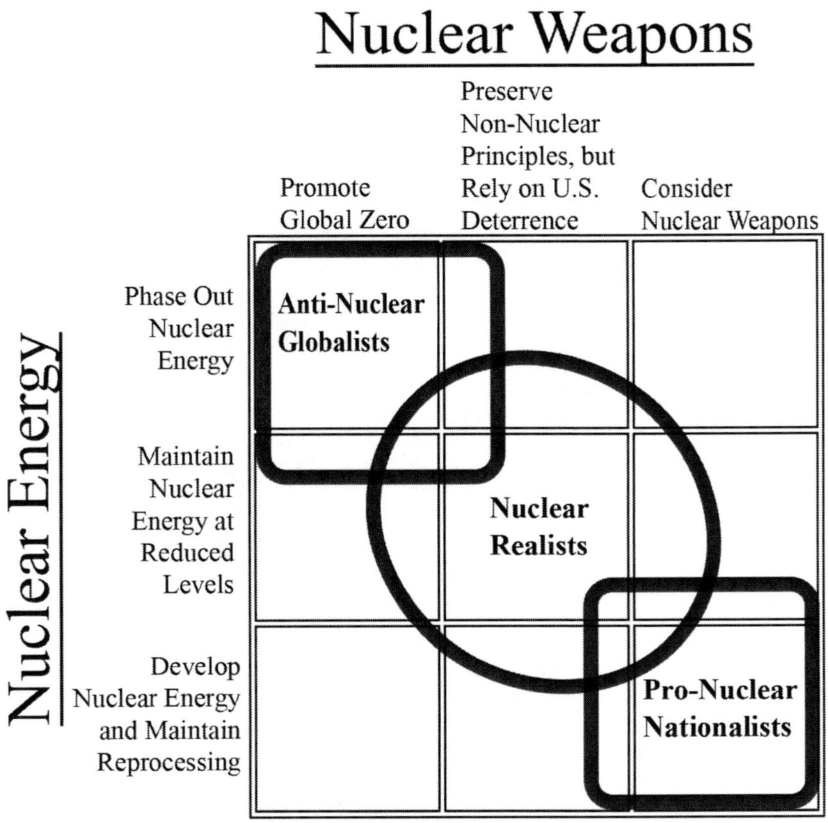

Figure 4.1.

around different policy preferences regarding nuclear energy and nuclear weapons, as shown in figure 4.1.

Anti-Nuclear Globalism

The opinion cluster of anti-nuclear Globalism encompasses those who seek to break apart the bargain that was struck in the 1950s between conservatives who promoted atomic power and progressives who opposed nuclear weapons. After Fukushima, these voices demand consistency in their nation's non-nuclear policy by not only forgoing nuclear weapons, but also by weaning from atomic power and promoting more actively global nuclear disarmament. In short, they advocate a "double zero": zero nuclear energy and zero nuclear weapons. In their view, the Fukushima disaster was a culminating event after a series of nuclear mishaps. They blame the collusive "nuclear village" for not heeding warnings from experts and for preventing the development of a truly independent regulatory regime to insure safety. But for them, more effective public regulation is not the answer. They argue that the safety of atomic power cannot be guaranteed in a country like Japan that is so vulnerable to earthquakes, tsunamis, and volcanic eruptions.

Several former prime ministers have joined the campaign to end nuclear power in Japan.[31] After coming out against nuclear energy, Koizumi Junichirō encouraged Hosokawa Morihiro to run in the Tokyo gubernatorial election in February 2014 on an anti-nuclear platform. Although former Prime Minister Hosokawa lost the race by a large margin to Masuzoe Yōichi who was endorsed by the LDP and Kōmeitō in the governing coalition, Hosokawa and Koizumi teamed up in May 2014 to form the Japan Assembly for Nuclear Free Renewable Energy to encourage citizens groups against nuclear power. Koizumi has charged that the notion of nuclear power being safe, cost-effective, and clean is a "blatant lie." He argues that Japan cannot secure disposal sites for radioactive waste produced by nuclear plants and that reactors are not safe because earthquakes, tsunamis, and eruptions occur all over the country. Therefore, Japan should not restart its nuclear reactors and should abolish them as early as possible.[32] Kan Naoto, the prime minister at the time of the Fukushima disaster, goes even further: "The lesson of the Fukushima atomic power accident is that atomic power, which artificially creates plutonium and other radioactive materials, cannot coexist with human beings."[33]

Because the renewed push for nuclear energy in 2010 was motivated in part by a Globalist alarm about climate change, Japanese opponents of atomic power wrestle with the challenge of how to meet national energy needs while decreasing carbon emissions. Some have advocated a "simple life" with an emphasis on energy conservation and slow growth.[34] By restraining energy consumption and promoting renewable energy sources, Japan could phase out nuclear energy without burning more fossil fuels. According to a plan promoted by a group of anti-nuclear energy political leaders, including former Prime Minister Kan, Japan would reduce its energy demand by 20 percent by 2030 through conservation and increase its reliance on renewables from 10 percent in 2010 to 50 percent in 2030. Under this scenario, Japan could then reduce fossil fuels in the nation's energy mix from 65 percent in 2010 to 50 percent in 2030.[35]

The above plan resonates with some analyses of mainstream economic think tanks in Japan. For example, Kobayashi Tatsuo, a senior economist at the respected Japan Center for Economic Research (JCER), reached this conclusion: "Making efforts in both adopting new forms of energy and energy saving would be cheaper and more economically advantageous than continuing with nuclear power." He argues that the government underestimates by half the cost of nuclear power generation if one factors in the risk of future accidents and subsidies for power plant sites. Moreover, energy volume from renewable sources and conservation "can substitute for the total output of all existing nuclear power stations prior to the Fukushima Daiichi accident." But dramatically increasing the use of wind and solar power and of new energy sources like geothermal, biomass, and low head hydropower will require "massive deregulation."[36] A follow-up study by the JCER reports that rising energy costs after Fukushima has spurred energy efficiency and conservation and even economic growth, and it projects that energy consumption in Japan could decline 40 percent by the year 2050.[37] The JCER studies and others like it lend credibility to the view that Japan can abandon atomic power without dire economic consequences or the need to embrace an ascetic "simple life."

Although many of the post-Fukushima, anti-nuclear power groups have sprouted up separately from existing pro-nuclear disarmament organizations, both sets of groups tend to draw their support from the same political and social constituencies. There is much overlap among the political and opinion leaders who oppose atomic power and those who oppose nuclear weapons. Advocates for abandoning nuclear energy are committed to the three non-nuclear principles and refrain from talking about a nuclear weapons hedge. They also emphasize that Japan's pursuit of reprocessing and a full nuclear fuel cycle undermines the global nonproliferation regime.

There are, however, some differences regarding how vigorously to promote nuclear disarmament. Some opponents of nuclear energy continue to recognize the importance of the so-called "U.S. nuclear umbrella" for Japan's security. But others believe that Japan should actively support initiatives to reduce the role of nuclear weapons for global security. For example, Hiraoka Hideo, who served briefly as Justice Minister in the Noda cabinet and has been a member of an anti-nuclear power caucus in the DPJ, has also been a vigorous proponent of a nuclear weapons-free zone in Northeast Asia. Hiraoka was also one of the 204 Japanese Diet members who signed a June 2010 letter to President Obama urging the United States to declare that the "sole purpose" of nuclear weapons is to deter other states from using such weapons against the United States and its allies and stating that Japan would not seek nuclear weapons if the United States were to adopt such a policy.[38] The equivalent of a "no first use" doctrine, such a policy would significantly circumscribe the role of U.S. nuclear weapons in deterring security threats against Japan. Akiba Tadatoshi, another opponent of nuclear energy, has gone even further. During the August 2010 anniversary ceremony of the Hiroshima atomic bombing, Akiba—then mayor of Hiroshima—implored Japan to reject the U.S. nuclear umbrella.[39]

Pro-Nuclear Nationalism

For pro-nuclear Nationalists, energy security (i.e., the stable supply of energy at affordable prices) requires continuing nuclear power generation. In their view, the increasing demand for oil and the growing geopolitical risks in oil-producing countries, along with the rise of resource nationalism, are intensifying resource competition. Although shale gas and renewables may be promising as energy sources, Japan must keep nuclear power in its national energy mix. But the pro-nuclear Nationalists also emphasize international status and national security. For example, the Nationalist think tank, Japan Institute for National Fundamentals (JINF), emphasizes that Japan has "the world's highest level of manufacturing technology for nuclear reactors." Therefore, Japan has an international obligation to maintain this technological edge and help to improve the safety of nuclear power plants in both developed and developing countries. According to the JINF, "To abandon nuclear energy generation as an emotional reaction . . . would jeopardize the international position of Japan in Asia and in the world." The JINF also wants Japan to move ahead with reprocessing and development of fast-breeder reactor technology. A full nuclear fuel cycle with plutonium use is essential for Japan's energy security because "proven recoverable reserves of uranium" are estimated to be 100 years at the current rate of production.[40]

Regarding nuclear weapons, pro-nuclear Nationalist views range from a more explicit "nuclear hedge" all the way to development of Japan's own nuclear arsenal. As a more explicit "nuclear hedge," some advocate that the Japanese prime minister inform the United States that Japan should also acquire nuclear weapons if North Korea is going to have them. Others point out that China's modernization and expansion of its nuclear forces and the development of the South China Sea as a possible bastion for its ballistic missile submarines weaken U.S. extended nuclear deterrence with respect to Japan. Therefore, Japan must retain a nuclear option and counter Chinese efforts to undermine Japan's access to highly concentrated plutonium.[41]

Nationalists who advocate nuclear weapons make the following arguments: They insist that global nuclear disarmament is a pipe dream since none of the great powers with nuclear weapons will give them up. Moreover, Japan will never have the international status and influence commensurate with its economic power unless it possesses nuclear weapons. They also see nuclear weapons as the most effective way to guarantee Japan's security. For example, Tamogami Toshio, who became a Nationalist pundit after being fired as Chief of Staff of Japan's Air Self-Defense Force, asserts that Japan cannot assume that the United States will defend Japan forever. With China's power rising, Japan must acquire its own nuclear arsenal to protect itself. According to Tamogami, if Japan had nuclear weapons, China would not be approaching the Senkaku Islands as it does now. He claims that world peace is preserved through a nuclear balance of power. In his twenty-year plan for Japan to become a nuclear weapons state, Tamogami notes how Japan's nuclear energy technology and reprocessing capabilities will help. As a step toward Japanese nuclearization, he recommends that Tokyo first ask Washington to include Japan, as it does NATO allies, in U.S. "nuclear sharing" policy. If the United States balks at this request, then it would demonstrate that Japan does indeed need to move toward a more autonomous defense policy.[42]

Nuclear Realism

The third opinion cluster is nuclear Realism, though both the anti-nuclear Globalists and the pro-nuclear Nationalists would bristle at the implication that their views are somehow unrealistic. Anti-nuclear Globalists would insist that the nuclear Realists are overly optimistic about the safety of nuclear plants in a country with

Japan's geological features and too sanguine about the challenges of nuclear waste disposal. Similarly, pro-nuclear Nationalists would argue that their support for a stronger nuclear hedge and even a plan for nuclear weapons development is consistent with certain forms of realism such as "offensive realism." But nuclear Realism may appear more realistic than the other opinion clusters because it tends to reflect views that come closest to the policy perspective that was mainstream before Fukushima.

According to nuclear Realists, the Fukushima nuclear accident occurred because of circumstances that were beyond imagination (*sōteigai*), and therefore does not fundamentally challenge nuclear power as a safe source of energy.[43] To prevent future accidents, they agree that Japan's regulatory regime must be more robust and independent of the nuclear power industry. They also support a careful review of existing plants to insure their safety and the retirement of older plants. But nuclear Realists argue that nuclear power must continue to be a significant part of the nation's energy mix. Abandoning atomic power would raise the cost of electricity, impede economic growth, decrease energy security, and prevent a significant reduction of greenhouse gas emissions. They question the viability of plans promoted by the anti-nuclear movement regarding a dramatic increase in renewables and energy conservation.

Among nuclear power advocates, there is some variation in terms of how much of Japan's electricity should be generated by nuclear plants. Ōmae Kenichi, a well-known international business consultant, calculates that about half of Japan's fifty-four nuclear plants could be restarted and generate about 15–20 percent of Japan's electricity demand.[44] Others advocate a higher percentage for nuclear power by constructing safer "third-generation plus" reactors like those being built in the United States and China.[45] The Institute of Energy Economics (IEE), which has close ties with the Ministry of Economy, Trade, and Industry's (METI's) Agency of Natural Resources and Energy, prefers that 25–30 percent of electricity be generated by nuclear plants in 2030.[46] With the retirement of old reactors, this projection would require constructing new plants.

Some nuclear Realists explicitly support continuing Japan's nuclear fuel cycle program. For example, Kaneko Kumao, who served as the first director of the Nuclear Energy Division in Japan's Foreign Ministry, insists that maintaining the nuclear fuel cycle program and building a second reprocessing plant at Rokkashō are essential to enable Japan to buttress the international nonproliferation regime as well as deal with the problem of nuclear waste. He argues that whereas China, Russia, and France are exempt from mandatory IAEA safeguards, Japan's fuel cycle program is under stringent IAEA inspections and safeguards. For Japan to abandon reprocessing would weaken the implementation of nonproliferation requirements in the global nuclear market because it would remove, according to Kaneko, "a faithful supporter" of the nonproliferation regime. Given many Asian countries will inevitably pursue nuclear energy programs, Japan could contribute to nonproliferation by becoming a regional nuclear fuel cycle center whereby spent fuels from other Asian countries would be reprocessed under tight IAEA safeguards and inspections in Japan.[47] In short, Kaneko turns on its head the anti-nuclear argument that Japan's fuel cycle program undermines nonproliferation.

While supporting the three non-nuclear principles, nuclear Realists stress the importance of U.S. extended deterrence, especially in the context of North Korea's development of nuclear weapons and ballistic missiles and China's military modernization and assertiveness. Although they might embrace the goal of global nuclear disarmament in the abstract, they are concerned that a reduction of U.S. nuclear weapons might weaken the nuclear umbrella. For example, former Japanese ambassador to the United Nations Satoh Yukio writes, "Japan's adherence to the Three Non-Nuclear Principles depends largely, if not solely, upon the credibility of the Japan-U.S. Security Treaty, or more specifically, that of the United States' commitment to defend Japan from any offensive action, including nuclear threats." While endorsing in principle the vision of a world free of nuclear weapons, Satoh warns that Japanese concerns about the credibility of extended deterrence could heighten if the United States unilaterally moved "to redefine the concept of nuclear deterrence and to reduce dependence upon nuclear weapons in providing deterrence."[48]

In addition to asking Washington to maintain its security commitment to Japan, nuclear Realists advocate that Japan enhance defense cooperation with the United States, invest more in ballistic missile defense, and upgrade its conventional air and maritime defense capabilities. Unlike the pro-nuclear Nationalists, the Realists refrain from supporting the development of nuclear weapons because such a move is likely to damage relations with the United States, as well as undermine Japan's civil nuclear industry. But some in this opinion cluster justify Japan's atomic energy programs as a way of preserving a "nuclear hedge." For example, Ishiba Shigeru, who served as defense minister and secretary-general of the LDP, stated that nuclear energy programs enable Japan to develop nuclear weapons quickly. During his tenure as the defense minister in 2012, Morimoto Satoshi echoed this view by supporting nuclear power because it helped deterrence and gave Japan a nuclear weapons option.[49] A "nuclear hedge" not only serves as a useful signal against Japan's potential adversaries (e.g., North Korea and China), but also gives the United States a strong incentive to keep robust its extended deterrence for Japan.

Public Opinion and Party Politics

After the Fukushima disaster, public support for nuclear power plummeted. In a Japanese government survey conducted in 2009, 59.6 percent of the respondents supported either active promotion (9.7 percent) or cautious promotion (49.8 percent) of nuclear power, while 18.8 percent favored keeping nuclear power at existing levels and 16.2 percent preferred abandoning nuclear power either in the future (14.6 percent) or as quickly as possible (1.6 percent). But according to a poll by the pro-nuclear power newspaper *Yomiuri Shimbun* taken in April 2011 soon after Fukushima, those who wanted to increase nuclear power dropped to 10 percent, while 46 percent supported nuclear power at current levels, 29 percent advocated a decrease, and 12 percent wanted a complete end to nuclear power.[50]

Opposition to nuclear power has been steady since then. A November 2014 survey by the public broadcasting corporation NHK showed that 30 percent favored ending nuclear power completely and 37 percent supported reducing reliance on this energy source. Only 21 percent supported maintaining atomic power at current levels, and only 3 percent thought that use of nuclear energy should be increased.[51] According to a February 2013 *Asahi Shimbun* survey, 46 percent opposed using nuclear power for electricity generation, while 37 percent supported this. Two years later public views had not changed. A February 2015 *Asahi* survey showed that 44 percent opposed nuclear power while 37 percent supported it. The public has been evenly split regarding the economic consequences of not restarting nuclear power plants. According to a July 2014 *Asahi* survey, 42 percent thought that not restarting the plants would have a negative effect, while 43 percent disagreed. Nevertheless, a September 2014 *Asahi* poll indicated that 57 percent still opposed the restarting of nuclear power plants while only 25 percent favored restarting them.[52] Public support for phasing out nuclear power has remained strong. According to a July 2015 *Asahi* survey, 16 percent support an immediate adoption of zero nuclear power and 58 percent favor a move to zero nuclear power in the near future. Only 22 percent opposed abandoning nuclear power.[53]

Opinion polls conducted by other organizations, however, suggest that public attitudes may be malleable. According to a November 2014 survey conducted by the Japan Atomic Energy Research Organization (JAERO), 31.5 percent felt that Japan can develop economically without nuclear power, while 18.5 percent disagreed. But 48.0 percent stated that they could not say one way or another, suggesting that some in this group could be persuaded to support atomic power for economic reasons. When JAERO asked about the effectiveness of nuclear power for stopping global warming, 32.1 percent thought nuclear power to be effective while 15.0 percent did not. But 51.8 percent could not say one way or another.[54]

Although Japanese citizens have become increasingly concerned about potential security threats to Japan, public support for the three non-nuclear principles continues to be remarkably robust. According to an *Asahi Shimbun* survey conducted in February-March 2014, 82 percent of the respondents favored keeping the non-nuclear principles, while 13 percent supported reexamining them.[55] This poll shows a slight increase in support for the three principles compared to a similar survey conducted by *Asahi* back in October 1988. In the 1988 poll, 78 percent supported maintaining the three non-nuclear principles, while 14 percent favored reexamining the principles and 8 percent gave either another or no response.[56] In short, despite the dramatic changes in the security environment from 1988 to 2014, public support for a policy of nuclear restraint has been resilient.

The above survey data indicate that after Fukushima public attitudes about nuclear power have shifted toward the anti-nuclear Globalism opinion cluster. The center of gravity of popular opinion now lies at the intersection of the anti-nuclear and nuclear Realist perspectives, thereby undercutting the pre-Fukushima policy consensus. By contrast, public opinion regarding nuclear weapons has not changed, and sentiments against nuclear power could even buttress popular support for the three non-nuclear principles. But how this change and continuity in public opinion will actually affect policy will depend on developments in party and electoral politics.

Despite the groundswell of public opposition to restarting nuclear reactors, the pro-nuclear energy LDP paradoxically defeated the ruling DPJ in a landslide during the December 2012 general election. Three factors explain this outcome. First of all, the LDP soft-pedaled its stance on atomic power. In its election platform, the LDP did not explicitly advocate nuclear energy, but rather stated that it would aim to establish an economic and social structure that would not have to rely on nuclear power and would decide about restarting all of the nuclear reactors within three years.[57] Second, the votes of citizens who opposed nuclear power were divided among at least four parties, which gave the LDP an advantage in the 300 single-member districts. Although the DPJ election manifesto declared that "all available policy resources will be mobilized to achieve zero dependence on nuclear power by the decade of the 2030s,"[58] the party's ambiguous stance while in power hampered its ability to attract the bulk of the anti-nuclear vote. After announcing in September 14, 2012, its commitment to do its utmost to abolish nuclear power plants during the 2030s, the Noda cabinet several days later appeared to walk away from this pledge after intense opposition from the business sector.[59] Finally, during the December 2012 election, voters tended to place greater priority on issues like economic policy, taxes, and welfare and social security rather than nuclear energy, and therefore they appeared quite willing to vote for the LDP, even if they did not support nuclear power.[60]

A similar dynamic operated in the July 2013 House of Councillors and December 2014 House of Representatives to yield decisive victories for the LDP despite the Abe government's support for restarting nuclear reactors. This was also the case in the February 2014 special gubernatorial election in Tokyo that attracted national attention because former Prime Minister Hosokawa ran on an anti-nuclear platform with a strong endorsement from former Prime Minister Koizumi. Although the Tokyo election was touted as an informal referendum on atomic power, Masuzoe Yōichi, who was endorsed by the LDP and Kōmeitō, sidestepped the nuclear issue by claiming that he supported reducing dependence on nuclear power. Masuzoe won the Tokyo governorship by a huge margin, and Hosokawa came in third by splitting the anti-nuclear vote with runner-up Utsunomiya Kenji.[61]

On the issue of nuclear weapons, compared to the general public, support for the three non-nuclear principles among members of the National Diet appears to have softened. For example, according to a *Mainichi Shimbun* survey of members of the House of Representatives, those who believe that "depending on the international situation, Japan should consider arming itself with nuclear weapons" increased from 7 percent after the August 2009 election to 29 percent after the December 2012 election. At the same time, Representatives who responded that "Japan should never consider possessing nuclear arms in the future" decreased from 87 percent in 2009 to 61 percent in 2012.[62] This shift, however, probably reflects the landslide victory of the conservative LDP that ousted the more centrist DPJ from power, as much as increasing concerns about Japan's security environment.

Japan's nuclear weapons option has never been tested explicitly at the election polls. Given the overwhelming public support for the three non-nuclear principles, no established political party has risked its electoral prospects by openly challenging these principles during an election campaign, even though individual politicians might personally favor considering the nuclear weapons option. One recent barometer for how nuclearization might resonate in the Japanese electorate is Tamogami's political career after he was fired from the Air Self-Defense Forces after writing a nationalistic essay on World War II. Tamogami, a proponent of the nuclear weapons option, ran in the February 2014 Tokyo gubernatorial election. He surprisingly received about 611,000 votes (12.4 percent of the votes cast), which was more than twice the predicted amount.[63] Some commentators saw this as evidence of rising nationalism, especially among the youth. During the December 2014 House of Representative elections, Tamogami challenged the incumbent Ōta Akihiro, a former leader of the pacifist Kōmeitō, in the 12th district of Tokyo. Frustrated that the Kōmeitō, as the LDP's coalition partner, has been the main brake on the Abe government's effort to develop a more robust defense policy, Tamogami sought to defeat Ōta by running as a candidate of the nationalistic leaning Party of Future Generations (*Jisedainotō*). Ōta won handily with 41.6 percent of the vote in the single-member district, and with 18.5 percent of the vote, Tamogami came in last out of four candidates. Tamogami's electoral fortunes indicate that there is a nontrivial group of voters who might be receptive to calls for Japan to exercise its nuclear weapons option, but at this point, this nationalist message is not close to making a major electoral breakthrough.

IMPLICATIONS FOR POLICY

How is the evolving debate about nuclear issues affecting concrete policy in Japan? The following section will examine four salient policy areas: (1) nuclear power and safety, (2) international promotion of nuclear energy, (3) reprocessing and the nuclear fuel cycle, and (4) nuclear weapons, deterrence, and disarmament.

Nuclear Energy and Safety

After the Fukushima disaster, the Japanese government led by the DPJ restructured the regulatory framework to separate "regulation" from "utilization" in an attempt to sever the collusive ties that had existed between regulators and the nuclear industry. The newly established Nuclear Regulation Authority (NRA), which replaced the ineffective Nuclear and Industrial Safety Agency (NISA), was granted more autonomy from government nuclear policymakers. The various regulatory activities that had heretofore been dispersed across multiple agencies were now concentrated in the NRA, which was housed in the Ministry of the Environment. NRA moved to strengthen the regulatory regime by developing countermeasures against severe accidents, introducing back-fit systems, establishing a forty-year operational time limit for nuclear reactor facilities, and enhancing a nuclear emergency preparedness system.[64]

Initially, the Japanese government allowed seventeen out of the fifty-four nuclear reactors to operate, but Prime Minister Kan Naoto declared in July 2011 that he did not believe that reactors could operate safely in Japan given the risks of earthquakes and tsunamis. By May 2012, all fifty reactors (which excluded the four reactors in Fukushima) ceased operation, and the government insisted on a two-step stress test to insure their safety before they could go back on line.[65] Japan had to make up the shortfall by increasing the use of fossil fuels for electric power generation from 60 percent in FY 2010 to 89 percent in FY 2012, with natural gas going up from 31 percent to 47 percent.[66]

After Kan's resignation in August 2011, Noda Yoshihiko succeeded him as prime minister and initially appeared to maintain Kan's anti-nuclear stance. In September 2012, the Noda cabinet announced a policy to phase out nuclear energy by the end of the 2030s. But as noted earlier, Noda backtracked after intense opposition from the business community. His METI minister, Edano Yukio, reversed the Kan government's decision to suspend new reactor projects by giving the green light on the construction of three new reactors.[67]

The DPJ electoral defeat in December 2012 brought the pro-nuclear LDP back into power, and Prime Minister Abe signaled his commitment to revive nuclear power to much applause from the utilities and the nuclear industry. With the government now unabashedly for nuclear power, the utility companies in July 2013 submitted applications to the NRA to restart ten reactors. In April 2014, the Abe cabinet approved a new Strategic Energy Plan that excluded any reference to phasing out nuclear power. Instead it only declared that dependence on nuclear power would be lowered to the extent possible through energy saving, renewable energy, and more efficient thermal power generation. While clearly stating that the government will proceed with the restarting of nuclear power plants, the report shied away for articulating the amount of electricity that would be generated by nuclear power.[68] In July 2015, however, METI issued its "Long-Term Energy Supply and Demand Outlook," which projected nuclear power would generate 20–22 percent of the electricity in the year 2030.[69] This projection would be considerably less than the 28.6 percent of electricity generated by nuclear plants before Fukushima.

For reactors to restart, they have to be approved by the NRA and local governments. In November 2014, two Kyushu Electric Sendai reactors in Kagoshima Prefecture cleared these two hurdles, and the first reactor began commercial operation in September 2015 and the second in November 2015.[70] There are now more than a dozen reactors in the queue for approval, but this is likely to be a slow process. Although the central government favors an early restart of most of the reactors, the NRA has autonomy in reviewing and approving the applications in order to insure safety. And then much depends on local politics. Moreover, even after NRA and local authority approval, the actual restart can be delayed or even blocked because of litigation brought by residents and anti-nuclear activists in local communities. For example, the NRA approved the restarting of two Kansai Electric Takahama reactors in Fukui Prefecture at the end of 2014. In April 2015, however, the Fukui

District Court cited safety concerns in issuing an injunction to block the restart of these reactors.[71] But after another presiding judge of this district court lifted the injunction in December 2015, one of Takahama reactors finally resumed operation in January 2016, and the second reactor restarted a month later.[72] Nevertheless, the path to commercial operations has not been smooth. A radioactive water leak and a malfunction near a transformer at one of the Takahama reactors not only prompted an emergency shutdown, but also has revived public concerns about safety.[73] Then in March 2016, the Ōtsu District Court of Ōtsu Prefecture, which borders Fukui Prefecture, issued an injunction to suspend operations of the Takahama reactors. The court questioned whether the nuclear power plant could withstand a powerful earthquake and tsunami comparable to the one that struck the region in 1586.[74] Therefore, although the anti-nuclear power movement suffered a major blow with the return of the LDP to power, it can still frustrate the government's plans by using the judicial system and mobilizing public sentiment against atomic power.

International Promotion of Nuclear Energy

Although the nuclear industry is embattled within Japan, it is very much an internationalized sector. All of the major Japanese nuclear companies have partnerships with foreign firms: Mitsubishi Heavy Industries with AREVA of France, Toshiba with Westinghouse, and Hitachi with General Electric. At the governmental level, the United States and Japan have promoted cooperation in the nuclear sector. In June 2006, President George W. Bush and Prime Minister Koizumi Junichirō issued a Joint Nuclear Energy Action to advance R & D cooperation in such areas as fast reactor and fuel cycle technology, nuclear waste management, and safeguards and physical protection. In November 2009, President Barack Obama and Prime Minister Hatoyama Yukio broadened the bilateral agenda to include the role of nuclear energy in promoting a clean energy economy. The two leaders also expanded R & D cooperation to include advanced simulations for enhanced seismic safety of nuclear plants, waste vitrification, development of sensors for under sodium inspection of liquid metal fast reactors, and gas-cooled reactor technology.[75]

Given the close relationship that had developed between Japan and the United States on nuclear energy, the Obama Administration became alarmed when the Noda cabinet initially announced a phasing out of nuclear power. U.S. officials warned their Japanese counterparts that a Japanese withdrawal from the nuclear business would jeopardize the improvement of nuclear safety, U.S.-Japan corporate ties, and the promotion of nuclear-related technology. They were so troubled that they even hinted that the United States might have to review the 1988 bilateral nuclear agreement that permitted Japan to reprocess spent fuel.[76] In short, U.S. external pressure worked in tandem with domestic lobbying from business leaders to steer the DPJ government away from abandoning nuclear power.

While domestic opposition complicated the restarting of reactors and put on hold an ambitious plan to construct more nuclear plants, the Abe government actively promoted nuclear cooperation with foreign countries and, as a consequence, overseas sales of nuclear plants and technology. Many of Japan's recent bilateral nuclear agreements were signed prior to Fukushima: with Russia in 2009, Kazakhstan in 2010, Jordan in 2010, and South Korea in 2010. Moreover, the agreements with Kazakhstan, Jordan, and South Korea, as well as the one with Vietnam (see chapter 7 for more on Tokyo-Hanoi nuclear cooperation), signed after Fukushima in 2011, were sealed under DPJ governments. But after his return to power, Prime Minister Abe quickly signed nuclear agreements with the United Arab Emirates and Turkey in 2013 and has engaged in serious negotiations with India toward the goal of a bilateral nuclear deal (see chapter 3 for more on Tokyo-New Delhi nuclear cooperation).[77]

The public, however, has been uneasy about exporting nuclear plants abroad while opposing the restarting of reactors in Japan. According to an April 2014 *Asahi* public survey, 51 percent opposed the Abe cabinet's policy of promoting the export of nuclear power plants to foreign countries, while only 29 percent supported this policy.[78] Ogata Sadako, the former U.N. High Commissioner for Refugees and one of the most respected Japanese voices in international affairs, expressed misgivings about helping to build nuclear facilities in foreign countries when Japan had not been able to manage such facilities safely at home.[79]

Closed Nuclear Fuel Cycle and Nonproliferation

Japan's pursuit of a closed nuclear fuel cycle, which appeared to promise a self-sustaining energy source, has been plagued with problems. First, the fast-breeder reactor program has suffered from cost overruns and delays, and its commercialization proved to be economically unattractive as uranium supplies became abundant and relatively inexpensive. Second, the Monjū fast-breeder reactor experienced serious accidents in 1995 and 2010, and the reactor has not been in operation for much of its twenty-year history. Because of these setbacks, the government in 1997 decided to use mixed-oxide (MOX) fuel, a mix of uranium oxide and plutonium oxide, in light-water reactors. According to preliminary government-business plans, this "plu-thermal" fuel option was to be utilized by sixteen to eighteen reactors by 2015, while the commercial use of fast-breeder reactors would not happen until at least 2050.[80]

The DPJ government's initial decision to phase out nuclear power by 2040 further complicated the fuel cycle program. If existing reactors were eventually going to be decommissioned and additional reactors would not be built, then how would Japan address the problem of disposing of spent fuel that would keep accumulating during the gradual phase-out period? The issue pitted the central government and Aomori Prefecture, the host of the Rokkashō reprocessing plant, against each other. Rokkashō village and Aomori Prefecture had heretofore agreed to store spent fuel from reactors on the condition that the fuel would eventually be reprocessed at Rokkashō. But if the government were moving away from nuclear energy and closing down Rokkashō, nearby local communities demanded that the spent fuel be removed from Rokkashō.[81]

The Japan Atomic Energy Commission in its spring 2012 review of the nuclear fuel recycling program identified three options: (1) reprocessing of all spent fuel, (2) a combination of reprocessing and direct disposal, and (3) direct disposal of all spent fuel. Neither the first nor the third option is attractive. Reprocessing all of the spent fuel would be too expensive and would contradict a policy of phasing out or at least reducing nuclear power. Direct disposal of all spent fuel raises the daunting problem of finding suitable disposal sites. Therefore, the commission suggested that the second option of a mix of reprocessing and direct disposal might be best, given its flexibility with comparatively fewer negative implications.[82]

Prime Minister Abe's intention to revive nuclear power has mitigated the domestic dilemma of Japan's nuclear fuel cycle program somewhat, but the long-term problem of disposing nuclear waste remains. Although its high cost makes reprocessing economically unattractive, the LDP-led government has decided to move ahead with the restart of the Rokkashō reprocessing plant. In reaffirming support for reprocessing and the use of retrieved plutonium, the 2014 Strategic Energy Plan articulated the objectives of "effective utilization of resources and reduction of the volume and harmfulness of high-level radioactive waste."[83]

This decision, however, could have negative implications for nonproliferation.[84] In the very least, it could encourage other countries, like South Korea, to insist that they should also be allowed to reprocess spent fuel to mitigate the disposal problem.[85] The other risk is a growing stockpile of plutonium. Japan already owns about 46 tons of plutonium of which 10 tons are currently stored in Japan, with the rest in the United Kingdom and France.[86] Although Rokkashō is under strict IAEA inspections and regulations, there are concerns that Japan has enough plutonium to develop thousands of nuclear warheads; and both nuclear Realists and pro-nuclear Nationalists in Japan recognize that this stockpile makes Japan's nuclear weapons hedge credible.

Japan's reprocessing program was dealt a severe blow in November 2015 when the NRA recommended that a new operator for the Monjū fast-breeder reactor be found to replace the existing operator, the Japan Atomic Energy Agency (JAEA). The NRA declared that JAEA "did not have the capability to operate [Monjū] safely." If the government cannot find a new operator that satisfies the NRA, then the nuclear facility could be decommissioned.[87]

Nuclear Weapons, Extended Deterrence, and Disarmament

Public support for the three non-nuclear principles remains robust with very little change over time. According to an *Asahi* opinion survey conducted in February 2014, 82 percent favor maintaining these principles.[88] This strong public support for forgoing nuclear weapons along with formidable domestic institutional barriers against nuclearization make it highly unlikely that Japan will become a nuclear-weapons state despite the uncertainties in the security environment.[89]

Nevertheless, because of the rise of China's and North Korea's nuclearization, the Japanese security policy community has become increasingly concerned about the credibility of U.S. extended deterrence. Therefore, while continuing to forgo nuclear weapons or even an explicit nuclear weapons hedge, the Japanese government reflects the dominance of Realist thinking in the way it soft-pedals support for global nuclear disarmament. To reassure Japan, the United States launched with Japan the bilateral "extended deterrence dialogue" (EDD).[90] This concern about extended deterrence initially caused Japan to be reluctant about signing the Joint Statement on the Humanitarian Consequences of Nuclear Weapons proposed by New Zealand, Switzerland, and other countries. Japan finally signed the statement in October 2013 after the phrase "all approaches and efforts toward nuclear disarmament" was included.

At the same time, a number of prominent politicians (e.g., LDP leader Ishiba Shigeru) and security policy experts (e.g., former Defense Minister Morimoto Satoshi) have talked openly of preserving a latent nuclear weapons capability through nuclear energy programs. There is also serious discussion within the LDP about the acquisition of offensive or counteroffensive strike capabilities against possible nuclear missile sites of potential adversaries. These statements notwithstanding, the Japanese government under Abe's leadership has focused on strengthening defense cooperation between Japan and the United States and relaxing somewhat the constitutional restrictions on such cooperation. By facilitating seamless coordination between Japan and the United States during a variety of contingencies ranging from peacetime to gray-zone tensions to high-intensity conflicts, Tokyo seeks to deter aggressive behavior by potential adversaries like North Korea and China.

There remains, however, a substantial gap between the nuclear Realism that is dominant in government circles and the anti-nuclear Globalism that galvanizes the public. Despite concerns about an increasing security threat voiced by Japanese security analysts, the public appears much more sanguine. For example, according to public surveys conducted by NHK, the percentage of Japanese who feel that U.S. nuclear deterrence or the so-called "American nuclear umbrella" is necessary for Japan's security has actually declined. In June 2010, the NHK poll found 20.8 percent believed that the American nuclear umbrella is necessary for Japan's security both at present and in the future, while 34.8 percent felt that it was unnecessary. According to the June 2015 NHK survey, only 10.3 percent stated that the American nuclear umbrella is necessary both now and later, while 48.9 percent responded that U.S. nuclear deterrence is unnecessary.[91]

CONCLUSION

Japan's nuclear policy has entailed forgoing the development of nuclear weapons and relying on U.S. extended deterrence on the one hand and vigorously pursuing a civilian nuclear energy program on the other. This policy stood on a political bargain and consensus that emerged over time and that represented an amalgamation of pacifism, an "allergy" toward nuclear weapons, political and security realism, concerns about energy security, and techno-nationalism. The pursuit of nuclear energy became institutionalized in a collusive network known as the "nuclear village," involving elite economic bureaucrats, conservative politicians, utility companies, nuclear vendors, and industrial consumers of electricity. The commitment to the three non-nuclear principles regarding nuclear weapons resulted from a contentious political and public debate about national security policy.

The consensus behind atomic energy frayed somewhat during the 1980s because of nuclear accidents in Japan as well as in the United States and the Soviet Union. Concerned about safety, local communities resisted hosting the construction of nuclear facilities. In response, the so-called "nuclear village" invigorated its public relations campaign to convince the public about both the economic necessity and the safety of nuclear power; and they softened local community opposition to nuclear plants through generous subsidies and the promise of steady jobs. Because of changes in the security environment, questions were also raised about the policy of forgoing nuclear weapons and relying on U.S. extended deterrence. But advocates of a reconsideration of the three non-nuclear principles and proponents of a more proactive nuclear disarmament policy tended to cancel each other out; and as a result, the non-nuclear principles remained intact. By the beginning of the 2010s, Globalist arguments about the challenge of climate change combined with Realist arguments about energy security to support a substantial expansion of nuclear-generated electricity. Moreover, a combination of Globalism and Realism steered Japan to embrace—albeit cautiously, as a long-term goal—President Obama's Prague Declaration of a world free of nuclear weapons.

The Fukushima disaster along with growing security concerns about North Korea and China reframed the nuclear policy debate. The Globalist rationale for more nuclear power in order to reduce greenhouse gas emissions lost its persuasive power, and anti-nuclear Globalism ascended with an agenda of abandoning nuclear power while continuing to eschew nuclear weapons. Nationalists, on the other hand, insisted that Japan needed to maintain a robust nuclear energy program to preserve its technological edge and a convincing nuclear weapons hedge. Some even supported taking concrete steps toward nuclear weapons development as a hedge against an aggressive China and a security retreat by the United States. Realists struck a middle ground by emphasizing nuclear regulatory reform and safety, a restart of reactors after stringent safety reviews, and continued reliance on U.S. extended nuclear deterrence while promoting seamless defense cooperation with the United States.

The center of gravity of this new debate was located between the anti-nuclear Globalists and nuclear Realists, with the pro-nuclear Nationalists the clear outliers. The sharp increase in public opposition to nuclear power compelled the Japanese government to revamp the nuclear regulatory regime, enhance safety measures, and subject all reactors to strict reviews before they could be restarted. Even proponents of nuclear energy were forced to acknowledge that Japan would inevitably depend less on nuclear-generated electricity and not meet the ambitious nuclear power targets articulated in June 2010. At the same time, despite widespread concerns about the proliferation risks of a closed fuel cycle, the government was unwilling to abandon reprocessing, in large part because of the local political complications this would pose for the storage of spent fuel. The public's anti-nuclear sentiment also reinforced Japan's commitment to the three non-nuclear principles, and to address the security challenges posed by North Korea and China, security Realists focused their attention on upgrading conventional defense capabilities and tightening the alliance with the United States, rather than seriously considering a nuclear weapons option.

NOTES

1. This chapter's author translated Japanese language works cited in this section. "'Hikaku san gensoku' ni kansuru Satō Shushō no Kokkai tōben" [Prime Minister Satō's answers in the National Diet regarding the three non-nuclear principles], University of Tokyo Institute of Oriental Culture, January 30, 1968, http://www.ioc.u-tokyo.ac.jp/~worldjpn/documents/texts/JPSC/19680130.O2J.html.

2. Benjamin K. Sovacool and Scott Victor Valentine, *The National Politics of Nuclear Power: Economics, Security, and Governance* (London: Routledge, 2012): 106–107.

3. Ibid., 109.

4. Richard J. Samuels, *The Business of the Japanese State: Energy Markets in Comparative and Historical Perspective* (Ithaca, N.Y.: Cornell University Press, 1987): 228–56.

5. Sovacool and Valentine, *National Politics of Nuclear Power,* 107–109, 119; Yuki Tanaka and Peter Kuznick, "Japan, the Atomic Bomb, and the 'Peaceful Uses of Nuclear Power,'" *Asia-Pacific Journal* 9, issue 18, no. 1 (May 2, 2011), http://japanfocus.org/-Yuki-TANAKA/3521.

6. Daniel P. Aldrich, "Networks of Power: Institutions and Local Residents in Post-Tōhoku Japan," in *Natural Disaster and Nuclear Crisis in Japan: Response and Recovery after Japan's 3/11*, ed. Jeff Kingston (London: Routledge, 2012): 132. On the historical roots of Japan's anxiety about resource vulnerabilities, see Eric Dinmore, *A Small Island Nation Poor in Resources: Natural and Human Resource Anxieties in Trans–World War II Japan* (Ph.D. dissertation, Princeton University, Princeton, N.J., 2006).

7. Eiichi Katahara, "Japan's Plutonium Policy: Consequence for Nonproliferation," *Nonproliferation Review* 5 no. 1 (Fall 1997): 56–57, http://cns.miis.edu/npr/pdfs/kataha51.pdf.

8. Akimoto Kenji, *Genshiryoku suishin no gendaishi* [Contemporary history of the promotion of atomic power] (Tokyo: Gendai Shokan, 2014): 183–208; Jeff Kingston, "Japan's Nuclear Village," *The Asia-Pacific Journal* 10, issue 37, no. 1 (September 10, 2012), http://www.japanfocus.org/-Jeff-Kingston/3822.

9. Daniel P. Aldrich, *Site Fights: Divisive Facilities and Civil Society in Japan and the West* (Ithaca, N.Y.: Cornell University Press, 2008): 120–132; Aldrich, "Networks of Power," 131–133; Sovacool and Valentine, *National Politics of Nuclear Power*, 121–123; Honma Ryū, *Genpatsu kōkoku* [*Atomic power advertisements*] (Tokyo: Aki Shobo, 2013).

10. Mike M. Mochizuki, "Japan Tests the Nuclear Taboo," *Nonproliferation Review* 14 no. 2 (July 2007): 305, http://cns.miis.edu/npr/pdfs/142mochizuki.pdf; *Bō'ei Handobukku 2014* [*Defense handbook 2014*] (Tokyo: Asagumo Shinbunsha, 2014): 627–629.

11. Taka Daitoku, "The Construction of a Virtual Nuclear State: Japan's Realistic Approach to an Emerging Nuclear Nonproliferation Regime, 1964–1970," in *Uncovering the Sources of Nuclear Behavior: Historical Dimensions of Nuclear Proliferation*, eds. Andreas Wenger and R. Popp (Washington, D.C.: Georgetown University Press, forthcoming); Kurt

M. Campbell and Tsuyoshi Sunohara, "Japan: Thinking the Unthinkable," in *The Nuclear Tipping Point: Why States Reconsider Their Nuclear Choices*, eds. Kurt M. Campbell, Robert J. Einhorn, and Mitchell B. Reiss (Washington, D.C.: Brookings Institution, 2004): 222.

12. "Peace Prize Winner Sato Called Nonnuclear Policy 'Nonsense,'" *Japan Times*, June 11, 2000.

13. Campbell and Sunohara, "Japan: Thinking the Unthinkable," 223.

14. Yuri Kase, "The Cost and Benefits of Japan's Nuclearization: An Insight into the 1968/1970 Internal Report," *Nonproliferation Review* 8 no. 2 (Summer 2001): 55–68, http://cns.miis.edu/npr/pdfs/82kase.pdf.

15. Etel Solingen, *Nuclear Logics: Contrasting Paths in East Asia and the Middle East* (Princeton, N.J.: Princeton University Press, 2007): 63–65; Maria Rost Rublee, *Nonproliferation Norms: Why States Choose Nuclear Restraint* (Athens: University of Georgia Press, 2009): 66–68; Daitoku, "Construction of a Virtual Nuclear State."

16. Ōta Masakatsu, *Nichi-Bei "Kaku" Dōmei: Genbaku, Kaku no kasa, Fukushima* [*Japan-U.S. "nuclear" alliance: Atomic bombs, nuclear umbrella, and Fukushima*] (Tokyo: Iwanami Shoten, 2014): 44–92; Yuki Tanaka and Robert Wampler, "Nuclear Noh Drama: Tokyo, Washington and the Missing Nuclear Agreements," *The National Security Archive* (Washington, D.C.: George Washington University, October 13, 2009), http://nsarchive.gwu.edu/nukevault/ebb291.

17. Katsuhisa Furukawa, "Japan's Nuclear Option," in *Over the Horizon Proliferation Threats*, eds. James J. Wirtz and Peter R. Lavoy (Stanford, Calif.: Stanford University Press, 2012): 18–19; and Sugita Hiroki, *Kenshō: Hikaku no sentaku* [Investigation: Non-nuclear choice] (Tokyo: Iwanami Shoten, 2005): 69.

18. Endō Tetsuya, *Nichi-Bei Genshi Ryoku Kyōtei (1988) no seiritsu kei-i to kongo no mondaiten* [The process for concluding the Japan-U.S. nuclear agreement (1988) and issues hereafter] (Tokyo: Nihon Kokusai Kenkyūjō, 2010); Motoya Kitamura, "Japan's Plutonium Program: A Proliferation Threat?," *Nonproliferation Review* 3 no. 2 (Winter 1996): 7–8, http://cns.miis.edu/npr/pdfs/kitamu32.pdf; Tomotsugu Shinsuke, "1970 Nendai no Beikoku kaku fukakusan seisaku to kaku nenryō saikuru seisaku" [U.S. nonproliferation policy during the 1970s and nuclear fuel cycle policy], *Ningen Kankyōgaku Kenkyū* 7 no. 2 (December 2009): 117–124; Charles S. Costello III, "Nuclear Nonproliferation: A Hidden but Contentious Issue in US-Japan Relations during the Carter Administration (1977–1981), *Asia Pacific: Perspectives* 3 no. 1 (May 2003): 1–5, https://www.usfca.edu/uploadedFiles/Destinations/Institutes_and_Centers/pacificrim/perspectives/docs/v3n1/app_v3n1_costello.pdf.

19. Peter Dauvergne, "Nuclear Power Development in Japan: 'Outside Forces' and the Politics of Reciprocal Consent," *Asian Survey* 33 no. 6 (June 1993): 581–583, http://www.politics.ubc.ca/fileadmin/user_upload/poli_sci/Faculty/dauvergne/AsianSurvey1993.pdf; Richard J. Samuels, *3.11: Disaster and Change in Japan* (Ithaca, N.Y.: Cornell University Press, 2013): 113; Aldrich, *Site Fights*, 139–145.

20. Nihon no Shōrai o Kangaeru Kai (NPO-IOJ), *Naikakufu Daijinkanbō Seifu Kōhōshitsu ni yoru "Genshiryoku ni kansuru yoron chōsa" no matome* [Summary of the "public opinion surveys regarding atomic power" by the Government Public Information Department of the Cabinet Office Secretariat], (Tokyo: NPO Innovation of Japan, January 6, 2011): 12–13, 16, http://ioj-japan.com/xoops/download/poll.pdf.

21. Kitamura, "Japan's Plutonium Program," 9.

22. Mochizuki, "Japan Tests the Nuclear Taboo," 312–314.

23. "Kakkoku to no shinrai ni kizu: 'Hikaku tenkan mo' Seifu shunō hatsugen" [Injury to trust with various countries: Government leader statements about "even a change in non-nuclear"], *Asahi Shinbun*, June 3, 2002; Howard W. French, "Taboo against Nuclear Arms Is Being Challenged in Japan," *New York Times*, June 9, 2002, http://www.nytimes.com/2002/06/09/world/taboo-against-nuclear-arms-is-being-challenged-in-japan.html.

24. "'Kaku no Kasa Fuyō' 55%: 'Genbaku no hi' mae ni zenkoku yoron chōsa" ["Nuclear umbrella unnecessary" 55%: National public opinion survey before "Atomic Bomb Day"], *Hiroshima Peace Media Center*, July 16, 2001, http://www.hiroshimapeacemedia.jp/abom/01abom/yoron/yoron.html.

25. Kumao Kaneko, "Japan Needs No Umbrella," *Bulletin of the Atomic Scientists* 52 no. 2 (March/April 1996): 48–50.

26. Mochizuki, "Japan Tests the Nuclear Taboo," 308–309; Rublee, *Nonproliferation Norms*, 69–76; Solingen, *Nuclear Logics*, 76–79; T. Ogawa, "Implementation of the Additional Protocol in Japan," IAEA-SM-367/6/06, International Atomic Energy Agency, http://www-pub.iaea.org/MTCD/publications/PDF/ss-2001/PDF%20files/Session%206/Paper%206-06.pdf.

27. Mochizuki, "Japan Tests the Nuclear Taboo," 307, 319.

28. Nihon no Shōrai o Kangaeru Kai (NPO-IOJ), 17.

29. Naikakufu Seifu Kōhō Shitsu, "Genshiryoku ni kansuru Tokubetsu Yoron Chōsa no gaiyō" [Summary of "Special Public Opinion Survey regarding Atomic Power"], Yoron Chōsa—Naikaku Fu, November 26, 2009, 2, http://survey.gov-online.go.jp/tokubetu/h21/h21-genshi.pdf.

30. Ministry of Economy, Trade and Industry, Japan, *The Strategic Energy Plan of Japan* (Summary), Ministry of Economy, Trade and Industry, June 2010, http://www.meti.go.jp/english/press/data/pdf/20100618_08a.pdf. See also John S. Duffield and Brian Woodall, "Japan's New Basic Energy Plan," *Energy Policy* 39 no. 6 (2011): 3741–3749.

31. They include Prime Ministers Hosokawa Morihiro (August 1993–April 1994), Murayama Tomiichi (June 1994–January 1996), Koizumi Junichirō (April 2001–September 2006), Hatoyama Yukio (September 2009–June 2010), and Kan Naoto (June 2010–September 2011).

32. Izumi Sakurai, "Koizumi: It's Up to Abe to Make the Move toward a Nuclear-Free Society," *Asahi Shimbun Asia & Japan Watch*, November 12, 2013, http://ajw.asahi.com/article/behind_news/politics/AJ201311120095; "Anti-Nuke Crusade Symbolizes Paradigm Shift in Cooperation," *Asahi Shimbun Asia & Japan Watch*, May 9, 2014, http://ajw.asahi.com/article/views/vox/AJ201405090024; "Koizumi Points to Mt. Ontakesan Eruption as Another Reason to End Nuclear Dependence," *Asahi Shimbun Asia & Japan Watch*, September 30, 2014, http://ajw.asahi.com/article/behind_news/politics/AJ201409300025.

33. Kan Naoto, "'Genpatsu zero' no ketsui" [Decision for "Atomic Power Zero"] (Tokyo: Nanatsumori Shokan, 2014): 3.

34. Samuels, *3.11: Disaster and Change in Japan*, 128–130.

35. Kan, "'Genpatsu zero' no ketsui," 58–59.

36. Tatsuo Kobayashi, "Energy Saving and Renewable Energy Development Less Costly than Sticking with Nuclear Energy," Japan Center for Economic Research, December 27, 2011, http://www.jcer.or.jp/eng/research/pdf/pe%28kobayashi20120130%29e.pdf.

37. Tatsuo Kobayashi et al., "Energy Consumption Down 40% in Fiscal 2020 by Economic Structural Changes: Energy Conservation Is a Growth Engine," Japan Center for

Economic Research, November 4, 2014, http://www.jcer.or.jp/eng/research/pdf/141104_policy%28eng%29.pdf.

38. Masami Ito, "Diet Members Send Obama Nuclear Letter," *Japan Times*, February 20, 2010, http://www.japantimes.co.jp/news/2010/02/20/national/diet-members-send-obama-nuclear-letter.

39. "Hiroshima to Reject Nuclear Umbrella," *Asahi Shimbun*, August 3, 2010, http://www.asahi.com/english/TKY201008020255.html.

40. "Japan Should Not Abandon Nuclear Power Generation," Japan Institute for National Fundamentals: Policy Proposal, November 11, 2011, http://en.jinf.jp/suggestion/archives/1036.

41. Tomiyama Yasushi, "'Nippon ni mo kaku opushion ari' to ittara yoi," [Good to say "Japan has the nuclear option, too"], Chokugen—Kokka Kihon Mondai Kenkyūjō [Plain Speaking—Japan Institute for National Fundamentals], February 18, 2013, http://jinf.jp/weekly/archives/9958; Yuasa Hiroshi, "Kaku no opushion wa hōki dekinai" [Nuclear option cannot be abandoned], Chokugen—Kokka Kihon Mondai Kenkyūjō, March 31, 2013, http://jinf.jp/weekly/archives/12496.

42. Tamogami Toshio, *Nippon kaku busō keikaku: Shin no heiwa to jiritsu no tame ni* [Plan for Japan's nuclear armament: For true peace and autonomy], (Tokyo: Shodensha, 2013): 182–206.

43. Samuels, *3.11: Disaster and Change in Japan*, 123–128.

44. Ōmae Kenichi, *Nihon no Ronten 2015–16* [Japan's issues of debate 2015–16] (Tokyo: Purejidento Sha, 2014).

45. Tetsuo Yuhara, "The Fukushima Nuclear Accident and a Long-Term Energy Vision for Japan," Canon-Ifri Paper Series No. 6, Canon Institute for Global Studies, April 2012, http://www.canon-igs.org/en/column/energy/20120412_1331.html.

46. Yanagisawa Akira, Aoshima M., and Ito K., "Toward Choosing Energy Mix," Institute of Energy Economics, Japan, January 16, 2015, http://eneken.ieej.or.jp/data/5887.pdf.

47. Kaneko Kumao, "Japan Cannot Abandon Her Nuclear Fuel Cycle!," *kanekokumao.jp*, November 26, 2012, http://kanekokumao.jp/publication-en/95.html.

48. Satoh Yukio, "Reinforcing American Extended Deterrence for Japan: An Essential Step for Nuclear Disarmament," *AJISS Commentary No. 57*, February 3, 2009, https://www2.jiia.or.jp/en_commentary/200902/03-1.html.

49. Samuels, *3.11: Disaster and Change in Japan*, 124.

50. "Shinsai taiō 'Dai-renritsu o' 64%— Honsha zenkoku yoron chōsa kekka," [Response to earthquake disaster—64% "for grand coalition"—result of newspaper's national survey] *Yomiuri Shimbun*, April 4, 2011, 9.

51. "'Sendai Genpatsu to enerugī ni kansuru chōsa' Tanjun shūkei kekka" [Simple aggregate results of "Survey regarding Sendai Nuclear Power Plant and Energy"], November 10, 2014, https://www.nhk.or.jp/bunken/summary/yoron/social/pdf/141110.pdf.

52. Asahi Shimbunsha Seron Chōsa: Shitsumon to kaitō" [*Asahi Shimbun* public opinion survey: questions and responses], *Asahi Shimbun*, February 19, 2013, 5; February 17, 2015, 4; July 29, 2014, 4; and September 8, 2014, 3.

53. "Sendai genpatsu no saikadō 'yokunakatta' 49%—Asahi yoron chōsa," [Restart of Sendai nuclear power plant—49% say "was not good"—Asahi public opinion survey] *Asahi Shimbun Digital*, August 25, 2015, http://www.asahi.com/articles/ASH8S5G03H8SU-ZPS007.html.

54. Japan Atomic Energy Relations Organization (JAERO), *Genshi Ryoku Riyō ni kansuru Seron Chōsa* [Public opinion survey regarding the use of atomic power] (Tokyo: JAERO, 2015): 80–82, http://www.jaero.or.jp/data/01jigyou/pdf/tyousakenkyu26/4.pdf.

55. "Shitsumon to kaitō: Kempō—Nit-Chū-Kan 3-kakoku yoron chōsa" [Questions and responses: Constitution and Japan-China-South Korea three-country opinion survey], *Asahi Shimbun*, April 7, 2014, 11.

56. "Shitsumon to kaitō: Bō'ei mondai—Asahi Shimbun yoron chōsa shōhō" [Questions and responses: Defense issues—*Asahi Shimbun* opinion survey full report"], *Asahi Shimbun*, November 6, 1988, 11.

57. Jimintō, "Jūten seisaku 2012: Nihon wo torimodosu" [Important policies for 2012: To recover Japan], 23, http://jimin.ncss.nifty.com/pdf/seisaku_ichiban24.pdf.

58. "Democratic Party of Japan's Manifesto: 2012 Making Decisions to Get Things Moving," Democratic Party of Japan, November 26, 2012, 8, https://www.dpj.or.jp/english/manifesto/manifesto2012.pdf.

59. Hiroko Tabuchi; "Japan, under Pressure, Backs Off Goal to Phase Out Nuclear Power by 2040," *New York Times*, September 20, 2012, A9, http://www.nytimes.com/2012/09/20/world/asia/japan-backs-off-of-goal-to-phase-out-nuclear-power-by-2040.html; Jeff Kingston, "Nuclear Power Politics in Japan, 2011–2013," *Asian Perspective* 37 (2013): 508.

60. "Datsu genpatsu no min-i no zure: 2012 Shū-In sen" [Gap with popular will: 2012 House of Representatives election], *Chūnichi Shimbun*, December 17, 2012.

61. "Masuzoe Wins Tokyo Gubernatorial Election in Landslide," *Asahi Shimbun Asia & Japan Watch*, February 9, 2014, http://ajw.asahi.com/printout/behind_news/politics/AJ201402090030.

62. "More Lawmakers Want Japan to Consider Nuclear Arms Option," *Mainichi Shimbun*, August 8, 2013.

63. "600,000 Votes for Tamogami May Signal Rise of Ultraconservatives," *Asahi Shimbun Asia & Japan Watch*, February 11, 2014, http://ajw.asahi.com/printout/behind_news/politics/AJ201402110060.

64. Nuclear Regulation Authority, Japan, "Nuclear Regulation for People and the Environment," accessed online January 15, 2015, https://www.nsr.go.jp/english/e_nra/nsr_leaflet_English.pdf.

65. Kingston, "Nuclear Power Politics in Japan, 2011–2013," 501–504; "Japanese Reactor Shutdowns Threaten Power Shortages," *World Nuclear News*, May 20, 2011, http://www.world-nuclear-news.org/RS-Japanese_reactor_shutdowns_threaten_power_shortages-2005114.html; "Two-Step Stress Tests for Japanese Plants," *World Nuclear News*, July 11, 2011, http://www.world-nuclear-news.org/RS-Two_step_stress_tests_for_Japanese_plants-1107115.html.

66. Yanagisawa Akira, ed., "Economic and Energy Outlook of Japan for FY 2014: Japan Is Reaching a Crucial Moment," Institute of Energy Economics, Japan, December 20, 2013, 13, http://eneken.ieej.or.jp/en/press/press_131220.pdf.

67. Kingston, "Nuclear Power Politics in Japan, 2011–2013," 502–505.

68. Ministry of Economy, Trade, and Industry, *Strategic Energy Plan*, Ministry of Economy, Trade and Industry, April 2014, 24, http://www.enecho.meti.go.jp/en/category/others/basic_plan/pdf/4th_strategic_energy_plan.pdf.

69. Keizai Sangyō Sho, *Chōki enerugī jukyū mitōshi* [Long-term energy supply and demand outlook], Ministry of Economy, Trade and Industry, July 2015, http://www.meti.go.jp/press/2015/07/20150716004/20150716004_2.pdf.

70. "Japan Nuclear Update," Nuclear Energy Institute, November 5, 2015, accessed online March 4, 2016, http://www.nei.org/News-Media/News/Japan-Nuclear-Update.

71. U.S. Energy Information Administration, "Japan Plans to Restart Some Nuclear Plants in 2015 after Fukushima Shutdown," U.S. Energy Information Administration, February 11, 2015, http://www.eia.gov/todayinenergy/detail.cfm?id=19951; "Japan Court Halts Restart of Two Reactors in Blow to Nuclear Sector," *Asahi Shimbun Asia & Japan Watch*, April 14, 2015, http://ajw.asahi.com/article/behind_news/social_affairs/AJ201504140032.

72. "Court Lifts Injunction against Restarting Takahama Nuclear Reactors, *Asahi Shimbun Asia & Japan Watch*, December 24, 2015, http://ajw.asahi.com/article/0311disaster/fukushima/AJ201512240047.

73. "Clearly, Kansai Electric's Priority Must Be Clarifying Cause of Reactor Shutdown," *Asahi Shimbun Asia & Japan Watch*, March 3, 2016, http://ajw.asahi.com/article/views/editorial/AJ201603030046.

74. "After Injunction, Work Starts to Halt Reactor at Takahama Plant," *Asahi Shimbun Asia & Japan Watch*, March 10, 2016, http://ajw.asahi.com/article/behind_news/social_affairs/AJ201603100005.

75. Jane Nakano, "Civilian Nuclear Energy Cooperation between the United States and Japan," in *The New Nuclear Agenda: Prospects for US-Japan Cooperation*, ed. Yuki Tatsumi (Washington, D.C.: Stimson Center, February 2012): 45–46.

76. Naoya Yoshino, "Nuclear Power Safety Deal Keeps Japan Firmly in U.S. Tech Camp," *Nikkei Asian Review*, December 3, 2014.

77. Center for Strategic and International Studies (CSIS), *Japanese Nuclear Policy Background Paper* (Washington, D.C.: CSIS, November 2014): 25–26, http://csis.org/files/publication/140912_CSIS_Japan_Background_Paper.pdf.

78. "Asahi Shimbunsha yoron chōsa: Shitsumon to kaitō" [*Asahi Shimbun* public opinion survey: Questions and responses], *Asahi Shimbun*, April 22, 2014, 4.

79. "Nihon no genpatsu yushutsu ni gimon—Ogata Sadako JICA Rijichō' [Doubts about Japan's exports of nuclear power plants—JICA President Ogata Sadako] *Asahi Shimbun*, March 24, 2012, 1, 15.

80. Masako Toki and Miles Pomper, "Time to Stop Reprocessing in Japan," *Arms Control Association* 43 no. 1 (January/February 2013), http://www.armscontrol.org/act/2013_01-02/Time-to-Stop-Reprocessing-in-Japan%20.

81. Ibid.

82. Tatsujiro Suzuki, "Japan's Nuclear Spent Fuel Management Challenges" (presentation, CNS Virtual Science Challenge for Youth e-Conference, March 7, 2013), http://cns.miis.edu/activities/pdfs/130320_virtual_science_challenge_conference_2nd_japan_tatsu_suzuki.pdf.

83. Ministry of Economy, Trade, and Industry, *Strategic Energy Plan*, April 2014, 53.

84. For an excellent critique of Japan's reprocessing program, see James M. Acton, *Wagging the Plutonium Dog: Japanese Domestic Politics and Its International Security Implications* (Washington, D.C.: Carnegie Endowment for International Peace, 2015).

85. Toki and Pomper, "Time to Stop Reprocessing in Japan."

86. CSIS, *Japanese Nuclear Policy Background Paper*, 22.

87. "NRA Seeks New Operator for Monjū Reactor; Decommissioning Likely," *Asahi Shimbun*, November 14, 2015, http://ajw.asahi.com/article/behind_news/social_affairs/AJ201511140031.

88. "Shūdanteki ji'eiken, kōshiyōnin hantai 63% sakunen yori zōka Asahi Shimbunsha yoron chōsa," [Collective self-defense, opposition to approval of its exercise 63%—increase

compared to last year—Asahi Shimbun public opinion survey], *Asahi Shimbun*, April 7, 2014, 1.

89. Jacques E. C. Hymans, "Veto Players, Nuclear Energy, and Nonproliferation: Domestic Institutional Barriers to a Bomb," *International Security* 36 no. 2 (Fall 2011): 154–189.

90. Richard J. Samuels and James L. Schoff, "Japan's Nuclear Hedge: Beyond 'Allergy' and Breakout," in *Strategic Asia 2013–14: Asia in the Second Nuclear Age*, eds. Ashley J. Tellis, Abraham M. Denmark, and Travis Tanner (Washington, D.C.: National Bureau of Asian Research, 2013): 247–248.

91. Nishi Kumiko, "Genbaku tōka kara 65 nen—kienu kaku no kyōi" [65 years since the dropping of the atomic bombs—the nuclear threat that does not disappear], *Hōsō Kenkyū to Chōsa* October 2010, 73; "Takahama genpatsu no saikadō ni kansuru chōsa (2015 nen 10 gatsu) tanjun shūkeihyō," *NHK Online*, October 2015, http://www.nhk.or.jp/bunken/summary/yoron/social/pdf/151017.pdf.

5

South Korea

Intense Debates and Global Nuclear Ambitions

Scott Snyder

Despite the long-standing extended deterrence relationship with the United States, South Korea's orientation on nuclear-related issues has not been entirely consistent with geopolitical factors. From the 1950s, a strong Realist impulse initially drove the country's pursuit of nuclear energy toward energy independence and energy sovereignty. Subsequently, possession of nuclear technology enabled the country's conservative leadership in the 1970s to launch Nationalist efforts to develop nuclear weapons under cover of the nuclear energy program, in part as a result of growing doubts about the reliability of the United States as an alliance partner.

Then, under U.S. pressure, the ROK (South Korea) gave up the nuclear military option in exchange for civil nuclear technology and the reaffirmation of the nuclear umbrella. South Korea's acceptance of this bargain and continued development of its nuclear energy sector have facilitated the rise of a Globalist orientation that has been sustained by a number of important technical and economic bureaucracies that seek to integrate South Korea into the global nuclear sector as a central player. On the questions of peaceful nuclear energy development and nuclear weapons acquisition, pragmatic assessments of South Korean vulnerabilities, needs, limits, and self-interest have tipped the balance between Realist-Nationalist and Globalist-Realist objectives as a decisive influence in South Korea's nuclear debates.

In the aftermath of South Korea's grand bargain affirming pursuit of nuclear energy while forgoing nuclear weapons acquisition, there have been two main drivers that have influenced the parallel evolution of South Korean debates on both the peaceful application of nuclear power and the debate over acquisition of nuclear weapons to match development of nuclear weapons in North Korea. First, the liberal Globalist tendency that has supported adoption of nuclear energy has enjoyed high levels of support in a democratizing South Korea, but the emergence of civil society groups in the aftermath of Fukushima has begun to oppose South Korea's ambitious plans for building nuclear reactors. The issue of nuclear waste disposition has added another domestically contentious impediment to the rapid development of nuclear power. Second, at the same time, North Korea's nuclearization and the restimulation of a small but growing Realist-Nationalist set of views calls for a rethinking of nuclear weapons strategy. Arrayed against this view is a Realist minority opinion that wants to reintroduce U.S. nuclear weapons on South Korean soil as an alternative to indigenous development of weapons. One current uniting factor across the spectrum is the muted but growing ROK demand that it receive the same reprocessing privileges from the United States as its fellow American ally Japan. Thus, the center of gravity in the nuclear debate in South Korea is a blended Globalist-Realist outlook that is underpinned by the U.S.-ROK alliance.

In this chapter, I review three consequential debates South Korea has experienced related to the development of its nuclear sector: (1) South Korean decisions to pursue nuclear energy as a means by which to achieve energy self-sufficiency and recent debates over the extent to which nuclear energy should be emphasized as a component of South Korea's overall energy mix; (2) nuclear weapons acquisition and South Korea's accompanying debate of options in response to North Korea's nuclear weapons acquisition; and (3) South Korea's debate over spent fuel management and its impact on negotiations with the United States over future nuclear cooperation between the two nations. As part of this discussion, I will trace debates of respective branches and constituencies within the government on these issues, since the outcomes of many of these debates have primarily been influenced through those discussions. Finally, I will trace the impact of South Korean public views and debates on each of these three aspects of South Korean nuclear policy.

THE SIGNIFICANCE OF SOUTH KOREA'S NUCLEAR ENERGY CHOICES

South Korea's choice to develop nuclear energy as a critical component of its overall energy strategy has deep roots in the country's own lack of indigenous energy resources and its resulting struggles with energy security and near total dependence on imports to meet its energy needs. The priority that South Korea places on nuclear energy is reflected in a 2014 White Paper on Nuclear Energy, which emphasizes the importance of the nuclear energy development given the high price of oil, climate change, and the need to ensure stable sources of electric power in the face of growing electricity demand within South Korea.[1] The Ministry of Trade, Industry, and Energy (MOTIE)'s seventh basic power supply plan for the period up to 2029, released in July 2015, envisions construction of two additional nuclear power reactors and elimination of four coal-fired plants as part of South Korean commitments to reduce greenhouse gas emissions.[2] The conclusion of the White Paper has deep roots as a strategic commitment that South Korea made in the 1950s to prioritize nuclear energy as part of a strategy to address its chronic energy dependency. As part of this strategy, South Korea has gradually mastered the construction processes involved in developing production of nuclear energy using indigenous companies, materials, and design, but still relies on the United States to supply critical parts within these reactors.

This section will discuss (1) South Korea's unique and vast energy requirements; (2) the ROK government's early efforts to jump-start the country's nuclear energy program; (3) the early linkages between nuclear energy knowledge and considerations about whether to build an indigenous nuclear arsenal; (4) the role of the Korean public and technical community in nuclear debates, especially after the 2011 accident at Japan's nuclear reactors in Fukushima; and (5) the current priority of nuclear power in South Korea's broader energy strategy.

South Korea's Vast Energy Needs

The development of South Korea's nuclear industry is part of the nation's economic success story, and a subtheme of South Korea's broader modernization story is its nuclear development. South Korea is the world's thirteenth-largest economy,[3] measured by purchasing power parity, and sixth-largest exporter,[4] but it is also the ninth-highest energy-consuming country and the fifth-largest oil importer.[5] South Korea trails only China and Japan as East Asia's third-largest crude oil importer, and the Korea Gas Corporation (KOGAS) is the largest corporate buyer of liquefied natural gas (LNG) in the world.[6]

South Korea's 2014 total primary energy supply (TPES) shows wide diversification of energy sources, with oil constituting 39 percent of South Korean energy supply, coal at 31 percent, natural gas at 16 percent, nuclear at 13 percent, and renewables at 1 percent.[7] The contribution of oil to South Korea's total energy supply has declined from 52.6 percent in 2000, while the use of natural gas, coal, and nuclear energy have gradually increased since that time. This combination of energy sources makes South Korea dependent on imports of oil from the Middle East (Saudi Arabia at 34 percent of Korean oil imports, followed by Kuwait at 16 percent, UAE at 11 percent, Qatar at 10 percent, Iraq at 8 percent, and Iran at 5 percent), coal (primarily from Australia, Indonesia, Russia, China, and the United States), and LNG (35 percent from Qatar, 14 percent from Indonesia, 10 percent from Oman, 10 percent from Malaysia, and 5 percent from Russia).[8] South Korean imports of oil, coal, and natural gas from international markets represent over four-fifths of South Korea's total energy supply. As nuclear power is South Korea's only indigenous source of energy production in a supply mix characterized by relative dependence on foreign energy sources, it is easy to understand why promotion of nuclear power has been an important component of South Korea's energy security policies.[9]

As of 2016, South Korea had twenty-five reactors that produced approximately one-third of its electricity, with construction plans for an additional eleven plants.[10] This rate of production would also make South Korea one of the top five nuclear energy–producing countries in the world. Following a successful $20 billion bid in

2009 for a contract to construct four nuclear reactors in the United Arab Emirates (UAE), South Korean planners raised hopes of capturing as much as a 20 percent share of the global market for nuclear technology, although this target appears to be overoptimistic.[11]

What Drives South Korea to Pursue Nuclear Energy?

Even prior to the establishment of the Republic of Korea (ROK, or South Korea) in 1948, it was widely known that the southern part of a divided Korean peninsula was rich in agricultural resources, while the northern part of the Korean peninsula had a manufacturing advantage and a rich natural resource base that included coal and uranium resources. The relative energy dependency of South Korea compared to its northern neighbor was a sensitive issue for South Korea during and after the Korean War as it sought to marshal resources necessary for its economic recovery. Given the acute vulnerability South Korea felt as an energy importer—exacerbated by its wartime experience and ongoing competition with the North—it is not surprising that South Korea would be interested in securing an autonomous means of energy production that might ameliorate its dependency on imported energy supplies. Thus, the acquisition of a nuclear energy production capacity was clearly regarded as an essential step in South Korea's postwar recovery and development.

As a result, South Korea laid the initial foundations for development of its nuclear energy sector with the passage of its Atomic Energy Act in 1958, even while it was still struggling to recover from the aftermath of the Korean War. With assistance from the United States under the Atoms for Peace program, South Korea established an experimental nuclear reactor that laid the foundations for a larger investment in the sector. To support the development of the nuclear energy sector along with the construction of the research reactor, in 1959 South Korea established its first nuclear research institute, the Korea Atomic Energy Research Institute (KAERI), under a newly created Office of Atomic Energy within the Ministry of Education. For over a decade in the 1960s, by operating an experimental reactor provided by the U.S.-based General Atomics, KAERI researchers gained experience with nuclear R & D, operation of nuclear research facilities, and training and education of specialized nuclear personnel.

As part of South Korea's emphasis on development under Park Chung-hee, the ROK government decided to pursue construction of a nuclear reactor through meetings convened by the Office of Atomic Energy together with the Ministry of Trade and Industry and the Korea Electric Power Company (KEPCO), as well as other government-controlled energy companies in South Korea. As South Korea's reactor efforts moved from the experimental phase to the commercial phase, the institutional lead for activities in support of the construction of a commercial reactor were undertaken by KEPCO. Park established the Nuclear Power Oversight Committee in 1965 and the Ministry of Science and Technology in 1967, strengthening the governmental commitment to developing nuclear power by committing to the construction of a 600,000-kilowatt reactor at a time when South Korea's total energy consumption was only one million kilowatts. These became the main organizations supporting the development of South Korea's nuclear energy infrastructure. Based on its experience with the research reactor, South Korea contracted for the construction of the Kori-1 and Kori-2 U.S. reactors from Westinghouse in a build-operate-transfer (BOT) arrangement in 1972. The completion of reactor construction in 1978 enabled nuclear to be part of South Korea's energy mix for the first time, and South Korea rapidly built additional reactors at Ulchin and Wolsong through the 1980s. During this period, nuclear energy came to be seen as an attractive means to reduce cost and increase availability of electricity, a way to reduce reliance on energy imports, and an opportunity to strengthen South Korea's capabilities in nuclear technology.

As part of South Korea's effort to build a nuclear energy sector, the South Korean government quickly established a bureaucratic and industrial infrastructure to promote the development of South Korea's nuclear energy production capability, including having South Koreans taking every opportunity to learn the technologies, construction, and operation of nuclear reactors.[12] To gain this knowledge, the state supported the education at U.S. universities of individuals who brought back with them critical knowledge and the establishment of professional ties with other leaders in the sector. The development of this capacity required the creation of a network of university programs to teach nuclear engineering at the undergraduate and graduate levels, including at Seoul National University, the Korea Advanced Institute of Science and Technology (KAIST), and Hanyang University. These institutions produced graduates that became the mainstay supporters and builders of the technical knowledge required to work at KAERI and other public- and private-sector institutions developing South Korea's nuclear energy sector.

The fundamental strategic decision in the 1960s to commit resources to support development of the nuclear sector were thus primarily motivated by Realist aims to establish a measure of self-sufficiency in energy production, which in turn was critical to South Korea's national development. In the nuclear sector, these Realist aims were in turn supported by an American Globalist vision for peaceful use of nuclear energy that in turn enabled the development of a South Korean nuclear sector. In its initial phase of development, there is little evidence of an overt Nationalist motivation for South Korea to develop a nuclear weapons capability, in part because during this period, South Korea remained almost wholly dependent on the United States to meet its defense needs.

Key Actors in the Maturation of the Nuclear Energy Program

Within the government of the 2016 ROK, the most relevant ministries in charge of energy policy remain the Ministry of Trade, Industry, and Energy (MOTIE) and the Ministry of Science, ICT, and Future Planning (MSIP), with the Ministry of Foreign Affairs (MOFA) taking the lead on international negotiations and participation in nuclear-related international organizations. MOTIE has traditionally taken the lead on setting broader energy strategies, including representing private-sector stakeholders, while the MSIP has succeeded the Ministry of Education, Science, and Technology's (MEST's) previous role of guiding the development of South Korea's nuclear industry by supporting nuclear R & D. The Nuclear Safety and Security Commission (NSSC), which was established in 2011 as an agency independent of the MEST, is the main body responsible for nuclear safety regulation.[13]

MOTIE, known as the Ministry of Knowledge Economy (MKE) from 2008 to January 2013, is responsible for South Korea's strategic economic planning. MOTIE identifies four primary objectives of its energy strategy: (1) to manage the national energy supply, (2) to promote overseas energy development projects, (3) to implement environmentally responsible growth policies, and (4) to combat climate change.[14] MOTIE works closely with the Korea Trade Investment Promotion Agency (KOTRA), the Korea International Trade Association (KITA), and other business groups to represent private-sector interests within the government. MOTIE formerly provided state direction to South Korea's overall industrialization strategy. Historically, close government-business ties have existed between the South Korean government and quasi-private energy-sector companies that were responsible for implementing the national energy policy. However, South Korea's democratization and industrial restructuring, facilitated in part by changes in corporate governance and deregulation following the Asian financial crisis in 1997–1998, have provided South Korean companies with greater autonomy in their decisions. Furthermore, these reforms have resulted in somewhat greater private-sector competition in various South Korean energy sectors, under the oversight of specialized regulatory authorities such as the Korea Electricity Commission (KOREC) and the Korea Institute of Nuclear Safety (KINS).

The major quasi-state company involved with procurement and implementation of South Korea's nuclear energy sector is the Korea Electric Power Corporation (KEPCO). KEPCO serves the Nationalist objective of pursuing energy self-sufficiency and acts as the umbrella and hub for South Korea's energy generation and supply capabilities, working across sectors and prioritizing procurement, infrastructure development, and energy efficiency. KEPCO operates as a holding company overseeing a variety of companies that have specific subspecializations, most notably including Korea Hydro and Nuclear Power (KHNP), KEPCO Nuclear Fuel (KEPCO NF), and KEPCO Engineering & Construction (KEPCO E&C). KHNP, which operates both nuclear power plants and hydropower plants, supplies approximately one-third of the nation's electrical power. KEPCO NF is concerned primarily with nuclear fuel design and production, with the aim of localizing nuclear fuel and gaining technological independence. KEPCO E&C's main objective is to secure independent technologies for designing and building nuclear power plants.

The Nuclear Energy–Nuclear Weapons Nexus under Park Chung-hee

Increasingly worried about U.S. capacity and willingness to defend South Korea from the more powerful North, Park Chung-hee capitalized on the knowledge transfer resulting from efforts to build nuclear plants and attempted to use this knowledge as part of efforts to procure equipment that could also be used to make a nuclear weapon.[15] In 1970, Park established the Agency for Defence Development and the following year created a high-level coordinating committee known as the Weapons Exploitation Committee (WEC) that was placed under the direct control of the Blue House through the Office of the Second Presidential Secretary for Economic Affairs (OSPSEA).[16]

South Korean acquisition efforts in Canada and France ultimately tipped off the United States to Park's dual aims. The United States blocked South Korea's nuclear weapons development efforts and forced changes to the program that ultimately resulted in ensuring that South Korea's nuclear development would be for peaceful purposes.[17] As part of South Korean efforts to ensure that its program would not have a covert component, the Chun Doo Hwan administration reduced budgets for the program and cut direct-reporting relationships with the Blue House, placing KAERI and all other components of South Korea's civilian nuclear program under the oversight of the MEST.

As another form of reassurance that South Korea had ended its nuclear weapons development efforts, the country took steps to join international regimes, such as the Convention on the Physical Protection of Nuclear Material in 1982 and the Nuclear Suppliers Group in 1995, as well as a 1991 pledge not to pursue enrichment and reprocessing and the signing of an inter-Korean denuclearization agreement accentuating this vow in 1992. Although the International Atomic Energy Agency (IAEA) raised questions regarding unauthorized laboratory experiments with lasers involving uranium enrichment that occurred around 2000, these experiments were investigated by the IAEA as part of South Korea's accession to the IAEA Additional Protocol in 2004. As a result of its accession to monitoring under the Additional Protocol, South Korea earned a clean bill of health and certified its compliance with international nonproliferation standards.[18]

The Role of the Korean Public and Technical Community

The ROK government has been sensitive to the need to maintain public support for its nuclear energy program, particularly as its objective of achieving national energy security objectives have been challenged by local not-in-my-backyard (NIMBY) concerns. Accordingly, the government has established the Korea Nuclear Energy Promotion Agency (KONEPA) to educate the Korean public on the benefits associated with nuclear energy production, and it closely monitors public opinion through regular surveys on nuclear-related issues. As a government agency responsible for promoting positive views toward nuclear energy production, these surveys may have a pro-government bias, but they provide a source of consistent data over time that clearly reflects fluctuations in Korean public attitudes toward nuclear energy and that appear to have had some influence on South Korea's development of its energy policy.

KONEPA's surveys since the mid-1990s show strong Korean public support for nuclear energy, with well over 85 percent consistently responding that South Korea needs to develop nuclear energy.[19] A majority of Koreans show appreciation for the contribution of nuclear energy to the nation's development and economic prosperity, while environmental and climate change benefits are important to a plurality of Korean respondents. In fact, the data from KONEPA in a December 2013 survey showed the greatest consensus in support of that need when compared to support in the previous two decades. However, since 2009 there has been a decline in Korean public support for continued construction of nuclear reactors; almost half of Koreans surveyed in December 2013 stated that the current number of nuclear reactors is sufficient to meet Korean energy needs, evidence of fading public support for continued expansion of nuclear plants. Interestingly, flagging Korean public support for nuclear industry expansion predates the March 2011 Fukushima disaster. According to data from KONEPA in a December 2015 survey, among those who support continued expansion of the nuclear sector, 85.1 percent advocate doing so to meet increased South Korean energy demand. Among those who want to decrease nuclear facilities, the dominant concerns are safety oriented, revolving around concerns that radioactive material may leak from plants, concerns about storage of spent fuel, and concerns about frequent accidents and breakdowns at nuclear plants.[20]

Impact of Fukushima on the Korean Public's Views of Nuclear Energy

While South Korean respondents to their government's polling on nuclear issues show strong recognition of the needs and benefits of nuclear energy, the effects of Japan's March 2011 triple disaster at Fukushima (as also discussed in chapter 4) have had a clear impact on Korean public perceptions of the safety of nuclear power. Prior to Fukushima, KONEPA surveys show that over 53 percent of Koreans viewed the nuclear energy sector as safe, but that perception dropped by almost 20 points to around 35 percent in a poll taken in December 2011 before recovering to almost 40 percent in December 2013.[21] The December 2015 poll shows a steady recovery, with 41 percent of Koreans viewing the nuclear energy sector as safe.[22]

An online survey by the Korea Federation for Environmental Movements (KFEM) conducted in February 2012, one year following the Fukushima incident, showed that 35 percent of those surveyed agreed with the government's plans to expand nuclear plants, while 65 percent were against them. Coming from a nongovernmental survey, the results may be skewed toward a more critical view of government policy, but are nonetheless significant as a measure of the extent of public distrust of established sources on nuclear issues. This survey also showed high mistrust of government and media information regarding the nuclear energy sector, with only 4 percent showing trust in government sources and 10 percent showing trust in media coverage of nuclear energy. A June 2012 survey by the Busan branch of KFEM showed that the percentage of people who expressed anxiety about plans to expand nuclear plants near Busan increased from 44 percent in 2011 to 75 percent in 2012.[23]

KFEM is one organization among a loose coalition of civil society actors that has attempted to mobilize periodic opposition to South Korean government plans to expand nuclear energy or to protest against measures that might have negative ramifications for public health or well-being. The level of protest activity against state-supported nuclear activities varies according to the issue and remains poorly organized, but anti-nuclear activities appear to be gaining greater traction in the southeastern portion of the country near Busan, where nuclear industrial activities are most concentrated. As a result, it will be more difficult in the future for the government to approve new plants, or even to gain life-extension approvals for some early model reactors as they come to the end of their initially approved terms for authorized use.

Contemporary Debates on the Future of Nuclear Power in South Korea

The turn in public opinion toward a more cautious attitude regarding nuclear energy has had an effect on South Korea's plans to expand the sector. In its first basic energy plan produced in 2008, the ROK government projected an ambitious expansion of the nuclear sector with a target of providing 41 percent of South Korean electricity needs by 2030, but this target has shifted to the goal of providing 29 percent of South Korean electricity needs by 2035. The revision in the target scraps plans for the development of new nuclear reactor sites, but will allow all previously approved projects to proceed. This plan still would represent an expansion of capacity for the nuclear sector from 21 gigawatts (GW) in 2012 to a projected 43 GW in 2035, but it would represent only a slight increase in the overall mix of nuclear energy compared to other energy sources and compared to ambitious plans charted in 2008.[24]

The downward revision in plans for new nuclear plants is directly connected to the drop in public support rates for construction of new nuclear plants following Fukushima that is evident in the KONEPA polling explained above and a scandal involving improper installation of substandard cables in nuclear power plants without proper certification. The 2014 Second Basic Energy Plan explained its new approach to nuclear energy policy as one that emphasizes safety, an effective regulatory process, anti-corruption and strengthened security-based oversight and management of nuclear plants, and prioritized exports of nuclear energy parts and equipment.[25]

"NUCLEAR SOVEREIGNTY": SOUTH KOREA'S RESPONSE TO NORTH KOREA'S NUCLEAR ARSENAL

A second controversial nuclear debate within South Korea has involved the question of whether South Korea might decide that the threat it faces from an unchecked North Korean nuclear weapons development program is so severe that South Korea has no choice but to pursue its own nuclear deterrent to match that of the Democratic People's Republic of Korea (DPRK, or North Korea). North Korea's nuclear tests in 2006, 2009, 2013, and 2016 and its accompanying efforts to develop miniaturized missile development capabilities that could be used on mid-range missiles have emerged as a source of threat to South Korea. This threat has engendered an active debate in Seoul over how to best counter the prospect that North Korea would employ nuclear blackmail by threatening Seoul with nuclear devastation were its requests for economic and/or political benefits not answered. Each of North Korea's nuclear tests has stimulated a more active debate in Seoul about how to respond to North Korea's nuclear capabilities. The debate has included efforts by the United States to reassure South Koreans regarding the credibility of American extended deterrence commitments to the defense of South Korea against North Korea's nuclear blackmail.

Looming over this discussion are two historical data points that have influenced both South Korean thinking on nuclear issues and U.S. views of the current South Korean debate. First, as mentioned in the previous section, for American nonproliferationists, Park Chung-hee's effort in the mid-1970s to develop a nuclear weapon capacity stands out as a significant leadership decision, and the tensions surrounding South Korea's nuclear acquisition effort also serve to underscore for South Koreans the idea that nuclear acquisition is directly at odds with maintenance of the U.S.-ROK alliance. South Korea's nuclear breakout efforts at the time were framed by doubts about the credibility of the U.S. security commitment to Asia, and the South Korean economy lagged behind that of North Korea, in contrast to the situation in 2016. South Korea's "nuclear breakout" attempt, nonetheless, looms large in the minds of American nonproliferation specialists as a prior expression of South Korean intent to secure a nuclear weapons option in response to its own security needs and despite the establishment of the Nuclear Nonproliferation Treaty (NPT).

A second event important to South Korea's nuclear weapons debate is Seoul's denuclearization commitments to forgo reprocessing and enrichment as part of a 1992 Joint Inter-Korean denuclearization declaration.[26] Despite North Korea's clear abandonment of the agreement, some American specialists resent any sign that South Korea might back away from the commitments made in that lapsed declaration. Some Realist South Koreans have quietly argued that the failure of the denuclearization declaration underscores the fact that the South Korean commitments made under that agreement were ill-considered. South Korean Globalists, on the other hand, argue that the South Korean economic and reputational costs involved in pursuing a nuclear capability are prohibitive.

The Impact of DPRK Nuclear Breakout on South Korean "Nuclear Sovereignty" Debates

North Korea's successive nuclear tests have renewed and sharpened a debate within South Korea over the question of whether to acquire a nuclear weapons capability.

North Korea's tests and the South Korean debate over how to respond have also stimulated U.S. efforts to provide reassurance to South Korea regarding its commitment to extended nuclear deterrence to ensure South Korea's security, further underscoring the close relationship between South Korea's perceived need for nuclear weapons capability and assurances provided in the context of the U.S.-ROK alliance. Within South Korea, the debate revolves around three main positions: (1) a Nationalist push for "nuclear sovereignty," (2) a blended Globalist-Realist approach centered on U.S. tactical nuclear weapons, and (3) a Globalist perspective that embraces U.S. extended deterrence commitments.

The Nationalist View: Pursuit of "Nuclear Sovereignty"

Following North Korea's nuclear tests, a small but growing group of conservative Nationalists has argued that the only way to respond to North Korea's nuclear advantage is for South Korea to match Pyongyang by acquiring its own nuclear capability. The response of these conservatives is strongly shaped by the perceived need to prevail in the legitimacy and power competition with North Korea. For this group, it is both unacceptable for North Korea to have a nuclear capability superior to South Korea's own and for North Korea to potentially be able to use that capability to threaten South Korea. The only acceptable solution is for South Korea to get out from under the thumb of U.S. pressure and reclaim its "nuclear sovereignty."

Following North Korea's first nuclear test in 2006, only a small number of Korean conservatives such as former ambassador Park Keun openly made the call for South Korea to pursue a nuclear weapons option. There are several reasons why the South Korean response to this option in 2006 might have been relatively muted. First, the uncertainty of being in uncharted territory and the desire to observe the Bush administration's handling of the issue at the UN Security Council may have induced a cautious response among Korean conservatives. Second, the media's willingness to run strongly conservative responses under the administration of the progressive President Roh Moo-hyun may have been weakened by a desire not to come into direct conflict with the Blue House. Third, the size of the yield of North Korea's first test suggested that the test was only partially successful, leaving some ambiguity regarding whether North Korea could be considered a full-fledged nuclear state.[27] Fourth, the Bush administration responded relatively quickly with strong reassurances regarding the U.S. commitment to defend South Korea's security from both conventional and nuclear threats.

In the aftermath of North Korea's 2009 test, however, voices in favor of a South Korean nuclear weapons option were heard more clearly, most notably through conservative columnist Cho Gap-je, who called on South Korea to regain its "nuclear sovereignty" from the United States.[28] The publication of this column would not have been inhibited by political constraints under the conservative administration of Lee Myung-bak, especially in the context of a stalemated inter-Korean relationship. Cho's column fueled an active debate within South Korea on possible options for responding to North Korea's nuclear capability, although the number of supporters of a nuclear sovereignty option for South Korea remained low.

A Blended Globalist-Realist View: The U.S. Tactical Nuclear Reintroductionists

A second view that has emerged within South Korea following North Korea's 2009 test has been that the United States should reintroduce tactical nuclear weapons to South Korea. The most prominent advocate of this view was National Assemblyman and Asan Institute for Policy Studies Founder Chung Mong-joon. Chung's arguments are notable both because they drew a response from U.S. officials and because he has made his case both domestically and internationally at the Carnegie Endowment's biennial nuclear conference, but supporters of this school of thought in South Korea are also relatively limited.[29]

Chung Mong-joon's call for reintroduction of tactical nuclear weapons grabbed headlines, but perhaps more notable was his accompanying suggestion that if this option is unfeasible, South Korea should be allowed by the United States, China, and the international community to step outside of the NPT and gain a nuclear option to deter threats from the North. This line of argument shades more closely toward the Nationalist "nuclear sovereignty" camp, but was presented more as a veiled threat of consequences should China and the United States fail to exert more visible efforts to take the North Korean threat more seriously.

This suggestion has historical resonance as the United States had kept tactical nuclear weapons on U.S. bases in South Korea from the mid-1960s until the end of the Cold War. However, those tactical nuclear weapons were removed by the early 1990s as part of an adjustment in U.S. nuclear policy following the end of the Cold War by President George H. W. Bush in 1991. The removal was supported by ROK President Roh Tae Woo. However, the idea of reintroducing tactical nuclear weapons into the South Korean theater—an idea that emerged in South Korea in response to North Korea's nuclear development—has been widely regarded among American specialists and government officials as a nonstarter.

Nonetheless, the impulse behind the recommendation that the United States reintroduce tactical nuclear weapons into the Korean theater could be seen as a vote of confidence for the U.S.-ROK alliance and for the credibility of the United States as South Korea's main security guarantor. This could be true even while some Koreans call for more active U.S. efforts to provide assurances to South Korea of the credibility of its extended nuclear deterrence commitments. At the same time, Chung's proposal reveals the deep level of concern in South Korea over the possibility of being subjected to North Korean nuclear blackmail, the need for an effective counterstrategy, and the expectation that the alliance with the United States can generate the deterrent capabilities to hold North Korean nuclear brinkmanship in check. There is also rising general concern among the South Korean public about the implications of China's rise, a trend that has occurred in tandem with rising levels of public support for the U.S.-ROK alliance.[30]

The Globalist View: Embracing U.S. Extended Deterrence Commitments

The dominant response among South Koreans following North Korea's second and third nuclear tests has been to continue to depend on U.S. alliance commitments. This response believes that the costs of leaving the NPT would outweigh the benefits by placing South Korea on a track toward international economic and political isolation that might in the end resemble the direction the DPRK has already taken. On the one hand, it is positive that, in its attempt to level the political playing field, most South Koreans have resisted the temptation to match North Korea's nuclear acquisition and instead have relied on the South's status as a mainstay within international nonproliferation norms. On the other hand, to the extent that the DPRK is able to sustain its program despite international pressure, there is increasing risk that support for this point of view will erode within South Korea. The other significant factor influencing the degree of support for this Globalist view is the necessity that the United States continue to both doctrinally and rhetorically underscore its commitment to extended nuclear deterrence by maintaining the forces and capabilities necessary to signal to South Korea the credibility of this commitment.

In this respect, it is notable that the United States has made significant efforts following each of North Korea's nuclear tests to reinforce the credibility of extended deterrence. On her first visit to Japan and South Korea less than a week following North Korea's October 9, 2006, nuclear test, Secretary of State Condoleezza Rice unequivocally stated that the United States was committed to defend South Korea against North Korean nuclear threats when she strongly affirmed "America's defense commitment to our allies, our commitment to the defense of our allies, because an attack on or threats to our allies are, in effect, like attacks and threats on the United States."[31] Weeks following North Korea's second nuclear test in 2009, the United States included a strong statement on nuclear deterrence as part of the U.S.-ROK Joint Vision Statement,[32] which represents one of the strongest statements to date outlining a comprehensive vision for alliance-based partnership between the two countries. In addition, the United States has stepped up joint coordination on deterrence prior to and following North Korea's third nuclear test in February 2013 through the establishment of a joint Extended Deterrence Policy Committee process involving deepening U.S.-ROK consultations and coordination on operational issues related to extended deterrence.

SOUTH KOREAN DEBATES ON NUCLEAR WASTE DISPOSAL

A third critical debate has emerged within South Korea in recent years over the issue of how to provide for permanent disposal of spent fuel from South Korea's reactors. As nuclear energy production has expanded in the country, so has its need for interim storage facilities capable of handling low-, intermediate-, and high-level nuclear waste. However, the establishment of these storage facilities has been met by rising levels of local opposition within communities identified as potential sites for storage of nuclear waste, due primarily to health concerns and negative reputational consequences to the local area that might derive from hosting a nuclear waste site.

Local opposition to the placement of nuclear waste disposal sites became more difficult for the central government to overcome following South Korea's democratization that increased local governments' ability to resist central directives. Korean democratization also provided greater space for public debate and media coverage over nuclear energy and its potential impact on the environment.[33] As a result, the Korean government faced real difficulties from the mid-1980s to the early 2000s in winning acquiescence from a series of local constituencies whose areas had been identified as potential sites to host a low- and intermediate-level waste (LILW) repository. Local opposition garnered widespread media coverage, putting other localities on guard.

The central government finally gained traction in managing the issue with the passage of a Special Act on Support for Areas Hosting Low and Intermediate Level Radioactive Waste Disposal Facility in March 2005. This legislation provided assurances that local government acceptance of an LILW repository would exempt that locality from hosting a high-level nuclear waste (HLW) storage facility and established a local referendum that created a competition to gain access to central government subsidies, rather than a situation in which the central government was compelling localities to accept the facility. In November 2005, the city of Gyeongju was selected as the host for the LILW repository after 89.5 percent of its voters approved of hosting the site.[34] The Korea Radioactive Waste Management Corporation was established to construct and manage the site, which began operations in December 2010.[35]

While this process constituted a success story for the ROK government in its efforts to establish a site for storage of low- and intermediate-level nuclear waste, it also illustrates the high level of public sensitivity within the South Korean public regarding the nuclear waste issue. In light of American failures to establish a permanent site for storage of nuclear waste, it should not be surprising that given South Korea's relatively small size and population density, this issue would be a very sensitive one among the Korean public.

As a result, a major ongoing nuclear debate in South Korea revolves around the question of how to identify an acceptable solution regarding permanent storage of spent fuel. The gravity of this problem has been underscored by the fact that South Korea's existing, temporary, on-site nuclear storage facilities are likely to fill up by 2019.[36] Given the relatively short timeline for maxing out existing on-site storage options and the public sensitivities regarding either transport or temporary off-site storage of nuclear materials, it is easy to see that the issue of spent nuclear fuel storage is one of the most pressing and publically debated issues facing South Korea's nuclear energy program.

Making Sense of South Korea's Spent Fuel Debate

The current debate over spent fuel options in South Korea has at least three dimensions: (1) the ongoing need for a permanent repository for storage of HLW and the resulting public sensitivities and debate regarding where in South Korea such a site might be established; (2) the debate among specialists in South Korea over the best available options to address its spent fuel needs; and (3) the ramifications for U.S.-ROK negotiations of a revised nuclear cooperation agreement to replace the current one, specifically the ROK request to the United States to allow pyroprocessing as one of South Korea's main options for managing disposition of spent fuel.

Given the long and complicated experience of the ROK government in winning local support for establishment of the LILW repository, it is clear that public opinion will play an important role in the viability of any recommended course of action. Furthermore, public input in a transparent and open process is the only feasible road to managing a successful outcome for South Korean efforts to arrive at an effective solution to the problem of interim storage of spent fuel from nuclear reactors. Toward this end, MOTIE has established a Public Engagement Commission on Spent Fuel (PECOS) consisting of eighteen members from a wide range of backgrounds. This commission is responsible for distilling technical options for addressing the spent fuel issue, as well as receiving public input on the issue for consideration by the government. Given the challenges associated with this issue, it is hard to imagine that overcoming differences will be easy, and it remains to be seen whether PECOS can identify a viable way forward that can satisfy both South Korea's technical needs and the wide range of stakeholders in the issue.[37]

The second dimension of the debate in South Korea is the technical debate among specialists over preferred options for addressing the spent fuel issue. The dominant preference among technical specialists at KAERI has been the pursuit of a technique called "pyroprocessing," through which plutonium is separated from other radioactive elements in the spent fuel, leaving plutonium potentially available for reuse in a fast-breeder reactor. This option has been highly promoted within the bureaucracy and has even been the main focus of a joint U.S.-ROK ten-year spent fuel study launched in 2012 at Argonne National Laboratory involving Korean and American specialists. However, the U.S. government position on pyroprocessing since at least 2009 states that the technique is a form of reprocessing and has been discouraged due to its potential for producing fuel for use in a nuclear weapon.[38] One of the objectives of the joint research project, therefore, is to determine whether adequate safeguards can be designed to prevent the products of pyroprocessing from being misused for nuclear weapons.

The third dimension of the debate over spent fuel management is directly related to the question of whether South Korea might someday win "advanced consent" to alter nuclear fuel in form or content. Currently, U.S. permission is required in advance before altering all U.S. fuel or fuel used in reactors that contain made-in-the-USA nuclear components. This issue was a major sticking point in negotiations over the renewal of the U.S.-ROK nuclear cooperation agreement and is widely seen as necessary both for South Korea to address its spent fuel problems and as South Korean engineers push for development of next-generation, fast-breeder reactors that presumably would then be used for domestic purposes as well as for export.

South Korean debate over each of these issues is a highly technical one that is driven primarily by the technical community, although it clearly has wider implications both for the South Korean government and for the public. The motivations surrounding the technical debate are particularly difficult to discern with absolute certainty. The main motivation is undoubtedly a Realist-Globalist one whereby scientist and engineers seek to maximize their potential to make technical contributions toward national development of new energy solutions that contribute to the global good. However, there remain suspicions that some South Korean scientists, especially those who cut their teeth in the 1970s, may still have Nationalist aspirations to ensure that South Korea has at least a "shadow" nuclear weapons capability. At the least, South Korea's industrial development has long benchmarked Japan's technical accomplishments and industrialization in many sectors, including the nuclear sector.

THE POLITICAL ECONOMY OF SOUTH KOREA'S NUCLEAR ESTABLISHMENT

Within South Korea's nuclear bureaucracy, there is a relatively close-knit group of scientists that has grown up working together in the same field and now manages the country's nuclear-related, public- and private-sector institutions. These scientists are linked by common professors and school ties, making informal interaction across institutional lines fairly easy. This technocratic constituency has become a powerful pro-nuclear industry force within South Korea's bureaucracy, and it has led South Korea to push for further development and export of its nuclear plant designs, technology, and capabilities. Other parts of South Korea's bureaucracy rely on South Korea's nuclear engineers for their specialized expertise on nuclear matters—their technical capabilities and knowledge, but there is also some wariness that technocratic arguments that may serve the needs of the sector before serving the national interest, and technocratic elites must win political support and budgets for their projects to proceed. For instance, since this technocratic group has a primary interest in the promotion and expansion of the nuclear industry, they tend to be less concerned about the impact of proliferation than their counterparts in the United States, where more specialists are trained on the nonproliferation side than on the nuclear energy industry side.

The Prominence of Nuclear Power in South Korea's Globalist-Realist Coalition

As South Korea's nuclear energy capabilities and export ambitions have grown, South Korean nuclear engineers have arguably evolved from Realist-Nationalist to now Globalist-Realist views. South Korea crossed the threshold into the international nuclear plant export market with its 2010 contract to build four reactors in the United Arab Emirates. Nuclear energy production has become symbolic of South Korea's homegrown capabilities and is both a source of pride and prestige as well as a tangible contribution to South Korea's ability to produce some of its energy indigenously without relying on external sources.[39]

The internationalization of South Korean nuclear interests and its desire to play a global leadership role in this area have enhanced the role of the Ministry of Foreign Affairs in the shaping of South Korea's nuclear policy, as the country liaises with international organizations, lobbies potential customers regarding the attributes of purchasing a ROK–model nuclear reactor, and negotiates nuclear cooperation agreements with both its customers and the United States. South Korea's foreign ministry not only mediates on behalf of Korean commercial interests around the globe, but it also actively conveys to the industry the importance of upholding international nonproliferation norms and incorporating these norms into international guidelines. In addition, the ministry has had to more actively "market" industry accomplishments as part of a presidentially led "sales diplomacy" on behalf of industry interests. Finally, the ministry has had to thread the needle between the interests of South Korean nuclear industry and technical expert communities and international nonproliferation norms and concerns. The simultaneous management of this last set of interests was particularly relevant to the foreign ministry's development of a unified inter-agency position during negotiations that led to the renewal of the nuclear cooperation agreement with the United States.

At the time South Korea won its first contract to export nuclear plants to the UAE in December 2009, press reports suggested that South Korea sought to build upwards of fifty nuclear plants around the world by 2030. In early 2015, South Korea won a contract to supply two SMART reactors to King Abdullah in Saudi Arabia and was targeting additional sales efforts to Vietnam (also discussed in chapter 7), Egypt, and the Czech Republic.[40] This goal, even if it proves to be unachievable, reflects the aspirations of South Korea's nuclear sector to gain a major share in the global nuclear power market. In this respect, South Korea's advancement into the nuclear energy market takes a page from South Korea's state-led development strategy, which has protected domestic markets in order to learn and develop technologies and design innovations that can subsequently be marketed to international consumers.[41] Following steel, automobiles, and smart phones, it appears nuclear plant exports are the latest high-technology item in the South Korean product catalog.

The Enduring Element of Nuclear Weapons in Korea's Nuclear Energy Debate

The development and cohesion of the main actors in the nuclear industry revolved initially around Realist impulses for energy self-sufficiency, but subsequently combined with Nationalist aims to strengthen South Korea's energy production and export capabilities and, some might allege, to become a virtual nuclear weapon state. South Korea has put all of its nuclear-related institutions under international safeguards inspections through the IAEA Additional Protocol in an effort to show its Globalist credentials. In addition, South Korea has become active as a resource for training in the safe and secure operation of nuclear equipment. On the occasion of Seoul's hosting the Nuclear Security Summit in March 2012, to advertise itself as a responsible and reliable player in nuclear energy production and nonproliferation, South Korea touted its contributions to nuclear safety and security at nongovernmental and industry side meetings held in conjunction with the summit.

In addition, South Korea's Nuclear Safety and Security Commission (NSSC), established in 2011, represents the evolution of domestic institutions needed to provide effective regulation and transparency to South Korea's nuclear industry. This process involves further enhancement of cooperation with the IAEA, both to allow inspections and to provide resources in support of strengthened nuclear safeguards. Given South Korea's growing interest in nuclear exports, a combination of Globalist and Realist motives have advocated for South Korea to contribute to the international regime for monitoring safety and security of nuclear programs in other countries. For instance, post-Fukushima, Seoul National University Professor Ha Young-sun proposed that South Korea, China, and Japan boost their technical cooperation and transparency on nuclear matters within the region, both as a means of imposing transparency and providing assurances among the three countries and as a way of leveraging availability of additional technical resources in a crisis.[42] This idea has been adopted as part of the Park administration's Northeast Asia Peace and Cooperation Initiative, which identifies nuclear safety as one of the specific functional areas on which it hopes to promote regional cooperation.[43]

As a next step in enhancing South Korea's competitiveness in the nuclear energy sector, South Korea has benchmarked other competitors, showing continued aims in an attempt to make inroads into the nuclear plant export sector. However, South Korea showed interest in enrichment and reprocessing of nuclear fuel as part of U.S.-ROK negotiations over the renewal of its bilateral nuclear cooperation agreement. This issue has raised old concerns that South Korea has Nationalist motives to acquire a latent or shadow nuclear capability. South Korea, on the other hand, has developed a Realist rationale for justifying South Korea's request for "advanced consent" from the United States to pursue these capabilities quite apart from the pursuit of nuclear weapons. South Korea argued that it is facing discrimination by the United States while other partners, such as Japan (see chapter 4 for more on this issue), India (also explored in chapter 3), and Europe have received permission (required under U.S. law) to reprocess and enrich U.S.-originated materials.[44]

After years of talks, U.S. and South Korean negotiators finally penned an agreement in April 2015. While the new framework does not allow the ROK to enrich uranium or reprocess spent fuel in the short term, it does not require South Korea to legally renounce its rights to those technologies.[45] Indeed, the agreement includes the possibility ROK enrichment could occur "in the future through consultations with the United States" and an option for spent fuel to be reprocessed by a third-party country that parties consider not at risk for proliferation. Further U.S.-South Korean nuclear cooperation outlined in the new agreement involves joint feasibility studies of new nuclear waste management options, including some in-country experiments on pyroprocessing.[46]

Nongovernmental Voices in South Korea's Future Nuclear Debate

Despite the efforts of South Korean industry to limit debate on nuclear energy, a more active set of nongovernmental voices critical of nuclear energy has emerged post-Fukushima. These writers point to trends among advanced countries, including Japan and Germany, toward decreasing dependency on nuclear energy and the potential for accidents, after a catastrophic event, that come at unbearably high costs.[47] South Korean anti-nuclear arguments have been boosted by a domestic scandal in which Korea Nuclear and Hydro Power Corporation (KNHP) officials were alleged to have procured substandard wiring and other components used in the construction of South Korean reactors, raising concerns and allegations that these components could represent a safety risk and the need for shutdown of these reactors until they were fitted with replacement parts that meet industry standards.[48] Under the leadership of Seoul City Mayor Park Won-soon, the Seoul Metropolitan Government has worked closely with civil society to launch a "One Less Nuclear Power Plant" campaign to reduce dependency on nuclear energy through the promotion of energy efficiency measures.[49] However, opposition from nongovernmental organizations (NGOs) and civil society institutions is still relatively weak, and anti-nuclear protests in South Korea are sporadic and suppressed. South Korea's public remains positive about the use of nuclear energy, although the level of support has dropped since 2011.

Although some small NGOs are concerned about the security risks connected to the nuclear sector, energy policy has not thus far attracted a great deal of activism or attention from the nongovernmental sector in South Korea. NGO influences on South Korea's energy policy have thus played relatively marginal roles, with the exception of local communities that have expressed concerns about the development of nuclear-related plants and waste facilities in their neighborhoods. However, the trend is toward greater public influence on governmental policy in the wake of the KNHP scandal.[50] In addition, the construction of electricity transmission towers in Miryang in Gyeongsang province has evoked serious protests from local residents concerned that they would have negative public health effects.[51] Moreover, the nuclear reactors South Korea first built in the 1970s will be coming to the end of their commercial lives, requiring reviews and decisions regarding licensing approval for life extensions, or alternatively, for decommissioning of the reactors. These decisions will be subject to close public scrutiny. The increasing influence of South Korean public opinion is most clearly evident in the decisions of South Korea's second energy plan to not pursue new nuclear power plant construction beyond plans already on the books.

CONCLUSION

South Korea's first nuclear debates over whether to invest in development of a nuclear energy sector were enabled by a Realist desire for energy self-sufficiency and cultivated by an American Globalist desire in the 1960s to expand global use of nuclear energy for peaceful purposes. The Realist decision to develop South Korea's nuclear energy sector has been so successful that South Korea has developed South Korean Globalist ambitions as a nuclear plant exporter.

The cultivation of a Realist South Korean nuclear establishment, however, took a Nationalist detour under Park Chung-hee's authoritarian dictatorship as a result of security concerns about the possibility of U.S. abandonment in the face of a stronger North Korea. In pursuing this objective, Park bypassed bureaucratic checks and put nuclear promotion on a financial fast track. The United States squelched the South Korean Nationalist turn toward nuclear weapons development in the 1970s, but that historical experience hangs over a second debate over nuclear weapons acquisition that has been revived by North Korea's emergence as a nuclear threat. The South Korean Nationalist strain of debate persists over how to respond to the prospect of North Korean nuclear blackmail against the South, but it is held in check by the considerable Globalist-Realist stake in nuclear energy production and exports, which would be put at risk if South Korea were to abandon international nonproliferation norms.

A third South Korean nuclear debate over nuclear waste disposal emerged in which South Korean technologists seek to utilize pyroprocessing and fast reactor development in tandem to achieve Realist and Globalist objectives of self-sufficient energy production and expanded international exports based on ambitious technical accomplishments. But South Korea's nuclear technologists have not been able to remove international concerns that such technological accomplishments might be shaped by the specter of Nationalist impulses to experiment once again with the development of nuclear weapons. Moreover, technologists must secure bureaucratic support both from within the government and from a South Korean public whose voices and sensitivities on nuclear issues have gradually increased along with South Korea's democratization and modernization. As a result, South Korea's nuclear debates have become more intense, but they have also become more Globalist in tone, even as the main driver for those debates continues to revolve around the basic Realist objective of pursuing energy sufficiency.

NOTES

1. Ministry of Trade, Industry and Energy and Korea Hydro and Nuclear Power Corporation, *2013 White Paper on Nuclear Power Generation*, administrative publication no. 11-1410000-00273-10 (Sejong City and Seoul: Republic of Korea, March 2014), 154, http://cms.khnp.co.kr/wp-content/legacy/pds/2013_whitepaper.pdf.

2. World Nuclear News, "South Korean Energy Plan Sees Two More Reactors," *World Nuclear News*, July 22, 2015, http://www.world-nuclear-news.org/NP-South-Korean-energy-plan-sees-two-more-reactors-2207154.html.

3. World Bank, "Gross Domestic Product 2014, PPP," The World Bank: World Databank, accessed online February 23, 2016, http://databank.worldbank.org/data/download/GDP_PPP.pdf.

4. CIA: World Factbook, "Country Comparison: Export," CIA: World Factbook, accessed online February 5, 2016, https://www.cia.gov/library/publications/the-world-factbook/rankorder/2078rank.html.

5. U.S. Energy Information Administration (EIA), "Korea, South Korea: International Energy Data and Analysis," U.S. Energy Information Administration, October 5, 2015, 1–3, accessed online February 2, 2016, https://www.eia.gov/beta/international/analysis_includes/countries_long/Korea_South/south_korea.pdf.

6. Korea Gas Corporation, *Annual Report: Creating KOGAS Energy for the Future*, Kogas, accessed online February 1, 2016, http://www.kogas.or.kr/en/main.action.

7. Korea Electric Power Corporation, "2015 Annual Report," Kepco, 62, accessed online February 1, 2016, http://home.kepco.co.kr/kepco/EN/C/htmlView/ENCBHP003.do?menuCd=EN030204. EIA, "Korea, South Korea: International Energy Data and Analysis," 4.

8. Ibid.

9. International Energy Agency, *Energy Policies of IEA Countries: The Republic of Korea 2012 Review* (Paris: International Energy Agency, 2012), 18, http://www.iea.org/publications/freepublications/publication/Korea2012_free.pdf.

10. World Nuclear Association, "Nuclear Power in South Korea," accessed online March 4, 2016, http://www.world-nuclear.org/information-library/country-profiles/countries-o-s/south-korea.aspx.

11. Ibid, 108.

12. Mark Holt, *U.S. and South Korean Cooperation in the World Nuclear Energy Market: Major Policy Considerations* (Washington, D.C.: Congressional Research Service, January 25, 2013), 1–3, http://www.fas.org/sgp/crs/row/R41032.pdf.

13. International Energy Agency, *Energy Policies of IEA Countries*, 103.

14. Suduk Kim and Yungsan Kim, "Energy and Environmental Policy in Korea," Energy Modeling Lab, Department of Energy Studies, Energy Systems Division, Ajou University, March 2, 2015, http://eml.ajou.ac.kr/suduk/teach/contents/sdkim_yskim.pdf.

15. The rise of these Nationalist motivations occurred in the following context: after the 1969 announcement of the Nixon Doctrine projecting U.S. troop drawdowns in East Asia; the 1972 Nixon visit to Beijing and the opening of U.S. relations with China; and the U.S. military quagmire and eventual defeat in Vietnam, followed by the 1976 election as president of Jimmy Carter, who had vowed during his campaign to remove U.S. troops from South Korea.

16. See Peter Hayes, *Pacific Powderkeg: American Nuclear Dilemmas in Korea* (Lexington, Mass.: Lexington Books, 1991), 203–205, 209–212; Sangsun Shim, "The Causes of South Korea's Nuclear Choices: A Case Study in Nonproliferation" (Ph.D. dissertation, University of Maryland, College Park, Md., 2008), 36–53, 200–213.

17. Don Oberdorfer and Robert Carlin, *The Two Koreas: A Contemporary History* (New York, N.Y.: Basic, 2013): 58.

18. Kyle Fishman, "IAEA South Korean Concerns Resolved," *Arms Control Today* (July/August 2008), https://www.armscontrol.org/2008_07-08/SouthKorea. For further details on South Korean nuclear decision-making, see Scott Snyder "South Korean Nuclear Decision Making," in *Forecasting Nuclear Proliferation in the 21st Century, A Comparative Perspective*, vol. 2, eds. William Potter and Gaukhar Mukhatzhanova (Stanford, Calif.: Stanford University Press, 2010), 158–181.

19. "2013 Nyun Wonjaryok kukmin insik josa kyulkwa" [Results of 2013 public survey on knowledge of nuclear power], *Korea Nuclear Power Promotion Agency*, accessed online May 26, 2014, results available in Korean at http://www.konepa.or.kr/new/main.asp.

20. "2015 Nyun Wonjaryok kukmin insik josa kyulkwa" [Results of 2015 public survey on knowledge of nuclear power], *Korea Nuclear Power Promotion Agency,* accessed online February 23, 2016, results available in Korean at http://konepa.or.kr/new/main.asp; the chapter's author was responsible for translations of Korean language works cited in this section..

21. "2013 Nyun Wonjaryok kukmin insik josa kyulkwa."

22. "2015 Nyun Wonjaryok kukmin insik josa kyulkwa."

23. As cited in Kim Myung-ja, *Korea's Atomic Energy Trilemma: Seeking Debates, Communication, and Solutions* (Seoul: Kachi Press, 2013), 171–173.

24. British Embassy Seoul, "South Korea Energy Policy: New Strategy Announced," Open to Export, January 2014, accessed online August 1, 2014, http://opentoexport.com/article/south-korea-energy-policy-new-strategy-announced.

25. Ministry of Trade, Energy, and Industry, *Je2-ch'a Eneoji gibbon kye-huik* [Korea energy master plan: Outlook and policies to 2035] (Sejong City, South Korea: Republic of Korea, January 2014), 89–95.

26. "Joint Declaration of South and North Korea on the Denuclearization of the Korean Peninsula," entered into force February 19, 1992, www2.law.columbia.edu/course_00S_L9436_001/North%20Korea%20materials/koreadenuclearization.htm.

27. Graham P. Collins, "Kim's Big Fizzle: The Physics Behind a Nuclear Dud," *Scientific American*, December 16, 2006, http://www.scientificamerican.com/article/kims-big-fizzle.

28. Cho Gap-je, "Bukhan-ui Haek-kaebal-eul To-un 'Naebu-ui Jeok-dul,'" ["Internal enemies" that helped North Korea's nuclear development], *ChoGabje.com*, June 23, 2009, http://www.chogabje.com/board/view.asp?C_IDX=27966&C_CC=BB.

29. M. J. Chung, "2013 Carnegie International Nuclear Policy Conference: Morning Keynote" (address, 2013 Carnegie International Nuclear Policy Conference, Washington, D.C., April 9, 2013), http://carnegieendowment.org/files/0409carnegie-day2-morning-keynote.pdf.

30. "66 Pct of S. Koreans View China's Rise as 'Military Threat': Poll," *Yonhap News Agency*, June 24, 2014, http://english.yonhapnews.co.kr/national/2014/06/24/90/0301000000AEN20140624003000315F.html.

31. Secretary of State Condoleezza Rice, "Interview with Hideo Yanagisawa of NHK News," October 19, 2006, http://2001-2009.state.gov/secretary/rm/2006/74744.htm.

32. The White House, Office of the Press Secretary, "Joint Vision for the United States of America and the Republic of Korea," The White House, June 16, 2009, http://www.whitehouse.gov/the_press_office/Joint-vision-for-the-alliance-of-the-United-States-of-America-and-the-Republic-of-Korea.

33. Kim Myung-ja, "Korea's Atomic Energy Trilemma," 182.

34. Caroline Peachey, "Korean Repository Realised," *Nuclear Engineering International*, July 22, 2014, http://www.neimagazine.com/features/featurekorean-repository-realised-4323899.

35. Herman Damveld and Dirk Bannink, "Management of Spent Fuel and Radioactive Waste: State of Affairs," *Nuclear Monitor* 746/7/8 (May 2, 2012): 17, http://www.nirs.org/mononline/nm746_48.pdf.

36. Meeyounjg Cho, "South Korea Needs New Facility for Spent Nuclear Fuel: Advisory Group," *Reuters*, June 11, 2015, http://www.reuters.com/article/us-nuclear-southkorea-spent-fuel-idUSKBN0OR0AT20150611.

37. Interview with Korean nuclear expert, Seoul, July 2014.

38. Choe Sang-hun, "U.S. Wary of South Korea's Plan to Reuse Nuclear Fuel," *New York Times*, July 13, 2010, http://www.nytimes.com/2010/07/14/world/asia/14seoul.html.

39. Korea Atomic Energy Research Institute, *50 Years of Nuclear Energy; 50 Years of Prosperity* (Seoul: Republic of Korea, Ministry of Education, Science, and Technology, April 9, 2009).

40. World Nuclear Association, "Nuclear Power in South Korea," accessed online October 1, 2015, http://www.world-nuclear.org/info/Country-Profiles/Countries-O-S/South-Korea/.

41. For more, see Jeffry Frieden, *Global Capitalism: Its Fall and Rise in the Twentieth Century* (New York: W. W. Norton, 2007).

42. Young-sun Ha, "Hankook-ui Ilbon dopki, Ilbon-ui Dongasia dopki," [Korea helps Japan, Japan helps East Asia], *Chosun Ilbo*, March 24, 2011.

43. Lally Weymouth, "South Korean President Park Geun-hye Answers Questions," *Washington Post*, May 7, 2013, http://articles.washingtonpost.com/2013-05-07/opinions/39090737_1_china-president-xi-jinping-northeast-asia.

44. Miles A. Pomper, "How to Unsnag US-South Korea Nuclear Negotiations," *Bulletin of the Atomic Scientists*, September 25, 2012, http://thebulletin.org/how-unsnag-us-south-korea-nuclear-negotiations; Toby Dalton and Alexandra Francis, "South Korea's Search for Nuclear Sovereignty," *National Bureau of Asian Research Asia Policy* 19 (January 2015): 115–136, http://www.nbr.org/publications/asia_policy/Free/AP19/AsiaPolicy19_Dalton_Francis_January2015.pdf.

45. Choe Sang-hun, "U.S. and South Korea Reach Revised Nuclear Deal," *New York Times*, April 22, 2015, http://mobile.nytimes.com/2015/04/23/world/asia/us-and-south-korea-reach-revised-nuclear-deal.html.

46. Jack Kim, "South Korea, U.S. Reach Deal to Revise Civil Nuclear Pact," *Reuters*, April 22, 2015, http://www.reuters.com/article/2015/04/22/us-southkorea-usa-nuclear-idUSKBN0ND08R20150422.

47. Haejung Kim, "Fukushima-ihu saegye wonjaryuk donghyang-gwa Hankook-ui wonjun junchek" [Global nuclear energy trend after Fukushima and Korea's nuclear energy policy], *Jaggasaegye* [Writers' world], 2012; Yoonnayoung Kim, "Wonjaryuk sago, daemun Kankook charae: hack opnun seasang-ul ggumggundah" [Nuclear energy accidents, next up is Korea: Dreaming of a world without nuclear energy], *Pressian*, March 11, 2012, http://www.pressian.com/article/article.asp?article_num=30120309200753.

48. Choe Sang-hun, "Scandal in South Korea over Nuclear Revelations," *New York Times*, August 3, 2013, http://www.nytimes.com/2013/08/04/world/asia/scandal-in-south-korea-over-nuclear-revelations.html?pagewanted=all&_r=1&.

49. "One Less Nuclear Power Plant 2012," Environmental Policy Division, Seoul Metropolitan Government, August 2013.

50. Choe Sang-hun, "South Korea Shuts Two Reactors over Faked Certificates," *New York Times*, May 28, 2013, http://www.nytimes.com/2013/05/29/world/asia/south-korea-turns-off-nuclear-reactors.html.

51. Hyo-sil Kim, "Miryang Residents in a Fight for 'Environmental Equality,'" *Hankyoreh*, October 7, 2013, http://english.hani.co.kr/arti/english_edition/e_national/606054.html.

6

Taiwan

Commitment to a Nuclear-Free Future

Robert Sutter and Yuan-Ming Alvin Yao

Taiwan in recent years has experienced a protracted and contentious national debate over the role of nuclear power in the country's energy supply, economic development, and overall modernization. The scope of the debate has involved a broad range of Taiwan government and society, including national economic and energy policy elites, local government leaders, legislators, media, and various nongovernment interest groups including environmentalists, businesses, and nuclear and consumer safety activists. The debate at times, notably in 2013–2014, became one of the most salient issues in the domestic politics of the highly competitive democratic system in Taiwan. It posed serious challenges for the ruling Nationalist Party (Kuomintang or KMT) administration of President Ma Ying-jeou and for the strongly oppositionist Democratic Progressive Party (DPP). The debate was resolved for the time being with a decision by the Ma government to reverse policy and order a stop to the completion of a controversial fourth nuclear power plant in 2014. The plant has been basically completed and in a storage or "moth-ball" status for the three years since then, at a cost well over $100 million USD.[1] The upsurge in Taiwan's nuclear power debate seen in 2013–2014 has deep roots. It followed past such episodes, notably an impasse over a decade ago between a newly elected DPP president and a KMT controlled legislature, which saw mass demonstrations and attempts to officially recall the president.

This chapter assesses the Taiwan nuclear energy debate against the background of Taiwan's debates over other aspects of nuclear developments relevant to this book, notably debates over Taiwan's indigenous development of nuclear weapons and the role the country has played and will play in international nuclear nonproliferation efforts. Taiwan has a long record of debate over the pros and cons of developing nuclear weapons. That debate has been quiet and seemingly resolved for many years amid government determination to pursue peaceful applications of nuclear energy rigorously consistent with international norms.[2] Taiwan also has a peculiar position in the international nuclear nonproliferation system, especially as a result of its loss of membership in the United Nations over forty years ago. China exerts strong efforts to keep Taiwan out of international bodies that regulate nuclear proliferation and other international bodies where statehood is a requirement for membership. The limited discussion and debate that takes place over Taiwan's role in the international nonproliferation regime is carried out among a small circle of specialists and other government and nongovernment elites in Taiwan.

A finding of this study is that the contentious Taiwan debate over nuclear energy involving a wide range of specialists, leaders, and other elites along with a broad range of popular representatives, has developed separately from the much less prominent and generally elite- and specialist-led discussions and debates on nuclear weapons development in Taiwan or Taiwan's role in international nonproliferation efforts (see table 6.1). However, the nuclear energy debate does reinforce a strong recent tendency in Taiwan to choose paths of future development that involve eschewing more nuclear material in Taiwan. This trend adds to circumstances that go against any possible revival of nuclear weapons development in Taiwan. And it adds to forces moving Taiwan to the forefront of international efforts to strengthen the global nuclear nonproliferation regime. Thus, in the context of this book, it may be concluded that the Nationalist school of thought is becoming less influential in the nuclear realm than might be expected (given Taiwan's divisive nationalism on other political issues), while growing Globalist and Realist thinking currently holds the center of gravity.

This chapter first reviews the background, status, and outlook of Taiwan's debate over developing nuclear weapons. It then reviews the limited discussion and debate over Taiwan's role in international nuclear nonproliferation regimes. Taiwan's salient debate over nuclear power is then assessed, followed by a review of findings relevant to the scope of this book and, finally, a section looking ahead to the future of Taiwan's nuclear debates and activities.

TAIWAN'S NUCLEAR WEAPONS DEVELOPMENT

Taiwan's clandestine development of nuclear weapons has a long history. The motives of Taiwan's leaders in developing nuclear weapons were widely seen as related to the security challenge Taiwan has faced with an ever more powerful adversary, the People's Republic of China (PRC) on the Chinese mainland and to fears that Taiwan's main support against China's power, the United States, would change policy and abandon Taiwan for the sake of accommodation with the PRC.[3]

What we know of Taiwan decision making at the time shows that a small group of Taiwan elites debated and decided to move ahead with a nuclear weapons program in the 1960s.[4] The group centered on the rulers of the then highly authoritarian KMT-ruled government, notably President Chiang Kai-shek and his son Chiang

Table 6.1. Review of Now-Moribund Nuclear Weapon Debates in Taiwan

Issues[1]	*Realists*	*Nationalists*	*Globalists*
1. Nuclear Weapons	High	High	Low
2. Safety Risks	Low	Low	High
3. View Adversary	Negative	Negative	Negative
4. Reprocessing Technology	Favor	Favor	Oppose
5. Spread Dual-Use Technology	Favor	Favor	Oppose
6 Export Nuclear Technologies	Favor	Favor	Oppose
7. Transparency	Low	Low	High
8. Openness to Nuclear Safety Regimes	Low	Low	High
9. Openness to Nonproliferation Regimes	Low	Low	High
10. Political Imperative for Proliferation	Strong	Strong	Weak
11. Role of Nuclear Weapons in Defense Strategy	Strong	Strong	Weak
12. Attitude on U.S. Nonproliferation Role	Unfavorable	Unfavorable	Supportive
13. Desirability of Role of Nongovernmental Groups	Low	Low	Medium

[1]Further details on the issues discussed in this chart:

1. Desirability of nuclear weapons in the country's security mix
2. Views reflecting level of concern on proliferation, terrorism, and safety risks of pursuing nuclear weapons
3. Views on adversary's pursuit of nuclear weapons capability
4. Attitude toward reprocessing technology (traditional and emerging technologies)
5. Attitude toward spread of dual-use technologies, especially enrichment and reprocessing
6. Views reflecting support or opposition regarding export of nuclear technology to entrant states and others
7. Acceptable level of transparency in the country's use of dual use technology, especially civil versus military nuclear program
8. Openness to formal and informal nuclear security and safety arrangements/regimes
9. Openness to formal and informal nuclear nonproliferation regimes
10. Attitude toward the country's political imperatives for nuclear proliferation
11. Preferred role of nuclear weapons in the country's defense strategy
12. Attitude toward U.S. leadership role in nonproliferation efforts
13. Desirability of nongovernmental groups influencing country's nuclear policy

Ching-kuo. The younger Chiang was the government's security overseer and was being groomed to take over the various government and party leadership responsibilities of the ageing "Generalissimo" Chiang—something he gradually did before and after the elder Chiang's death in 1975.[5]

The reasoning of the group centered on what this book would categorize as Nationalist and Realist schools of thought. While Taiwan seemed reasonably secure vis-à-vis the PRC in the 1960s, circumstances foreshadowed change adverse to Taiwan's security. Beijing was then in the throes of a very radical phase of domestic politics known as the "Cultural Revolution." The communist regime was capable of remarkably capricious and provocative actions. For example, Chinese anger at the British arrest of disruptive, pro-Maoist demonstrators in the then–British colony of Hong Kong in 1967 saw a mob of several thousand Chinese demonstrators surround the

British diplomatic mission in Beijing and set it afire, forcing British diplomats to flee the flames into the Chinese mob. For several weeks, many of the diplomats remained unaccounted for. At the same time, this often rabidly hostile Chinese regime successfully developed atomic nuclear weapons in 1964, and its sixth nuclear test in 1967 was a successful detonation of a hydrogen bomb.[6]

The United States at this time was becoming bogged down in a protracted and costly conflict in Vietnam, which soon added to strong domestic divisions weakening U.S. resolve to sustain the existing containment system against China on which Taiwan depended. Under these circumstances, Nationalist sentiment in Taiwan urged the country to take steps to develop its own power in order to deter China. Realist thinking was evident in the judgment that a domestically preoccupied America inclined to pull back from Asia would have less leverage to exert in order to coerce Taiwan against seeking nuclear weapons. Thus, Realist leaders in Taipei came to judge that the risks of going forward with the clandestine program seemed offset by the need for Taiwan to secure its integrity in the face of growing dangers from China.

Indeed, subsequent developments went against Taiwan's interests and reinforced the Nationalist and Realist rationale supporting Taiwan's clandestine nuclear weapons program. In late 1967, Mao Zedong called in the military to reestablish order in chaotic Chinese cities. Under de facto martial law, the military occupiers removed millions of disruptive Red Guards who had been engaged in leadership purges, antiforeign demonstrations, and often-intense, armed conflict with one another. The military rulers began the rebuilding of the Chinese Communist Party and Chinese government administrative structure that had been destroyed under Mao's direction in the previous two years. Meanwhile, tension and armed conflict along the Sino-Soviet border also rose, prompting some Chinese leaders to argue for outreach to the United States and the West. Seeking China's assistance in their intense rivalry with an ever more powerful Soviet Union, the United States and a number of Western leaders joined developing countries in seeking closer ties with China. A strong international majority supported China in the United Nation's vote in 1971 recognizing the PRC and expelling Taiwan. President Richard Nixon's visit to China in 1972 foreshadowed the official decision by President Jimmy Carter in late 1978 to end U.S.-Taiwan official relations, including the termination of the U.S. defense treaty with Taipei. The adverse trend for Taiwan continued even under the U.S. leadership of President Ronald Reagan, an ardent anti-communist and longstanding supporter of Taiwan. Reagan went further in accommodating China at Taiwan's expense by agreeing in 1982 to limit American arms sales to Taiwan.[7]

Secretly Seeking Nuclear Weapons and American Responses

In 1958, National Tsing Hua University bought Taiwan's first nuclear reactor from the General Electric Corporation; the reactor reached its first criticality in 1960. Available evidence indicates that the government did not launch a full-scale nuclear weapons effort until after China's successful nuclear weapons test in 1964. Taiwan leaders urged the United States to bomb PRC nuclear installations. Because the United States refused, Taiwan by 1967 began a $140 million program in a secretive defense ministry institute that led the way to nuclear weapons development for the country.[8] It is important to remember that Taiwan in those years was a very authoritarian state ruled by a KMT party organized along the same lines as Communist parties created by Vladimir Lenin. The loyalty of the government and party hierarchy and rank-and-file to Chiang Kai-shek and his very capable and often ruthless son Chiang Ching-kuo was strong. The KMT regime employed martial law, prevalent in Taiwan until 1988, to censor media, quell dissent, and suppress demonstrations. Under these circumstances, keeping a secret defense ministry program under wraps was not difficult.[9]

The weapons program involved procuring and operating a heavy-water reactor, a heavy-water production plant, a reprocessing research laboratory, and a plutonium separation plant. It began through legal means in accord with international obligations. As noted, the General Electric Corporation built Taiwan's first nuclear power plant, and Taiwan also sought abroad additional nuclear materials for so-called "research purposes." Germany sold Taiwan a nuclear reactor in 1967, and in the following years Taiwan bought nuclear equipment, supplies, and expertise from the United States, Germany, Canada, France, and South Africa.[10]

The Republic of China (ROC) government on Taiwan joined the Nonproliferation Treaty (NPT) in 1968, and the International Atomic Energy Agency (IAEA) began the process of negotiating a safeguards agreement to treat the ROC as a non-nuclear weapon state. As a result of Taiwan's expulsion from the United Nations, official IAEA contacts with the ROC ended. The United States, Taiwan, and the IAEA agreed in 1971 that an existing trilateral nuclear agreement would form the basis in the future for safeguards on Taiwan's nuclear program. Any nuclear materials that Taiwan acquired were treated as if they came from the United States and thus were to be subject to restrictions in their application under U.S. law. The United States thus became the ultimate legal guarantor of Taiwan's non-nuclear status, albeit facilitated by continued IAEA inspections.

The years following saw an expansion of secretive programs in Taiwan's defense ministry involving ostensible research reactors and reprocessing facilities. U.S. officials strongly suspected Taiwan efforts to produce plutonium. The Americans monitored Taiwan's nuclear efforts and put increasing pressure on the Taiwan government to renounce nuclear ambitions or risk losing U.S. support for its nuclear power program. In 1974, the Central Intelligence Agency (CIA) judged that Taiwan was working toward a nuclear weapons capability and would be able to produce a nuclear weapon in five years. The IAEA inspections became more rigorous. Various indications showed Taiwan was diverting nuclear material from spent fuel rods, possibly for reprocessing into material for nuclear weapons. Taiwan's new president, Chiang Ching-kuo, pledged to the United States in 1976 that Taiwan would not engage in nuclear reprocessing, but evidence of the latter caused the United States to insist that Taiwan dismantle suspicious facilities and return U.S.-supplied plutonium. Taiwan did so in the 1980s, seemingly putting to rest U.S. and other international concerns over Taiwan's nuclear weapons ambitions.

The Chiang government began the process of dismantling the authoritarian one-party rule of the KMT in the late 1980s. It seemed resigned to a future of contested elections and opposition politics, along with secret-ballot voting, a freer media, freedom to demonstrate, and greater accountability in government. Nonetheless, most forecast Taiwan's continued rule under a dominant KMT, with the opposition DPP remaining in a subordinate position.[11]

Under these more open and liberal circumstances in Taiwan's political system, the risks of exposure of the clandestine nuclear weapons program seemed higher. Nevertheless, the prevailing Nationalist sentiment and Realist calculus among Taiwan national security decision makers at the time supported a continuation of and indeed stepped up efforts to develop nuclear weapons. The precise deliberations of the Chiang regime at that time remain unclear; Chiang Ching-kuo was in declining health before his death in 1988. And there is little question about the Taiwan president's seemingly progressive intention in taking significant steps to insure a more liberal political order would prevail in Taiwan after his passing, notably by ending martial law in 1987 and tacitly allowing for the formation of a viable political opposition, formed in 1986 under the DPP.[12]

In this mix of circumstances reflecting Nationalist, Realist and Globalist tendencies among Taiwan's ruling elite (see table 6.1 for more on these debates), the Chiang government stepped up efforts to acquire nuclear weapons. The United States found out about these clandestine efforts when a well-placed U.S. informant in 1987 passed information to the United States about Taiwan's nuclear weapons ambitions. The American government concluded in 1988 that Taiwan was a year or two away from developing a nuclear weapon. Confronted with U.S. spy satellite data and under intense pressure from the United States, Taiwan agreed to terms that were designed to conclusively and verifiably end its nuclear weapons program. The conditions involved a written assurance from the newly installed Taiwan President Lee Teng-hui to President Ronald Reagan that Taiwan would end its nuclear program and close the research reactor suspected of nuclear weapons related activities. Taiwan also agreed to a ban on any research that could be applicable to weapons development. Overall, Taiwan's commitments went beyond those obligations required under the NPT.[13]

When the DPP won the presidency in 2000, it supported the position that Taiwan would not apply its scientific know-how in order to build nuclear weapons. The KMT returned to the presidency with Ma Ying-jeou winning the office in 2008 with a platform that opposed nuclear weapons acquisition.

Assessing Recent Discussion and Developments

Given the long record of repeated duplicity and deception by the Taiwan government regarding the development of nuclear weapons, and in light of the growing security asymmetry Taiwan faces in China's rapidly advancing power and influence over Taiwan, foreign and Taiwan specialists have repeatedly reexamined the calculus behind Taiwan's decision to foreswear nuclear weapons. While some Nationalist sentiment continues in Taiwan and urges the country to secure its future with its own nuclear weapons, the assessments see a Realist perspective dominating Taiwan decision making. Taiwan leaders are seen examining the costs and benefits of reviving a nuclear weapons program, concluding that the costs strongly outweigh any benefits for Taiwan. Meanwhile, the capacity of Taiwan to develop nuclear weapons continues to decline with the passing of technical experts, Taipei's adherence to ever more restrictive international control mechanisms, and Taiwan's vibrant and transparent democratic politics moving the island closer to a desired nuclear-free status.

Writing soon after leaving government for a few years, U.S. government Asian security expert Derek Mitchell focused in 2004 on what were and are widely seen as important and growing technical and practical obstacles to Taiwan's pursuing nuclear weapons. Taiwan at the time no longer had materials or reprocessing capabilities. The research reactor suspected of reprocessing had been dismantled under U.S. and IAEA supervision; spent fuel rods were under strict IAEA safeguards. All of Taiwan's nuclear power reactors were subject to regular careful inspection.[14]

Moreover, Taiwan leaders and popular opinion at the time reflected strong Globalist tendencies. Taiwan endeavored to go the extra mile to be seen as a constructive supporter of international norms in handling nuclear materials. Taiwan leaders agreed for Taiwan's nuclear programs to become the first "substantial" (defined as involving big power reactors) program that came to be subject to the additional protocol to the IAEA Safeguards Agreement instituted after the 1991 war in Iraq. The protocol expanded the range of items subject to the IAEA inspection from "nuclear material" to "nuclear related activity"—a much broader scope that severely restricts a nation's ability to cheat. The U.S. State Department and Taiwan participated regularly in the "Joint Standing Committee on Civil Nuclear Cooperation," a confidence-building measure that promotes health and safety in civilian use and includes elements that insure that Taiwan meets its obligations stemming from integration of U.S. nuclear components.

The Mitchell advisory was based on interviews in Taiwan and existing information that Taiwan in 2004 could reverse its recent policies and rebuild the capability to produce a nuclear weapon in one year. The constraints on Taiwan leaders' pursuing such a path included such costs as China's attacking Taiwan as it has promised to do if Taiwan develops nuclear weapons, Taiwan's open society and the questionable loyalty of the Taiwan military to support such provocative efforts by the newly ruling DPP government at that time, and strong opposition by the United States as well as by Japan and other countries Taiwan depends on for its well-being. Mitchell averred that there were strong Nationalist tendencies among the independence-minded wing of the DPP who might have been tempted to pursue a nuclear weapons program in order to support efforts to foster Taiwan's independence of China; but the risks and costs of adopting such an approach were overwhelming.

Six years later, in 2010, American-based nonproliferation specialists Monte Bullard and Jing-dong Yuan adopted a different line of assessment in coming up with conclusions similar to those of Mitchell. They recalled how the threat from China and concern with possible U.S. abandonment drove Taiwan leaders beginning in the 1960s to deal with the problems through Realist means involving domestic balancing— notably through the development of a nuclear weapons capability. They judged that the asymmetry in China-Taiwan military strength had grown steadily since that time, and Taiwan remained unsure of U.S. resolve in protecting Taiwan from China. But Taiwan no longer followed the past path of developing nuclear weapons. The authors used a multi-causal framework to explain Taiwan's recent security calculus. They showed that Realist views persist in Taiwan and judged that the costs of nuclear weapons development far outweighed the benefits.[15]

In particular, Taiwan's close interdependence with the United States, Japan, and other Western aligned countries committed to international nuclear nonproliferation norms meant that pursuing nuclear weapons could jeopardize Taiwan's economic foundations. Moreover, the thaw in Taiwan-China relations initiated under incoming–KMT President Ma Ying-jeou beginning in 2008 greatly strengthened Taiwan's already strong economic dependence on China, adding to reasons to avoid provocative actions surely to be strongly opposed by Beijing. China's threat to use force if Taiwan developed nuclear weapons remained as a credible deterrent to Taiwan's taking this action under prevailing circumstances. Meanwhile, the authors noted that Taiwan's technical abilities to develop nuclear weapons continued to erode with the passing of time. The vibrant, open, and politically competitive Taiwan society meant that taking such a controversial step as developing nuclear weapons would almost certainly become public knowledge and be used by the president's opponents both within the KMT and in the DPP to undermine his leadership.

The authors rightfully called attention to the new era of thaw in the cross-strait relations since 2008 and the significance of Taiwan's new strategy, resting on reassurance and accommodation, toward China. Though Beijing's military buildup targeting Taiwan has continued, the PRC leaders also have reciprocated Taiwan's moderation with accommodations on their side, resulting in a positive-sum dynamic that benefits both sides. President Ma's legacy rests fundamentally on managing this complicated process effectively; developing nuclear weapons appeared to have no place in this strategic calculus. Meanwhile, President Ma's approach also stressed Taiwan's international role as a responsible power, determined to support existing norms that foster regional and global stability. The importance of this Globalist thinking to the Ma government and internationally minded elite and the importance of public opinion inside Taiwan further argued against pursuing nuclear weapons. How far DPP presidential candidate Tsai Ing-wen might deviate from this moderate approach to China were she to win the presidential election in January 2016 remains to be seen. She promises to sustain cross-strait stability.

In 2012, Taiwan-based specialist Arthur Ding reinforced and updated earlier findings on why developing nuclear weapons is neither the subject of meaningful debate nor a likely option for Taiwan in the foreseeable future. Ding probed the thinking of the few Taiwan commentators who talk openly about developing nuclear weapons. Like Derek Mitchell, Ding suspected that some radical DPP members might consider developing nuclear weapons in pursuing their Nationalist goal of Taiwan independence. However, the evidence against such an option includes the continued erosion of Taiwan's capabilities to carry out such a nuclear weapons program; the stronger, domestic, anti-nuclear movement, which would very likely find out about any clandestine nuclear weapons program and mount enormous pressure against it; and powerful opposition from the United States. Ding underlined the importance of President Ma Ying-jeou's reassurance policies toward China as well as the United States, pledging to abide by international safeguards in seeking a nuclear weapons–free zone in the Taiwan area. Ding viewed Taiwan's existing cruise missile programs and possible ballistic missile programs, rather than a nuclear weapons program, as providing options favored by Nationalist opinion in Taiwan seeking indigenous weapons to deter China.[16]

Interviews conducted in Taiwan in 2013 showed that the Taiwan government still maintains that Taiwan has the ability to develop nuclear weapons, but it has disavowed the intention to do so and has placed Taiwan under strong international supervision in order to buttress the credibility of that stated intention.[17] Those interviewed reaffirmed the arguments of the assessments noted above as to why Taiwan would not pursue nuclear weapons under existing circumstances. Those interviewed were well aware of Taiwan's growing disadvantage vis-à-vis China's massive military buildup. They understood Taiwan's declining capacity to develop nuclear weapons and the latter trend's likely continuation with the island-wide pressure to move Taiwan to a nuclear-free status.

The sense of ambiguity in Taiwan over U.S. willingness to support Taiwan with advanced weapons and defend it in the event of Chinese attack remains strong. Notably, the remarkable public pledge of President George W. Bush in April 2001 that the United States would do "whatever it takes" to help Taiwan in the face of a Chinese attack was not repeated by him or President Obama.[18] Nonetheless, American military planning prepares to fight China in the event of an attack from the mainland against the island.[19] Against this background, the Ma government places heavy reliance on its ability to manage tensions with China within an overall policy of reassurance. The Chinese and American governments both support Ma's policy and the resulting thaw as positive cross-strait exchanges expand and advance. As seen in table 6.1, the overall situation argues against Taiwan moves toward developing nuclear weapons.

TAIWAN AND THE GLOBAL NUCLEAR NONPROLIFERATION REGIME

Taiwan's status as an isolated government officially unconnected with most international nuclear nonproliferation regimes prompts concerns that government elites in

Taiwan discuss and endeavor to deal with in generally pragmatic and accommodating ways along lines of the Globalist and Realist schools of thought defined in this study. Taiwan's past experience developing nuclear weapons, its longstanding civilian nuclear program, its technologically advanced economy and knowledgeable people, and its location along widely used international transit routes are salient factors added to the growing international concern since the end of the Cold War in the late 1980s to monitor more carefully and closely the transfers of nuclear materials and technology that rogue governments, non-state actors, or others could use in possibly disastrous ways. The combination of these considerations means that more stringent mechanisms have been seen as needed in order to insure that Taiwan is part of the solution and not part of the problem in the still porous international nuclear control regime.

Taiwan's Unique Contributions to the Nonproliferation Regime

The Taiwan government by and large has been cooperative in these efforts to establish more stringent controls in the international nuclear control regime. Interviews in Taiwan and other evidence underline a recent emphasis of the ruling KMT government of President Ma Ying-jeou to enhance Taiwan's international image and status as a "good citizen" in international affairs, seeking to sustain stability and development through adherence to international norms. Although the opposition DPP disagrees strongly with the Ma government on a host of issues, including major debates over the future of nuclear power in Taiwan, DPP representatives in Taiwan and abroad disclosed in interviews that the party is in accord with Ma's approach regarding the global nuclear nonproliferation regime, emphasizing Taiwan's responsible international behavior in line with world norms.[20]

The completion of an important atomic cooperation agreement between Taiwan and the United States in 2014 replaced an expiring forty-year bilateral nuclear agreement. It was hailed particularly for Taiwan's pledges to forego plutonium reprocessing or nuclear enrichment. It was accompanied by classified and unclassified U.S. government assessments to Congress backing President Obama's judgment that "over the past two decades, the authorities on Taiwan have established a reliable record on nonproliferation and on commitments to nonproliferation."[21]

Togzhan Kassenova of the Carnegie Endowment for International Peace has provided a special service among leading specialists in calling attention recently to Taiwan's peculiar position in the international nuclear nonproliferation regime that requires unusual means to insure Taiwan is not a source of dangerous proliferation.[22] As noted by Kassenova, Taiwan is one of the world's key suppliers and consumers of high-technology, dual-use goods, and technology. It has an active nuclear energy program that involves the flow of sensitive materials and technology. Taiwan is located along major sea lanes and is a major transit and transshipment hub. And there have been reported attempts—some successful—to smuggle weapons of mass destruction (WMD)–sensitive goods and technological knowledge out of Taiwan or across its territory.

Taiwan's "non-state" status, reinforced by rigorous monitoring and pressure by China, prevents the government from participating in multilateral nonproliferation treaties and export control systems, and it limits Taiwan's access to information and intelligence sharing from national and international security agencies. The United States and other states and international bodies, however, endeavor to create mechanisms that allow for more active and effective Taiwan participation in the substance of international nuclear nonproliferation efforts. They find the Taiwan government to be a willing partner, reflecting the commitment of the KMT government and the support of the opposition DPP to emphasize a Globalist perspective of international cooperation and accommodation with international norms.

Also, Taiwan officials and opposition politicians almost certainly consider, along the lines of the Realist school of thought, the pros and cons of Taiwan's recent emphasis on good citizenship abroad in terms of the degree of support Taiwan can expect from the world in the years ahead. The Ma government and the previous DPP government have demonstrated an increasing interest in enhancing Taiwan's international image in such ways, thereby strengthening Taiwan's "soft power" as it seeks international support in a still quite dangerous situation for Taiwan's future both vis-à-vis China and more broadly.[23]

As Kassenova points out, Taiwan's "non-state" status means that Taiwan is not a party to nonproliferation requirements such as U.N. Security Council Resolution 1540, which places a legal requirement on all U.N. member states to implement "appropriate" domestic controls to prevent WMD proliferation to non-state actors. Taiwan's loss of U.N. membership since 1971 in effect excludes Taiwan from membership in international nonproliferation treaties and Taiwan has lost the right to be considered for membership in multilateral exports control regimes (MECRs). Those regimes are the Australia Group, the Nuclear Suppliers Group, the Missile Technology Control Regime, and the Wassenaar Arrangement. Each of these groups adopts a list of items that should be controlled for nonproliferation purposes. They also provide forums for member countries to exchange information on emerging proliferation-sensitive products, suspicious actors, export license denials, and other proliferation-related factors.[24]

Meanwhile, Taiwan is unable to call on international law or its enforcement agencies, such as Interpol, in pursuing known proliferators and their agents. Taiwan can enact official extradition agreements only with the handful of states that still recognize the Republic of China. As a result, Taiwan is not able to extradite criminals from other nations, nor can other nations easily repatriate criminals from Taiwan. The strict curbs on Taiwan's official international interaction mean that it does not have access to intelligence available through Interpol's I-24/7 global police communications system that has been effective in dealing with international criminals and their activities.[25]

Taiwan's decision to end clandestine nuclear weapons development coincided with the greatly expanded scope and strengthened concern internationally to fortify international nuclear nonproliferation and WMD control regimes in the post–Cold War period. As is known, the scope and degree of concern with controlling the international flow of nuclear and other WMD materials and technology grew substantially after the terrorist attack on America and the U.S.-led war on terrorism during the George W. Bush administration. Against that background, the United States played a major role in getting Taiwan to conform to international norms by ending its nuclear weapons program, and the United States has continued to work actively in more recent years to strengthen Taiwan's strategic trade control system. Taiwan now participates in several U.S.-led nonproliferation initiatives, including the Container Security Initiative (CSI) and the Megaports Initiative. Taiwan's two main ports, Kaohsiung and Keelung, participate in the CSI, in which ports conduct screening of high-risk cargo before it departs for the United States. Kaohsiung port, one of the world's largest container ports and transshipment points, is a member of the Megaports Initiative that is designed to strengthen the capabilities of key ports to deter, detect, and interdict illegal radioactive and nuclear cargo.[26]

As noted above, in the absence of a standard IAEA agreement with Taiwan, the United States and the IAEA have shared responsibility for monitoring and enforcing nuclear safeguards in Taiwan. After the Taiwan government ended its clandestine nuclear weapons program, it undertook added safeguard responsibilities, notably in line with the requirements contained in the IAEA Model Additional Protocol of 1998. It also strengthened domestic proliferation controls. Taiwan's Foreign Trade Act provides the legal basis for strategic trade management; it includes provisions for penalizing those who intentionally or unwittingly engage in unauthorized transfers of goods and technology that can lead to WMD proliferation. A range of additional laws and regulations based on the foundation of the Foreign Trade Act provide detailed guidance on specific categories of strategic goods and technologies.[27]

While not a member of the MECRs, Taiwan has adhered to all four regimes by incorporating their respective control lists into Taiwan's national list. The Taiwan list and resulting enforcement have been seen by specialists as demonstrating a commitment to follow MECR guidelines closely and to take necessary measures to prevent proliferating states and non-state actors from acquiring sensitive goods and technologies.[28]

Togzhan Kassenova and other foreign specialists point to the necessity, among areas needing improvement in Taiwan's controls, for better control of so-called "intangible technology transfers," which involve transfers of sensitive technology by electronic means or technical discussions related to controlled technology with foreign citizens. They further note limitations in Taiwan controls over transit and transshipment of strategic cargo. Taiwan's detection, investigation, prosecution, and punishment of violators also are seen to need improvement as some violators have received only minimum punishments.[29]

That the Taiwan government has seen its interests well served by continuing to improve its record in embracing international norms regarding the international nuclear nonproliferation regime and other global regulations on the flow of WMD materials and technology was evident in deliberations between Taiwan and the United States over a renewal of the above-noted 2014 American agreement with Taiwan regulating U.S. nuclear cooperation with Taiwan.[30] Taiwan's adherence to such restrictions on reprocessing and enrichment went against the Nationalist viewpoint heard during interviews in Taiwan in 2013 that Taiwan should use reprocessing of spent fuel in order to allow for more Taiwan self-sufficiency in meeting its energy needs.[31] By accommodating the American preference and international norms consistent with high nonproliferation standards, Taiwan reinforced its positive international image, in line with the Globalist perspective of Taiwan elites. For those Taiwan leaders of a Realist mind-set, the purported Taiwan move benefitted Taiwan's close ties with the United States and other friendly countries, reassured Beijing of Taiwan's strategic intentions regarding possible development of nuclear weapons, and accorded with growing overall anti-nuclear sentiment among Taiwan voters. Taiwan's image building in this way was reinforced as neighboring U.S. ally South Korea remained very reluctant to abide by such a U.S.-sought restriction on nuclear reprocessing in South Korea.[32]

TAIWAN DEBATES NUCLEAR ENERGY

The quiescent discussion and debates among contemporary Taiwan government and nongovernment elites and specialists underlining a consensus in Taiwan today regarding the pros and cons of developing nuclear weapons and accommodating to international nonproliferation norms often stood in marked contrast to the recurring intense debates in Taiwan over whether or not to complete a long-delayed and almost-finished fourth nuclear power plant and the overall future of nuclear power in Taiwan. The scope of the latter debate goes well beyond policy elites and special interests. It involves mass demonstrations, widespread media coverage, and extensive public opinion polling.

The debate is highly politicized as the main political parties, the KMT and the DPP, have long been at odds over the fourth nuclear power plant at Lungmen (the Lungmen plant) and the future of nuclear power in Taiwan. The upsurge of the debate in 2013–2014 ended with the KMT government of President Ma Ying-jeou retreating in the face of popular protests led by a former DPP chairman against the plant. On April 27, 2014, the Ma government decided to halt construction on the almost-completed second reactor in the plant and to seal off the plant's first reactor once safety checks are completed.[33] As noted above, by early 2016 the entire fourth plant stood basically completed and had been in a storage or "moth-ball" status for three years, at a cost of over U.S. $100 million. To compensate for the lost electric power that was supposed to come from the new plant, Taiwan has increased its imports of expensive liquefied natural gas (LNG).[34] The prospects for actually using the plant remain poor, especially as the DPP's Tsai Ing-wen, an opponent of the plant, won the presidential election in January 2016.

The two main political parties repeatedly maneuver for advantage or to avoid negative repercussions on nuclear power and related issues, thereby adding to the salience of the highly charged nuclear power discussions and debates prevalent in recent years in Taiwan. The Taiwan debate on the Lungmen plant and the future of nuclear power in Taiwan has a long history, which is discussed in box 6.1. Over time, the debate has waxed and waned for various reasons. The level of concern during 2013–2014 was high, but arguably it was not as intense as the debates over the Lungmen plant after the DPP finally won control of the presidency in 2000 and considered ending the plant construction in line with its longstanding platform favoring an end to nuclear power in Taiwan.

BOX 6.1. THE CONTROVERSY OVER THE LUNGMEN NUCLEAR POWER PLANT, 2000

The Democratic Progressive Party long aligned with Taiwan's environmental and anti-nuclear power activists in its struggle for power against the dominant KMT. The 2000 presidential election saw DPP candidate Chen Shui-bian win less than 40 percent of the popular vote, but nonetheless he won the presidency against a divided opposition of a KMT candidate and a former KMT stalwart who had broken from the party. Chen faced a Legislative Yuan still dominated by the KMT and allied politicians. And his government apparatus, under KMT dominance for decades, was slow to follow his lead.[1]

Controversy over the Lungmen Nuclear Power Plant dominated the new president's first year in office. Environmental and anti-nuclear advocates within and allied with the DPP pushed President Chen to live up to his campaign promise to terminate the Lungmen nuclear plant that was undergoing construction. Following Chen's taking office in May 2000, a government committee reviewed the plans for the plant in June but could not reach a consensus on the project's fate, though a majority favored ending the project. Chen's prime minister, a former general and member of the KMT, favored building the nuclear power plant and indicated that he would resign if the project were cancelled. President Chen gave indications that he would support cancelling the project and a new economics minister in September recommended suspending the project; the prime minister resigned and was replaced by a prime minister who supported ending the Lungmen plant.

The KMT-led legislature argued that the plant could not be terminated without that body's approval. On October 27, 2000, Chen tried to find common ground in a meeting with the KMT leader; an hour after the meeting,

Chen's prime minister announced that he would end the plant on his own authority. The seemingly abrupt and likely poorly coordinated move embarrassed the KMT leader and steeled the KMT opposition in the legislature. The KMT-led opposition launched an unprecedented effort to unseat Chen through a presidential recall. Public and business confidence in Chen and his rule declined. Chen was forced to apologize in November. And he asked the Taiwan's high court, The Council of Grand Justices, to determine whether the Taiwan executive on its own authority could end the nuclear power plant project. The judges ruled in January 2001 that the decision required approval by the legislature. After negotiations between the prime minister and the legislature's leader in February, a compromise was reached to resume construction of the power plant.

The tortuous events over ten months of the Chen government's first year in office showed poor coordination and serious divisions among the DPP members and an enormous gap between the DPP and the KMT. The events raised serious and arguably lasting questions as to whether Chen and his associates had sufficient skills to manage contentious issues in Taiwan politics. Although the DPP gained some seats in the legislative elections in December 2001, President Chen still did not have sufficient support to push cancellation of the Lungmen Nuclear Power Plant project through the legislature. In particular, the DPP's main allied party in the legislature was directed by former President Lee Teng-hui who was known to oppose the decision to cancel the Lungmen Nuclear Power Plant.

NOTE

1. Steven Goldstein and Julian Chang, *Presidential Politics in Taiwan: The Administration of Chen Shui-bian* (Norwalk, Conn.: Eastbridge, 2008).

Nuclear Dissonance in Taiwan's Domestic Debates

Anti-nuclear sentiment in Taiwan revived strongly after the Fukushima disaster in Japan in March 2011. The DPP presidential candidate in the January 2012 election, Tsai Ing-wen, staked out a position against the President Ma Ying-jeou government's policy to complete construction and begin operation of the Lungmen plant. The candidate pledged that Taiwan under DPP rule would end nuclear power by 2025. President Ma favored a more gradual path involving terminating the three existing nuclear power plants once their normal length of operation concludes and seeking an end of nuclear power on the island in the much later future, presumably once the Lungmen plant reached the end of its operation. Tsai's position failed to generate support sufficient to close the gap between her and President Ma as the incumbent was reelected by a wide margin.[35]

Persistent controversy and repeated media reports of safety and operational shortcomings at the Lungmen plant and at already-operating Taiwan nuclear power plants combined with an initiative in February 2013 by the newly installed KMT Premier Jiang Yi-huah to propose a national referendum to decide the fate of the Lungmen plant. The result was a large increase in debate, including a March 9, 2013, rally that drew 200,000 protestors calling for the halt of the Lungmen plant.[36] By June, Chinese Nationalist Party (KMT) legislators, backed by the government, proposed a referendum with the question, "Do you agree that the construction of the Fourth Nuclear Power Plant should be halted and that it not become operational?" The referendum needed to be approved by Taiwan's legislature, the Legislative Yuan, and then voted on by the electorate.[37] In the event, the referendum remained to be approved by the legislature, amid extraordinary efforts by the DPP to block legislative action and apparent divisions within the KMT legislative caucus about the wisdom of going forward with the referendum at that time and amid the broader debate about nuclear power in Taiwan's future.[38]

The decision of former DPP leader Lin Yi-hsiung to begin a hunger strike on April 22, 2014, against the Lungmen plant revived public attention to this divisive issue following weeks of political turmoil in Taipei. In March 2014, crowds of protestors occupied the legislature and carried out other acts of civil disobedience in response to the KMT efforts to push through the legislature a controversial services trade agreement with China reached in 2013. The dissident movement—not affiliated with any political party but obviously opposed to Ma and KMT practices—became known as the "sunflower movement." Lin's hunger strike caused April 27 demonstrations urging a stop to the Lungmen plant. Against this background, President Ma faced strong divisions within the KMT about the Lungmen plant and overall public opinion ratings that at times were in the single digits. The result was the KMT decision on April 27 to halt the Lungmen plant.[39]

Despite the strong attention to the Lungmen plant and related issues by media, politicians, and various activists, it was commonly predicted in Taiwan that the referendum would fail due to insufficient votes. Taiwan's Referendum Act of 2003 requires a voter turnout of at least 50 percent of *eligible* voters. That means about 9.15 million votes are needed for a referendum to be considered valid, and of those who voted, at least 50 percent must vote "yes" to the question for it to be passed. Getting 50 percent of eligible Taiwan voters to vote in favor of referendum issues failed in 2004. Polls in Taiwan show that a majority supports deciding the fate of the Lungmen plant by means of a national referendum. Nevertheless, interviews with Taiwan anti-nuclear activists in May 2013 pointed to the Taiwan public's "fatigue" and disenchantment with what the public reportedly sees as politicians manipulating debates on sensitive issues like nuclear power for the sake of the narrow political interests of their parties. The dissatisfaction adds to reluctance to vote on what is widely seen as a "politicized" referendum, even though popular concern with safety and other issues at the Lungmen and other plants remains high.[40] Nevertheless, in 2014 President Ma decided to abandon the referendum option and end the controversy by shutting the fourth plant.

Features of the Nuclear Power Debate

Taiwan's energy sources rely heavily on imports. Taiwan imports more than 97 percent of its primary energy as of January 2014. In recent years, Taiwan's electric power came from coal (about 40 percent), with liquefied natural gas providing over 30 percent, nuclear power over 18 percent, hydropower 4 percent, oil 2.5 percent, and wind and solar .7 percent.[41] The output of Taiwan's three existing nuclear power plants, each with two nuclear power reactors, declined by 4 percent in 2012 due mainly to an extended shutdown during a refueling outage to obtain restart approval after replacement of seven damaged anchor bolts at one of the nuclear reactors. The discovery of the damaged bolts was among a litany of safety concerns highlighted by the media and activists warning that Taiwan's existing nuclear plants and the Lungmen plant are unsafe.[42]

Reviews of safety and other considerations in Taiwan's existing nuclear plants and the almost completed Lungmen plant were carried out by the Taiwan government in the wake of the Fukushima disaster and the enhanced public debate over nuclear power. The government reviews found a number of safety concerns, many of which were highlighted graphically by activists in the media and in videos over the Internet to demonstrate what were seen as serious dangers posed to Taiwan by nuclear power plants and related issues, notably the storage of nuclear waste. The government's Atomic Energy Council affirmed that these shortcomings were being addressed and they arranged to have international teams review a so-called "stress test" of Taiwan's three existing nuclear power plants and the overall safety of Taiwan's nuclear power system.[43]

Interviews in May 2013 with a variety of specialists and activists in Taiwan, including opponents and supporters of the KMT government, showed that the credibility of Taipower in dealing with nuclear matters and the credibility of the Atomic Energy Council, the government body that oversees Taipower's operation of nuclear matters, had been questioned repeatedly. As a result, despite ongoing efforts of the company and its oversight body to show the safety of Taiwan's existing and planned nuclear power, basic issues involving the location of the plants, their construction, their operations, and their arrangements for storage of nuclear waste remain subject to wide media and public debate with no resolution in sight.[44]

A senior government official with responsibility for nuclear energy and economic development echoed others in explaining that the Fukushima disaster influenced opinion in Taiwan against nuclear power for several reasons. For one, the Fukushima plant used U.S. technology similar to that used in Taiwan's three operating plants. Seeing such technology, viewed as high quality by Taiwan in the past, fail in Fukushima raised serious questions in Taiwan about the safety of the Taiwan plants.

Also, Taiwan opinion is said to regard Japanese industry as more cautious and more carefully regulated than that in Taiwan. The failure of Japanese oversight in Fukushima caused many in Taiwan to have more serious doubts about real or suspected shortcomings in Taiwan's operation and regulation of nuclear power plants. Such concern was exacerbated by the fact that the Lungmen plant wasn't being developed under the guidance of a U.S. company as the first three plants had been, and that reportedly poor coordination among the many different contracting entities hired by Taipower to build the Lungmen plant led to numerous reported structural-integrity issues and other shortcomings.[45]

Against this background, basic questions were raised. They involve (1) whether or not the Taiwan plants are located in zones vulnerable to tsunamis, earthquakes, or other seismic dangers; (2) whether or not the large populations near the nuclear power plants would have the ability to evacuate effectively in the event of nuclear disaster at one of the plants; and (3) whether or not the seemingly ad hoc decisions to store spent nuclear fuel from the nuclear power plants temporarily at reportedly crowded power plant sites because of the absence of a permanent storage site will lead to leakage from rusting storage barrels or other dangers to the Taiwan environment.[46]

Nevertheless, the support for the anti-nuclear movement remained limited in 2012, underlined by the failure of the DPP challenger in the January 2012 presidential election. Among the reasons cited by those interviewed in May 2013 for the revival of the nuclear energy debate was Taipower's request in 2013 for additional funds from a skeptical Taiwan legislature to finish the Lungmen Nuclear Power Plant, and then Premier Jiang Yi-huah's surprise call for a national referendum on the future of the Lungmen plant.[47]

According to officials, specialists and activists interviewed in Taiwan in May 2013, other elements influencing the recent debate in Taiwan include the following:[48] Without the Lungmen plant and other nuclear power plants, Taiwan's electricity will cost more. Nuclear power is more than four times less expensive than LNG-generated power and more than two times less expensive than coal-generated power.[49] Moreover, Taiwan people resist increased costs for electric power.[50] Efforts by the Ma government in 2012 to pass to consumers more of the actual cost of producing electricity proved to be widely unpopular. Thus, rather than resort to what one observer called "political suicide" in passing big electricity cost increases on to consumers, continuing plans for nuclear power may appear more acceptable, to at least some leaders. On the other hand, anti-nuclear activists assert that the Taiwan people and they will be willing to "turn up the air-conditioning thermostat" and take other steps to deal with more expensive electricity, provided Taiwan benefits from more rapid movement away from nuclear power. The low regard that some observers in Taiwan have of Taipower leads them to argue that the utility should be reformed in ways that would result, in the view of these critics, in considerable cost savings for Taiwan electricity consumers.

Nuclear power does not produce the greenhouse gases associated with coal and, to a lesser degree, LNG-produced energy. Thus, Ma government environmental officials have maintained that Taiwan faces more danger from climate change caused by production of greenhouse gases than it does from plans for continued use of nuclear power plants.[51] On the other hand, anti-nuclear advocates argue that reducing overall electricity consumption by curbing air-conditioning and other means is a better way to reach emission goals than continued use of nuclear power plants with what they view as their associated dangers.[52]

Advocates of continued use of nuclear power argue that such power generation makes Taiwan more secure in the face of a blockade or other curtailment of international supplies. Taiwan maintains only a ten-day supply of LNG and a sixty-day supply of coal for its power reactors; it has the nuclear materials it needs to keep its reactors going for one year.[53] Some advocates for nuclear power go further in regard to energy security and self-reliance, arguing for Taiwan to arrange reprocessing of its spent nuclear fuel for reuse in Taiwan reactors.[54]As noted in the second section of this chapter, the Taiwan government has foresworn reprocessing on Taiwan soil, though such reprocessing possibly could be done in the United States or some other location. Predictably, those opposed to continued use of nuclear power in Taiwan see the costs of these options outweighing perceived benefits.

COMPARING TAIWAN'S NUCLEAR DEBATES

The overall framework of analysis for this book affirms that the Nationalist, Realist, and Globalist schools of thought will act consistently on multiple issues and provide indicators of how and why Taiwan decision makers would be likely to move in one direction or another on key questions in international affairs. The framework of analysis seems to be valid when looking at the assessments of Taiwan's development of nuclear weapons and the global nuclear nonproliferation regime. Available assessments reviewed above show that on the question of nuclear weapons development in Taiwan, the Nationalist and Realist viewpoints among the KMT elite decision makers saw the merits of developing nuclear weapons in a period of ascendant Chinese nuclear weapons and American vacillation over security commitments to Taiwan. Decision makers with this viewpoint preferred the security of Taiwan's possessing nuclear weapons to the uncertainty of continued reliance on waning international support and uncertain American backing.

Those with a Realist viewpoint among the KMT elite almost certainly calculated the costs and benefits of Taiwan's clandestine nuclear weapons program, presumably concluding that the benefits outweighed the costs. As they pursued a secret program, the Taiwan leaders may have judged that the costs would be low if the program were not detected. However, they pursued an intensified clandestine effort to develop nuclear weapons in the 1980s, even after taking serious measures to ostensibly end the program in the face of strong U.S. pressures in the late 1970s and early 1980s. This suggests that they judged that the risks of being caught cheating again on existing commitments to the United States were offset by the perceived need to have nuclear weapons to preserve the security of their increasingly isolated state.

Throughout these years, the influence of KMT elite officials with a Globalist outlook that would work against the Nationalist and Realist views seemed weak. Perhaps such influence and officials didn't exist. Also during these years, the nascent environmental and related anti-nuclear power movement in Taiwan and the emerging opposition party, the DPP founded in 1986, still were marginal forces in a martial law order dominated by Chiang Ching-kuo and his KMT colleagues.

According to assessments by foreign and Taiwan experts since the 1980s and the end of the Cold War and authoritarian rule in Taiwan, the Realist perspective among government decision makers in the post–martial law order in Taiwan has been heavily influenced by circumstances that include eroding capabilities in Taiwan to produce nuclear weapons; threats of attack by China if Taiwan develops nuclear weapons; strong opposition from the United States, Japan, and other states on which Taiwan depends; and vocal opposition from within Taiwan's increasingly vibrant and transparent democracy. The thaw in cross-strait relations initiated by Taiwan President Ma Ying-jeou in 2008 adds to circumstances working against pursuing nuclear weapons in Taiwan. Under these conditions, those remaining Nationalists or Realists favoring Taiwan's acquisition of nuclear weapons are viewed as fringe elements. A common speculation is that strongly committed, pro-independence elements in DPP would be prone to consider developing nuclear weapons in order to preserve Taiwan's independence, but there is little publicly available evidence to support this view. Meanwhile, since the end of the Cold War and authoritarian rule in Taiwan in the late 1980s, both DPP and KMT leaders have adopted strongly Globalist views emphasizing that Taiwan's future rests heavily on international norms that Taiwan is committed to protect. Nuclear nonproliferation norms are part of this broad commitment.

Post–martial law Taiwan has featured growing opposition to nuclear power and nuclear waste in Taiwan that has focused heavily on safety concerns involving having potentially very dangerous nuclear power plants and nuclear waste sites in small, crowded, and seismically active Taiwan. The public debate has not focused on Taiwan's development of nuclear weapons, its adherence to international nuclear proliferation regimes, or its willingness to foreswear reprocessing nuclear fuel in Taiwan. However, the debate over nuclear energy has been so widespread and politically salient that it has compelled previously pro-nuclear energy advocates and politicians to reconsider their positions. One of the two KMT politicians seen in 2013 as most likely to be selected as the KMT candidate in the 2016 presidential election, Taipei Mayor Hau Ling-bin, changed his position and went against his party's leaders on the proposed referendum on the Lungmen plant. He argued in 2013 that the referendum should not be held and that construction of the plant should stop right away.[55]

The result is that both the DPP, long associated with the anti-nuclear power movement, and the KMT have increasingly moved in directions that will diminish the nuclear footprint in Taiwan. Both parties are now committed to goals of a nuclear-free Taiwan. Under these circumstances, consideration of the introduction of nuclear weapons in Taiwan or nuclear reprocessing there appears increasingly remote. The anti-nuclear sentiment also would appear to impel Taiwan leaders to advance their recent Globalist outlook in becoming one of the world leaders in fostering a strong global nuclear nonproliferation regime. The 2014 atomic cooperation agreement between Taiwan and the United States discussed earlier appears to underline this trend.

The utility of the Nationalist, Realist, and Globalist schools of thought is less effective when one assesses Taiwan's nuclear energy debates, as opposed to its nuclear weapons and nonproliferation debates. One reason is that the former debates have become so widespread among a variety of interest groups and constituencies in Taiwan. Another reason relates to the fact that the energy debates have become politicized—part of the highly partisan struggle for political leadership in Taiwan between the KMT and the DPP. Each political party continues to endeavor to bring constituencies and interest groups into its camp. The result is that in the recent debate focused on whether or not to begin operation of the Lungmen Nuclear Power Plant, one finds Nationalists, Realists, and Globalists on both sides of the debate.

The strong opponents of the Lungmen plant include DPP Nationalists like former DPP leader Lin Yi-hsiung, whose concern relates to the purity and integrity of their homeland, Taiwan, which they see challenged by the danger of potentially massive contamination from nuclear reactors and nuclear waste. The Globalists among the opponents of the plant feature advocates active in demonstrations who physically and virtually join hands with skeptics of nuclear power throughout the world. They work together with advocates in Germany, Japan, and other highly developed countries who are part of a recent world movement against nuclear power. The Realists among the opponents include some KMT dissenters from the Ma government push in 2013 to complete the Lungmen plant who endeavor to reconcile doing away with nuclear power with Taiwan's commitment to reduce its international carbon footprint and help slow global warming and climate change; and they address the purported rise in electricity costs that is forecast with the end of nuclear power. In both cases, they tend to stress greater energy conservation along with greater emphasis on renewal energy as an appropriate path.

By contrast, as noted above, some Nationalists among the supporters of the Lungmen plant argue that nuclear energy provides much more energy security for Taiwan than imported coal or LNG. Some favor reprocessing spent nuclear fuel in Taiwan as a way to enhance Taiwan's energy security. Globalists among the supporters of the plant include business people and economists and their allies in the KMT who argue that Taiwan needs dependable, affordable, and sufficient energy if it is to compete effectively in the global economy. In their view, conforming to the dictates of international economic globalization makes nuclear energy a cost-effective and necessary component in Taiwan's energy supply mix. As noted above, other Globalists, including environmental officials in the current government, see a trade-off in that using nuclear energy avoids increasing Taiwan's carbon footprint and allows Taiwan to meet commitments to curb greenhouses gases and resulting climate change.

Table 6.2. Review of Contemporary Issues and Questions in Taiwan's Nuclear Debates

Issues[1]	*Realists (Anti-Nuclear)*	*Realists (Pragmatic)*	*Nationalists (Anti-Nuclear)*	*Nationalists (Pragmatic)*	*Globalists (Anti-Nuclear)*	*Globalists (Pragmatic)*
1. Nuclear Weapons	Low	Medium	Low	Medium	Low	High
2. Safety Risks	High Concern	Medium Concern	High Concern	Medium Concern	High Concern	Medium Concern
3. View Adversary	Moderate Negative	Moderate Negative	Moderate Negative	Moderate Negative	Moderate Negative	Moderate Negative
4. Reprocessing Technology	Oppose	Some favor	Oppose	Some favor	Oppose	Some favor
5. Spread Dual-Use Technology	Oppose	Oppose	Oppose	Oppose	Oppose	Oppose
6. Export Nuclear Technologies	Oppose	Oppose	Oppose	Oppose	Oppose	Oppose
7. Transparency	Favor	Favor	Favor	Favor	Favor	Favor
8. Openness to Nuclear Safety Regimes	High	High	High	High	High	High
9. Openness to Nonproliferation Regimes	High	High	High	High	High	High
10. Political Imperative for Proliferation	Opposed	Opposed	Opposed	Opposed	Opposed	Opposed
11. Role of Nuclear Weapons in Defense Strategy	Opposed	Opposed	Opposed	Opposed	Opposed	Opposed
12. Attitude on U.S. Nonproliferation Role	Positive	Positive	Positive	Positive	Positive	Positive
13. Desirability of Role of Nongovernmental Groups	Medium	Some Low	High	Some Low	High	Low

[1]Further details on the issues discussed in this chart:

1. Desirability of nuclear energy in the country's energy mix
2. Views reflecting level of concern on proliferation, terrorism, and safety risks of pursuing nuclear power
3. Views on adversary's pursuit of nuclear energy and/or nuclear weapons capability
4. Attitude toward reprocessing technology (traditional and emerging technologies) in Taiwan
5. Attitude toward spread of dual-use technologies, especially enrichment and reprocessing
6. Views on export of nuclear technology to entrant states and others
7. Acceptable level of transparency in the country's use of dual use technology, especially civil versus military nuclear program
8. Openness to formal and informal nuclear security and safety arrangements/regimes
9. Openness to formal and informal nuclear nonproliferation regimes
10. Attitude toward the country's political imperatives for nuclear proliferation
11. Preferred role of nuclear weapons in the country's defense strategy
12. Attitude toward U.S. leadership role in nonproliferation efforts
13. Desirability of nongovernmental groups influencing country's nuclear policy

Realists among the supporters of the plant include those who take account of the loss to Taiwan's taxpayers and Taiwan's economy if the almost-completed plant is abandoned. These Realists are not convinced by the opponents' arguments that the loss of inexpensive electricity from nuclear power could be offset by conservation or renewable energy in Taiwan.

Against this background, listed in table 6.2 are the substantive issues and questions tracked in this volume and how Nationalists, Realists, and Globalists in Taiwan agree and disagree with one another and among themselves about these issues. The split in each group (Nationalist, Realist, and Globalist) is evident between those strongly committed to a nuclear-free Taiwan and those with more pragmatic leanings that give greater weight to other considerations.

CONCLUSION

Taiwan's debates over nuclear energy will continue, even with the halt in completing the Lungmen plant. The highly politicized debates of 2013–2014 were used by DPP candidates to mobilize support in island-wide elections that the DPP won by a landslide in November 2014. The KMT government was placed on the defensive. By reversing policy and halting the fourth plant, KMT leaders can try with better effect to portray the party as also moving Taiwan to a nuclear-free future, albeit in a less radical and disruptive manner than would be required by the DPP's commitment to a 2025 deadline. Of course, the debate over nuclear energy for now has waned in light of the January 2016 presidential and legislative elections that saw a landslide victory of the DPP's Tsai Ing-wen and her legislative colleagues who are opposed to nuclear power in Taiwan.[56]

The broad nuclear energy debates include so many groups and constituencies on the two opposing sides that the Nationalist, Realist, and Globalist framework used in this book is not very useful in assessing likely directions. As summarized in table 6.2, there are Nationalists, Realists, and Globalists in both camps, in more or less comparable numbers.

The Nationalist, Realist, and Globalist schools of thought effectively explain Taiwan's past decisions in pursuing clandestine acquisition of nuclear weapons, as well as Taiwan's growing commitment to Globalist norms regarding support for international nonproliferation regimes. One outcome of the broad nuclear energy debates summarized in table 6.2 is that there is important common ground among the three schools of thought that add to powerful circumstances that push Taiwan toward a nuclear-free future. Such circumstances add to reasons why Taiwan decision makers would eschew nuclear weapons development and strengthen Taiwan's growing interest in a position as a world leader in fostering and supporting rigorous, global, nuclear nonproliferation.

NOTES

1. "Taiwan's Fourth Nuclear Power Plant to be Mothballed in July," *CNA*, February 4, 2015, http://www.wantchinatimes.com/news-subclass-cnt.aspx?id=20150204000044&cid=1201.

2. Taiwan government representatives stress, among other evidence of Taiwan's strict adherence to peaceful uses of nuclear energy, that the government has been upholding the spirit of the NPT; has pursued a policy of "no development, no production, no acquisition, no storage, and no use" of nuclear weapons; and has worked closely with the International Atomic Energy Agency (IAEA) and the United States in assuring nuclear safeguards in Taiwan.

3. David Albright and Corey Gay, "Taiwan: Nuclear Nightmare Averted," *Bulletin of the Atomic Scientists* 54 (1998): 54–60.

4. See Derek Mitchell, "Taiwan's Hsin Chu Program: Deterrence, Abandonment, and Honor," in *Nuclear Tipping Point: Why States Reconsider Their Nuclear Choices*, ed. Kurt Campbell, Robert Einhorn, and Mitchell Reiss (Washington, D.C.: Brookings Institution, 2004), 293–313; Monte Bullard and Jing-dong Yuan, "Taiwan and Nuclear Weaponization: Incentives vs. Disincentives," in *Forecasting Nuclear Proliferation in the 21st Century*, vol. 2, *A Comparative Perspective*, ed. William Potter and Gaukhar Mukhatzhanova (Stanford, Calif.: Stanford University Press 2010), 182–204; Arthur Ding, "Will Taiwan Go Nuclear?" in *Over the Horizon Proliferation Threats*, ed. James J. Wirtz and Peter R. Lavoy (Stanford, Calif.: Stanford University Press, 2012), 33–46.

5. Jay Taylor, *The Generalissimo's Son: Chiang Ching-kuo and the Revolutions in China and Taiwan* (Cambridge, Mass.: Harvard University Press, 2000).

6. Harold Hinton, *China's Turbulent Quest* (New York: Macmillan, 1970).

7. Nancy Bernkopf Tucker, *Strait Talk: United States–Taiwan Relations and the Crisis with China* (Cambridge, Mass.: Harvard University Press), 29–128.

8. Mitchell, "Taiwan's Hsin Chu Program," 296.

9. Denny Roy, *Taiwan: A Political History* (Ithaca, N.Y.: Cornell University Press, 2003), 76–104.

10. This and the next two paragraphs are sourced from Mitchell, "Taiwan's Hsin Chu Program," 297–300.

11. Roy, *Taiwan: A Political History*, 152–182.

12. John Copper, *Taiwan: Nation State or Province*, 5th. ed. (Boulder, Colo: Westview Press 2009), 53.

13. Mitchell, "Taiwan's Hsin Chu Program," 300–301.

14. This and the next two paragraphs are based on ibid., 301–307.

15. This and the next two paragraphs are based on Bullard and Yuan, "Taiwan and Nuclear Weaponization," 182–204.

16. Ding, "Will Taiwan Go Nuclear?," 33–46.

17. The interviews were conducted by Robert Sutter in Taipei, May 20–24, 2013, with twenty-four officials, specialists, and advocates involved in Taiwan nuclear debates. Included were officials from the Atomic Energy Council, government economic and environmental officers, and officers from the Ministry of Foreign Affairs. Sutter thanks the officials from the Taiwan Economic and Cultural Representative Office in Washington and the Ministry of Foreign Affairs in Taipei for their help in arranging these official meetings and meetings with scholars, notably including some prominent former officials from the Chen Shui-bian government. Alvin Yao of the Prospect Foundation was interviewed by Sutter and assisted

in arranging meetings with members of the Foundation, opposition politicians, and activists opposed to the Lungmen power plan. Yao serves as the "in-country author" for this chapter, providing guidance on the project outline, reviewing the drafts of the chapter, and making very helpful and well-informed suggestions throughout the process.

18. "Bush Pledges Whatever it Takes to Defend Taiwan," *CNN*, April 25, 2001, http://edition.cnn.com/2001/ALLPOLITICS/04/24/bush.taiwan.abc/.

19. Thomas Christensen, *The China Challenge* (New York: W. W. Norton, 2015): 95–114.

20. Communications and interviews with DPP representatives, Washington, D.C. and Taipei, May 22–23, 2013.

21. Elaine Grossman, "Obama Sends Taiwan Nuclear Pact to U.S. Congress," *Global Security Newswire,* January 8, 2014, http://www.nti.org/gsn/article/obama-submits-taiwan-nuclear-trade-pact-congress/; The White House, Office of the Press Secretary, *Message to the Congress—Agreement for Cooperation Between the American Institute in Taiwan and the Taipei Economic and Cultural Representative Office Concerning Peaceful Uses of Nuclear Energy*, White House, January 7, 2014, http://www.whitehouse.gov/the-press-office/2014/01/07/message-congress-agreement-cooperation-between-american-institute-taiwan.

22. This section of the chapter relies heavily on Togzhan Kassenova, "A 'Black Hole' in the Global Nonproliferation Regime: The Case of Taiwan," *Nautilus Institute Workshop Forum 11-30,* September 8, 2011, http://nautilus.org/napsnet/napsnet-policy-forum/a-black-hole-in-the-global-nonproliferation-regime-the-case-of-taiwan/); Togzhan Kassenova, "Global Non-Proliferation and the Taiwan Dilemma," *Global Asia*, March 22, 2012, http://carnegieendowment.org/2012/03/22/global-non-proliferation-and-taiwan-dilemma/a5gh.

23. See notably President Ma's emphasis on Taiwan's soft power as one of its three lines of defense. Kwei-Bo Huang, "Taiwan's Incomplete Third Line of Defense for National Security," *Brookings Northeast Asia Commentary* 61, February 2013.

24. Kassenova, "A 'Black Hole' in the Global Nonproliferation Regime," 3.

25. Kassenova, "Global Non-Proliferation and the Taiwan Dilemma," 3.

26. Kassenova, "A 'Black Hole' in the Global Nonproliferation Regime," 4.

27. Ibid.

28. Kassenova, "Global Non-Proliferation and the Taiwan Dilemma," 4.

29. Togzhan Kassenova, "Strategic Trade Controls in Taiwan," *Nonproliferation Review* 17 (2010): 389–390.

30. Elaine Grossman, "U.S. May Land Key Asian Nuclear Trade Deals in 2013," *Global Security Newswire*, January 11, 2013, http://www.nti.org/gsn/article/us-could-secure-key-asian-nuclear-trade-deals-2013/.

31. Interviews conducted by Robert Sutter in Taipei, May 20–24, 2013.

32. Grossman, "U.S. May Land Key Asian Nuclear Trade Deals in 2013."

33. "Taiwan to Halt Construction of Fourth Nuclear Power Plant," *Reuters*, April 27, 2014, http://www.reuters.com/article/2014/04/27/taiwan-nuclear-idUSL3N0NJ08C20140427.

34. "Taiwan Seals Lungmen No. 1 Nuclear Reactor," *Taiwan Today*, July 1, 2015, http://www.taiwantoday.tw/ct.asp?xItem=232105&ctNode=421.

35. Andrew Jacobs, "Vote Holds Fate of Nuclear Power in Taiwan," *New York Times*, January 12, 2012, http://www.nytimes.com/2012/01/13/world/asia/nuclear-power-emerges-as-election-issue-in-taiwan.html.

36. "The Nuclear Power Debate Heats Up," *Taiwan Communiqué* (April/May 2013): 10–11.

37. Lee I-chia, "Bicyclists Start Referendum Drive," *Taipei Times*, June 6, 2013, 3.

38. Shih Hsiu-chuan et. al., "KMT Vows to Pass Referendum Proposal," *Taipei Times* August 2, 2013, 1.

39. Yu-Huay Sun, "Taiwan Ruling Party Concedes on Halting Nuclear Power Plant," *Bloomberg,* April 28, 2014, http://www.bloomberg.com/news/2014-04-28/taiwan-ruling-party-concedes-on-putting-nuclear-plant-on-hold.html.

40. Lee, "Bicyclists Start Referendum Drive"; Interviews conducted by Robert Sutter in Taipei, May 20–24, 2013.

41. David Yih-Liang Chan et al., "Energy Efficiency Benchmarking of Energy-Intensive Industries in Taiwan," (Abstract) *Energy Conversion & Management* 77 (January 2014): 216; *Status of Nuclear Power in Taiwan,* Taiwan Government Fact Sheet, May 2013 (provided to author during interviews in Taiwan, May 2013).

42. *Status of Nuclear Power in Taiwan,* 1; interviews with opponents of Lungmen plant, Taipei, May 22–23, 2013.

43. *Status of Nuclear Power in Taiwan,* 1; interviews with opponents of Lungmen plant, Taipei, May 22–23, 2013.

44. Interviews conducted by Robert Sutter in Taipei, May 20–24, 2013; "The Nuclear Power Debate Heats Up"; Alan Romberg, "Striving for New Equilibria," *China Leadership Monitor* 41 (2013): 14–17.

45. Interview with Taiwan government economic official, Taipei, May 21, 2013; interviews conducted by Robert Sutter in Taipei, May 20–24, 2013 and Washington, D.C., August 2013.

46. Interview with DPP legislator and staff aides, Taipei, May 23, 2013. The Taiwan authorities periodically have sought to export nuclear waste abroad, but thus far they have not secured a satisfactory arrangement.

47. Interviews conducted by Robert Sutter in Taipei, May 20–24, 2013.

48. Ibid.

49. *Status of Nuclear Power in Taiwan,* 1.

50. Romberg, "Striving for New Equilibria," 15.

51. Ibid., 16; interview with Taiwan government environmental officials, Taipei, May 21, 2013.

52. Interviews conducted by Robert Sutter in Taipei, May 20–24, 2013.

53. Interview with Taiwan government economic official, Taipei, May 21, 2013.

54. Briefing by Taiwan academic expert, Taipei, May 20, 2013.

55. Romberg, "Striving for New Equilibria," 15.

56. Min-Hua Huang, "The KMT's Landslide Defeat in the Nine-in-One Elections," *Brookings East Asia Commentary,* December 2014, http://www.brookings.edu/research/opinions/2014/12/08-taiwan-political-landscape-elections-huang; Julie Makinen, "Taiwan's Nationalist Party Looks to Be Heading to a Historic Defeat," *Los Angeles Times,* September 14, 2015, http://www.latimes.com/world/asia/la-fg-taiwan-election-20150914-story.html.

7

Vietnam

Nuclear Ambitions and Domestic Dynamics

Linda J. Yarr and Nguyễn Thị Thanh Thủy

In a unitary state such as Vietnam where open debate and channels of independent expression are limited, if not restricted, associating specific individuals or groups with discrete schools of thought is a challenging proposition. The Political Bureau of the Vietnam Communist Party (Politburo), operating by consensus and in private, has unlimited power to set policies carried out by the government, including plans for nuclear power. Nevertheless, these policy directions reflect the ideational frames emerging from the country's history, international environment, and resource endowment.

Vietnam's nuclear debates open a window to discern points of view that are at times discordant and at other times complementary. Despite some relative shifts in emphasis over time, the salient positions are Realist, Globalist, and to some extent Nationalist. While international factors have played a critical role in how Vietnam's nuclear program has evolved in recent times, these have been based more on Vietnam's desire for a positive global reputation than determined by other geopolitical considerations. This finding may come as somewhat surprising given Vietnam's war-torn past and nationalist ambitions—and persisting tensions with a rising China.

While chapter 8 explores the nuclear ambitions held by other ASEAN members, the scale and progress made on Vietnam's nuclear energy plans warrants its own chapter. This section seeks to reveal the fault lines of the debates over Vietnam's pursuit of nuclear power and gives greater texture to the influences on its foreign policy with respect to nuclear issues by (1) identifying the multiple drivers for Vietnam's nuclear energy program at the domestic and international levels; (2) discussing the country's nuclear landscape and the key actors and institutions; and (3) assessing the domestic debate on cost, safety, and nonproliferation issues in Vietnam.[1]

DRIVERS OF THE TURN TO NUCLEAR POWER

Vietnam entered the nuclear age as a recipient of nuclear materials and technology under the Atoms for Peace (AFP) program announced by President Dwight D. Eisenhower at the United Nations General Assembly in 1953. A major purpose of AFP was to share nuclear technology with countries around the world and harness the atom for peaceful uses, resulting in a nuclear research facility at Dalat in Vietnam. Built in 1960, the small reactor operated from 1963–1968 and was shut down by the United States as its war in Vietnam escalated. In 1980, under a nuclear cooperation agreement with Vietnam, the former Soviet Union rebuilt the facility to accommodate a larger research reactor that was commissioned in 1984.[2] Originally limited to research, medicine, and agriculture uses, the push to expand Vietnam's capability to build nuclear power plants came from both internal and external factors. Vietnam's long struggle with the United States to unify the country and border skirmishes with the Khmer Rouge–led Cambodia that propelled Vietnam to occupy that country left Vietnam with a faltering economy and deep diplomatic isolation.

In 1986, recognizing that a go-it-alone strategy was no longer viable in a globalized economy, Vietnam launched its economic reform policy known as *đổi mới*, thereby setting the stage for the country's spurt of industrialization and urban construction. The accelerated economic growth (averaging 6.12 percent from 2000 to 2014)[3] precipitated increased demand for energy. In addition to powering manufacturing, energy is needed for exploiting certain natural resources in Vietnam such as bauxite, a highly energy-intensive raw material to mine and refine into aluminum.[4] On the diplomatic front, Vietnam also sought to assure its neighbors and other powers of its benign intentions. These drivers—economic and diplomatic—influenced Vietnam's turn to nuclear power and reflect its prominent Realist and Globalist schools of thought.

Energy for Economic Growth

A rising middle-income country, Vietnam faces two imminent and linked challenges: (1) providing jobs for a growing and young population, and (2) meeting the increasing energy needs of industries slated to offer those jobs. Consequently, Vietnam has pursued multiple options to secure energy for its economic growth and has embraced nuclear power as a potential source. The growth rate of industrial production in Vietnam is expected to rise to 10.9 percent in 2017 and 10.5 percent by 2019.[5] Thus, Vietnam will require a plentiful and reliable supply of electrical power, which currently is generated by natural gas, hydropower, and coal. According to the most recent data from the U.S. Energy Information Agency, "Vietnam anticipates power demand to more than triple to 330 billion kilowatt-hours (kWh) between 2011 and 2020."[6] *The Economist* pegs energy demand to grow by up to 14 percent per year.[7] To meet this demand, in mid-2011 with direct approval by the prime minister, the Vietnamese government set out its National Power Development Plan for the 2011–2020 Period with the Vision to 2030, known as Power Master Plan VII.[8] In December 2014, the prime minister approved a Master Plan for NPP (nuclear power plant) Infrastructure Development to 2020. Under the Plan, the Atomic Energy Law will be revised and more considerations will be given to compensation for nuclear damage and relevant agencies' responsibilities for NPP safe operation and management of nuclear waste.[9]

According to one close analysis of Power Master Plan VII, the state aims to put "the first nuclear power plant into operation in 2020 and develop more nuclear power plants going forward with the hope that electricity generated from nuclear energy will account for about 10.1 percent of the total power output by 2030."[10] The Plan recognizes that Vietnam will reach the limits of its reliance on fossil fuels and hydropower. To shore up the mix of energy sources available, it designates nuclear power as a necessary addition to coal, natural gas, renewables, and hydropower.[11] The Vietnamese government maintains that nuclear power offers the steady base-load power that can be relied upon to provide electricity night and day, thereby attracting foreign investment for manufacturing, information and communication technology, and mineral refining. Although the energy plan calls for renewables to be included in the energy portfolio, government officials assert that most non-nuclear renewable energy systems are still too expensive and fail to offer a steady energy output.

Vietnam's quest to ensure energy security for economic growth aligns with a Nationalist view of energy independence and the prospect of knitting further national unity through expansion of a reliable electricity grid from north to south. This view, however, is tempered by the particular nature of the nuclear turn, which has been determined less by ideological factors than Vietnam's history might indicate.

International Motivations

Unlike the decision to build a new coal-fired power plant, the nuclear choice does not solely remain an issue of domestic purview restricted to cost-benefit analysis or site-selection debates. For Vietnam, the decision has major foreign policy salience, particularly as the country aims to reassure regional and global actors about its peaceful intentions, as well as its capacity to ensure the security and peaceful use of nuclear materials. In the initial stages of Vietnam's opening to the world economy in the mid-1980s, leaders sought to "make more friends and fewer enemies."[12] To overcome its diplomatic isolation and in pursuit of future development, Vietnam acceded to a host of treaties and conventions so that the country could take its place in the international community as a "responsible power." The country gained membership in international organizations such as the World Trade Organization, secured a turn as a member of the United Nations Security Council in 2008–2009, and served as chair of ASEAN in 2010. As discussed later in this chapter, Vietnam has proclaimed its resolve to foreswear the development of nuclear armaments by its membership in multiple regimes and accession to nonproliferation agreements. Even if Vietnam's nuclear capability is limited to supplying energy, there are risks that nuclear materials could be diverted for aggressive or terrorist purposes or result in an accident that could have far-reaching effects on neighboring countries. Furthermore, its choice of nuclear energy to displace fossil fuels aligns Vietnam with the developed nations' agenda on climate change negotiations. Thus the outlines of a more Globalist outlook emerged for policymakers despite the lack of any structural changes toward a more favorable geopolitical environment.

Indeed, as its northern neighbor China has begun to take a more assertive stance in regional affairs by laying claim to a large swath of the South China Sea, Vietnam has ramped up its diplomatic engagement to secure strategic partnerships with major powers. Having obtained a comprehensive partnership with the United States, it seeks an upgrade to a strategic partnership and to bring its concerns to the attention of the international community about aggressive actions by Chinese fishing vessels in areas Vietnam claims as its territorial waters, protests against the positioning of a Chinese oil-exploration rig off its coast, and China's creation of artificial islands with space for military facilities on reefs and rocky marine features.[13]

Although going down the nuclear power path opens the country to closer international scrutiny, Vietnam's leadership sees a number of foreign policy objectives that may be fulfilled.[14] First, acquisition of nuclear technology brings with it the prestige of belonging to the nuclear club and a more secure *place at the table* when new agreements or regimes on nuclear issues are proposed and discussed internationally. Second, the process of building nuclear capacity requires expanding the numbers of technicians, engineers, and nuclear physicists who are qualified to run and manage the nuclear program. This expansion of expertise could help to strengthen the country's science and technology assets. As Trần Chí Thành, head of the Vietnam Atomic Energy Institute (VINATOM) stated, "Vietnam needs to build a nuclear power program to have sufficient electricity for the country, and this process will also help boost the development of human resources in other sciences that eventually support the economic development."[15] Third, contracts for nuclear power plant construction and the education of nuclear scientists, engineers, and technicians offer the means to reward and deepen relationships with leading powers. Fourth, nuclear energy offers the potential to gain relative independence from the vagaries of global oil and gas prices and the risk of supply interruptions. Finally, a not inconsequential reason is to demonstrate to the international community the desire to join global efforts to mitigate emissions contributing to climate change.

This list of international objectives indicates that diverse schools of thought come into play: Nationalists seek to boost the country's capabilities in the nuclear sphere and secure energy self-sufficiency; Realists can use nuclear policy to cement relations with major nuclear powers; and Globalists demonstrate their willingness to follow international norms and join the treaties, regimes, and international organizations that promote nuclear safety, security, and safeguards as well as contribute to global measures to deal with climate change. Apart from energy independence, all other objectives are by and large consistent with Realist and Globalist viewpoints.

Choosing Russia as the builder of the first two nuclear power plants reaffirms and rewards Russia's historic role as a supporter of the Socialist Republic of Vietnam and current status as its strategic partner.[16] Moreover, with no outstanding issues of contention, such as the differences with the United States over human rights, cooperation with Russia on nuclear energy is likely to be free of political discord. Likewise, the selection of Japan to build the second two plants anchors a trade and commercial relationship with a country responsible for $24.7 billion in trade in 2012 and with Vietnam's leading supplier of official development assistance.[17] On the political level, Realists might judge Japan to be a possible counterweight to China since both Vietnam and Japan seek to check China's maritime ambitions. As noted in chapter 5, South Korea—also an important source of investment and trade with Vietnam—aspires to be Vietnam's third NPP supplier.[18]

MAPPING VIETNAM'S NUCLEAR DECISION-MAKERS AND OPINION SHAPERS

As we might expect under a centralized system, Vietnam's nuclear bureaucracy tends to hold fairly consistent views. In this section, we outline the main actors and process of nuclear decision-making, which is still a fairly recent development, with the Atomic Energy Law passed only in 2008. [19] Unlike more established nuclear states experiencing bureaucratic competition for resources and influence, it is too early to make such judgments in Vietnam's case. In a subsequent section, we explore the spectrum of opinion as it plays out domestically on specific nuclear issues, where it is possible to find more variation in thinking.

Government Actors in Vietnam's Nuclear Debates

Vietnam's nuclear program is run under strong civilian control through political and bureaucratic structures. The ultimate policy on nuclear issues is set by the Politburo, a subunit of the Central Committee, the highest authority of Vietnam's Communist Party system, which is elected by the Party National Congress. Nevertheless, the complex environment concerning policy-making on nuclear issues involves numerous authorities and sources of influence. The Politburo directly discusses, formulates, and approves nuclear energy policies, national targets, and specific projects. While Politburo debates are not made public, policy and implementation are shaped by a range of players with different areas of responsibility. As an official with the Ministry of Foreign Affairs explained, "The policy-making mechanism in Vietnam relied on consensus. . . . Sometimes they [important members of the policy-making structure] may have different views, but not to the extent of competing with each other. Each different standpoint could be argued differently, but eventually, all have to be agreed upon for any decision to be made."[20]

Once the Politburo sets the direction of a policy, the National Assembly, particularly through its Commission for Science, Technology, and Environment, deliberates and passes laws required to implement the policy. The 2008 Atomic Energy Law laid the foundation for nuclear power in Vietnam. A year later, the National Assembly approved the first nuclear power plant in Vietnam.[21] During the 2008 law's debate, the National Assembly's website reported that "most deputies highly valued the formulation of the draft law in the context of Vietnam's national industrialization and modernization process."[22] However, it also noted that certain deputies were reported to have raised concerns about safety and constituents' worries about nuclear accidents. They asked for assurances that leaders would ensure the safety of populations and the environment. Others wanted details about safe transport and disposal of radioactive materials.[23] These concerns were also raised and debated in the aftermath of the 2011 incident in Japan at Fukushima.[24]

The prime minister is responsible for approving any nuclear energy policies, national targets, and projects submitted by relevant administrative agencies.[25] The National Atomic Energy Council established in May 2013 counsels the prime minister and assists in steering and resolving interdepartmental issues related to development and application of atomic energy.[26] Within the Ministry of Science and Technology (MOST), the Vietnam Atomic Energy Commission (VAEC), also known as the Vietnam Atomic Energy Institute, carries out scientific research and development of nuclear technology and atomic energy and provides technical support to state management activities on atomic energy, radiation, and nuclear safety.[27]

Another agency housed within MOST, the Vietnam Atomic Energy Agency (VAEA), implements the state's management of research, application, and development activities for nuclear energy. Central to this effort is its work with the International Atomic Energy Agency (IAEA), which reviews and assesses the country's nuclear infrastructure and regulations. VAEA also plays a public relations role and disseminates information through the mass media and educational institutions. Furthermore, VAEA oversees the complex task of human resource development for the atomic energy field in Vietnam.

On nuclear safety policy, the National Nuclear Safety Council (NNSC) established April 2010 coordinates these issues and includes representatives from MOST (who serve as chair of the board), the Ministry of Industry and Trade (MOIT), the Ministry of Public Security, the Ministry of Natural Resources and Environment, and the Ministry of Defense. [28] The National Council for Science and Technology Policy (NCSTP) provides direct advice to the prime minister on these questions. The Vietnam Agency for Radiation and Nuclear Safety (VARANS), also under MOST, serves as the country's nuclear regulator.

The mass media—including *Vietnam Television*, *Voice of Vietnam*, and print publications—help to disseminate official information on these nuclear ambitions. VAEA develops public communication strategies to promote citizen acceptance of nuclear power plans and invites media outlets to report on events in the run-up to building the first nuclear power plant in Vietnam. However, the press does occasionally report on questions and concerns of local authorities about the siting of nuclear facilities, such as the opposition to a larger reactor at the Dalat Nuclear Research Center.[29] Local officials fear that the new location will diminish the town's appeal as a tourist destination.

External Actors in Vietnam's Nuclear Debates

Externally, the IAEA has wielded the dominant influence on Vietnam's nuclear policies and practices. Vietnam joined the IAEA in 1978 and was elected to the Board of Governors on three occasions. The IAEA has provided various forms of assistance to Vietnam, including technical cooperation, training, laboratory building, exchange of information, and the strengthening of Vietnam's IAEA safeguards.[30] In fact, it is likely that the advice proffered by IAEA Director General Amano Yukiya during his visit to Hanoi in January 2014 prompted the Vietnam government's decision to postpone the start of construction of the first NPP in Ninh Thuận.[31] Citing persistent safety concerns, further legal challenges, and slower than expected economic growth, Hoàng Anh Tuấn, director general of the Vietnam Atomic Energy Agency, announced in early 2015 another delay, with construction not likely to begin until 2019.[32]

A number of countries are important for Vietnam's nuclear ambitions. As of late 2013, Vietnam had signed bilateral nuclear cooperation agreements with Russia, China, France, India, Japan, the Republic of Korea, and Argentina, as well as the so-called "123 Agreement" with the United States.[33] Russia and Japan exercise significant influence on Vietnam's nuclear policy as the suppliers of the country's first two NPPs. Both Japan and the United States helped Vietnam draft the 2008 Atomic Energy Law.[34] In addition, Japan is assisting VARANS to develop a legal framework on nuclear matters, improve human resources in Vietnam's nuclear bureaucracy, and review amendments to some aspects of the Atomic Energy Law related to licensing, emergency response, and compensation for nuclear damage.[35]

FAULT LINES IN VIETNAM'S NUCLEAR DEBATES

Although Vietnam's leadership has maintained its commitment to proceed with plans for building nuclear power plants, it is not immune to criticisms expressed by nuclear experts inside and outside the country.[36] It has also not escaped questioning from local grassroots groups. In this section, we outline three main clusters of concerns: (1) the financial cost of a national nuclear energy program, (2) nuclear safety risks, and (3) nuclear nonproliferation.

Financial Costs in Realizing Vietnam's Nuclear Ambitions

To build the first plant in Vietnam, Russia extended a low-interest loan of up to $10 billion.[37] The size of this obligation raises questions about whether the anticipated level of economic growth and electricity will offset the cost of the investment. According to the World Nuclear Association, Vietnam Electricity (EVN), the largest public power company in Vietnam and an investor for building and operating the first two plants, would finance up to 25 percent of the money required, with the rest borrowed from the foreign country supplying the technology. EVN considers the financial burden acceptable based on the expectation of averaging 7–8 percent GDP growth per year.[38] However, the Asian Development Bank (ADB) was less sanguine on such growth, pegging the 2014 rate as 5.6 percent with an increase to 5.8 percent in 2015.[39] Nevertheless, the ADB's latest 2015 forecast has GDP growth to "edge up to 6.1 percent in 2015 and 6.2 percent in 2016, with FDI [foreign direct investment] an important driver."[40] The Organization for Economic Cooperation and Development projects Vietnam's average annual growth rate for the period 2014 to 2018 at only 5.4 percent.[41] These discrepant predictions only underscore the uncertainties and risks assumed by the government in undertaking the quest for nuclear power.

Before further consideration of the price tag, it is worth taking a look at the cost effectiveness of the nuclear industry worldwide. In their summary of the *World Nuclear Status Report for 2013*, nuclear industry analysts Mycle Schneider and Antony Froggatt contend,

> The nuclear industry is in decline, however, with global electricity generation from nuclear power plants down by a record 7 percent in 2012 . . . [whereas for renewables,] the costs of equipment also declined, making it possible to significantly increase renewable capacity. Three of the world's four largest economies [China, Germany, and Japan] now run on a higher share of renewables than nuclear power.[42]

This challenges the view expressed by the deputy director general of VARANS that wind and solar electricity generation are still too expensive for Vietnam to consider, leaving nuclear power as the cheapest alternative to fossil fuels.[43] Consequently, Vietnam's decision to borrow the money necessary to invest in nuclear power needs to be weighed against the opportunity cost of not investing more aggressively in renewables such as wind, solar, and tidal power, for which installation costs are decreasing.

Within Vietnam, there is no absolute consensus that assuming the cost of developing nuclear power is the best course of action. Phạm Duy Hiển, professor and former director of the Atomic Energy Institute, questioned whether the business generated by nuclear power would cover the expense of construction and expected that the state would have to end up subsidizing the price.[44] He also noted that Vietnam's energy productivity was low in comparison to other countries in the region. For example, Vietnam earns one dollar for every kWh consumed, whereas Thailand earns $2 and the Philippines and Indonesia earn close to $3 (see chapter 8 for more on ASEAN's nuclear plans). He further cautioned that as other renewables, especially wind power, become more advanced in the next several years, the price of nuclear power will be less competitive.

Additionally, Professor Phạm Duy Hiển observed that new foreign investment is moving to more energy-efficient industries rather than high-energy-consumption industries such as cement factories and steel mills.[45] As a result, electricity consumption for 2013 and 2014 has plateaued,[46] and is less than what was anticipated by some nuclear power advocates. In November 2013, *Thanh Niên News* reported that while "the government has enacted a green growth strategy, experts say other renewable resources—such as wind, solar, and biofuels—have not been given priority."[47] Moreover, "Vietnam has still not paid enough attention to energy efficiency," according to Trịnh Lê Nguyên, executive director of People and Nature Reconciliation, a conservation group based in the country.[48] This diverse range of opinions on the cost of a national nuclear energy program point to a more complex nuclear debate in Vietnam than its centralized unitary government may initially suggest.

Nuclear Safety Risks

The March 2011 nuclear plant disaster at Fukushima in Japan marked a new reference point for considering the safety factors in nuclear plant construction. As the disaster continues to impact Japan,[49] Vietnam's nuclear plans have come under additional scrutiny, and it is fair to compare Vietnam's current state of readiness to what went wrong at Fukushima. While prospective nuclear vendors assure potential buyers that they have learned the lessons from Chernobyl and Fukushima and have incorporated them in their designs,[50] it is not enough to just put in place the most up-to-date technology. Ultimately, an analysis of what went wrong at Chernobyl and Fukushima points above all to the impact of the human factor: the roles of regulatory bodies, decision-makers, management systems, and training of personnel.

The independent online news network *NucNet* reported on a study by Airi Rye and Najmedin Meshkati at the University of Southern California comparing the case of NPPs located in the Fukushima province with that of the Onagawa Nuclear Power Plant in Japan, which was also subject to the earthquake and tsunami, was of a similar design, and operated under an identical regulatory regime. The authors found that "their fates were very different."[51] In particular, the paper credits Hirai Yanosuke, vice president from 1960 to 1975 of the Tohoku Electric Power Company and operator of the Onagawa plant, with being "strict about the promotion of safety protocols, which led to the establishment of a strong safety culture."[52]

In light of these conclusions, how are Vietnam's plans to build nuclear power plants being viewed? Headlines around the world question, "After Japan, Where's the Next Nuclear Weak Link?"[53] and "Is Vietnam Ready for Nuclear Power?"[54] Dr. Nguyễn Khắc Nhẫn, former deputy director general of EVN, opposes Vietnam's nuclear program. Among his many reasons, three stand out: (1) nuclear power is both outmoded and more expensive than claimed by the industry; (2) whereas major catastrophes such as those at Chernobyl and Fukushima can be ascribed to human incompetence rather than technical failure, Vietnam has neither sufficient trained personnel nor a culture of safety to reassure the population of safety; and (3) widespread corruption could lead to inadequate construction materials and practices, increasing the risk of accidents. As Dr. Nguyễn Khắc Nhẫn put it, "I think that Vietnam must quickly and massively promote renewable energy, energy conservation, and energy efficiency rather than dangerously plunging into a nuclear program that will ruin and destroy the country."[55]

These challenges have not gone unnoticed by Vietnam's authorities. In an interview with the authors, Lê Chí Dũng, the deputy director general of VARANS, conceded that there is much to be done before nuclear power plants can be built and safely operated in Vietnam. He identified four key areas that require attention: (1) training adequate personnel; (2) developing a robust regulatory framework; (3) creating an independent, qualified, and authoritative regulatory body to carry out inspections and hold operators accountable for following regulations; and finally, (4) care in siting the plants to be built.[56] On the first issue, the IAEA has declared VARANS to be in serious need of rapid scaling up of experts. [57] This is essential, as Lê Chí Dũng stated, because more expertise is needed to prepare all the legal and regulatory documents necessary. Vietnam draws from Russian, Japanese, and American regulatory documents to adapt them to the Vietnam context. Lê Chí Dũng considered documents from the United States preferable to those from Japan or Russia, but those documents are voluminous, and many provisions reference numerous other regulations that need to be studied in turn. He said it took his agency six months to prepare for a one-week training program by experts coming from the United States.

On the other hand, IAEA model regulations are drafted to apply to many different countries, thereby lacking specificity. Lê Chí Dũng stated that since the legal and regulatory framework had to be in place before starting construction of the first nuclear plant, VARANS asked the government for at least a two- or three-year delay in the start-up. Shortly after this interview, it was reported that the Vietnamese prime minister "announced that the construction of the country's first nuclear power plant, the 4,400-megawatt Ninh Thuận 1—slated to begin this year [2014]—will be delayed until 2020 due to concerns about safety and efficiency."[58] Although the National Assembly of Vietnam originally expected construction to begin in 2014, Lê Chí Dũng noted in his interview that 2017 might be more realistic. In January 2015, Hoàng Anh Tuấn, director general of the Vietnam Atomic Energy Agency indicated construction on the country's first nuclear plant would not begin until 2019.[59]

In early 2015, in line with their official missions, government leaders met with relevant ministries, agencies, and leaders of Ninh Thuận to discuss new policies to be implemented through a Mechanism Plan to support socioeconomic development in areas impacted by NPP construction in Ninh Thuận. For example, the Ministry of Planning and Investment assesses how much capital is required for essential projects to be carried out in the province. The Ministry of Education and Training (MET) creates policies to incentivize nuclear power–associated higher education, vocational training in NPP construction, and management and operation, and for special attention to be given to students of Ninh Thuận. The Ministry of Health must work with the Ninh Thuận People's Committee to double the number of beds in the provincial hospital and upgrade its equipment to better serve the province's population and the experts who will be engaged in NPP construction. The Ministry of Transportation cooperates with the province's Committee and EVN to identify priorities for improving the traffic infrastructure, including the province's ports, so as to facilitate construction and operation of the NPP.

The Ministry of Industry and Trade was urged to complete a project on a "Mechanism, Special Policy for the NPPs-Building Project in Ninh Thuận" for submission to the prime minister. The project is expected to report on the management experiences of countries operating NPPs and propose mechanisms and policies to support socioeconomic development of areas to be affected by the NPP building and operation.[60] Since then, concern for having sufficient trained personnel to handle the nuclear facility was a key factor in the extended timetable.[61] In April 2015, Russian Minister of Industry and Trade Denis Manturov, who accompanied Prime Minister Dmitry Medvedev in his visit to Vietnam, confirmed that Russia would fulfill all of its commitments and obligations toward construction of NPP 1, agreed to lend Vietnam $8 billion USD to implement the project, and affirmed that construction would start no later than 2017, with the first machine group operated in 2020.[62]

Measures taken since the announcement that construction on the nuclear power plants would be delayed indicate that ensuring safety is a preeminent concern and the lessons of the Fukushima incident are being taken into account. On July 16, 2015, the People's Committee of Ninh Thuận Province announced the decision to adjust planning of the construction sites of the NPP 1 and NPP 2 in Ninh Thuận. Under the new planning, the site of the NPP 1 in Vĩnh Trường Village, Phước Dinh Commune, Thuận Nam District will be expanded to the southwest and deeper into the mainland. The plant will also be built more than 12 meters above the sea line. Meanwhile, the site of the NPP 2 in Thái An Village, Vĩnh Hải Commune, Ninh Hải District will be expanded as well, with the plant built 15 meters above the sea line. Earlier, on June 9, 2015, the prime minister approved a resettlement project to move 1,288 local households or 4,911 residents living in the selected sites of the two NPPs to new settlements, and the project is financed with 3,235 billion đong.[63]

On July 30, 2015, the Russian State Atomic Energy Corporation (ROSATOM) and EVN signed a framework agreement in Hanoi to implement the first phase construction of the first nuclear power plant, "Ninh Thuận 1." The parties approved the list of measures to be taken in the near future and determined the implementation timeframe.[64] The following month, in Hanoi, a three-day workshop on "Reliability of Electric Grid and Grid Interface for Nuclear Power Plants" was jointly held by the EVN Project Management Unit of Ninh Thuận Nuclear Power and the IAEA Nuclear Power Engineering Unit. The workshop was within the framework of an IAEA technical assistance project for Vietnam to develop infrastructure of nuclear power, phase III. At the workshop, Vietnamese and international experts discussed how to integrate the future NPPs into the current national electric grid, the particular characteristics of such integration, and other technical issues.[65] In 2013, Deputy

Minister of Science and Technology Lê Đình Tiến called on the IAEA and other countries to assist Vietnam with its comprehensive human resource development.[66]

Staff from Vietnam's nuclear agencies and power company are increasingly being sent abroad to study the experiences of other countries with nuclear power programs in place—especially Japan, Russia, South Korea, and France. In the aftermath of Fukushima, Ngô Đăng Nhân, then director of VARANS, suggested, "If we don't have professional engineers and experts, it will be very tough in case of problems." He continued that Vietnam needed "a legal framework with strict operation and maintenance regulations to make the nuclear power plants safe."[67] Despite this goal, the quality, cost, and extent of training plans have also come into question. Trần Đại Phúc, a consultant to the nuclear project in Ninh Thuận reportedly "expressed his disappointment about the program to prepare human resources for the nuclear power industry, saying that the results had been unsatisfactory."[68]

The question of human resource development has subsequently been addressed in a number of ways. In September 2014, fifteen EVN staff members completed the first two-year training course in nuclear power supported by Japan International Nuclear Energy Development (JINED) at Tokai University. A second course with participation of another nine EVN staff members slated to terminate in 2015 followed immediately thereafter. The participants will constitute the core personnel of NPP 2 in Ninh Thuận, working as managers and supervisors at construction sites of a Japanese contractor at the NPP 2. EVN requested the JINED and other Japanese counterparts to train 100 more Vietnamese students in nuclear power.[69]

In September 2014, the Electric Power University (EPU), the International Atomic Energy Agency (IAEA), and VAEA jointly organized a three-day international workshop in Hanoi on a Human Resource Development Plan for Nuclear Power Program. Participants included international experts from the IAEA, SE-Engineering Consulting Company (Germany), EDF Energy Company (a UK branch) and representatives from the MET, EVN, Vietnam Atomic Energy Institute (VINATOM), Vietnam Agency for Radiation and Nuclear Safety (VARANS), Ministry of Science and Technology (MOST), Ministry of Industry and Trade, the Hanoi University of Science and Technology, University of Natural Sciences in HoChiMinh City, and Dalat University. The purpose of the workshop was to share international experiences in human resource development, exchange information on present and future plans for human resource development, and discuss a national framework for human resource development in the field of nuclear power in Vietnam.[70]

The need to establish a legal framework raises another problem for Vietnam. Whereas carefully trained personnel are required to draft the regulations and anticipate eventualities, any framework is inadequate without the authority and strict independence to act without fear or favor. The Vietnamese agency charged with researching and putting such regulations in place, VARANS, is not completely independent, but rather comes under MOST, as does VAEA, which is charged with personnel training and technology support. Consequently, VARANS does not appear to have the standing to hold the government as a whole accountable for anything amiss in the process of procuring, staffing, and operating nuclear power plants. Moreover, Vietnam's lack of a truly independent press with a tradition of investigative journalism leaves scant checks on the safety of construction or operations of nuclear installations.

Questions have also been raised about the suitability of sites chosen for the first plants. Citing a paper by Lê Đại Diễn of the Nuclear Safety Center at the Institute for Nuclear Science and Technology in Hanoi, a *Congressional Research Service* report notes that "Vietnam's coast has been subject to tsunamis in the past, and one study suggests more investigation is still needed on seismic conditions and tsunami risk."[71] If an accident were to occur in the currently designated sites, the country could be cut in two, with transportation links between the northern and southern parts severed.[72]

Prompt and accurate information sharing among government agencies, nuclear power companies, and the public is a central component of nuclear safety. The European Nuclear Safety Regulators Group issued guidelines in March 2013 on how nuclear regulatory bodies ought to provide information to the public.[73] The document recommends, for example, developing a communications strategy, disclosing information in a timely manner, and promoting "a culture of openness and transparency within the [national regulatory organizations] so that all staff understand the importance of being transparent and of proactive engagement with all stakeholders."[74] And yet, in her assessment published in *Asahi Shimbun*, Nakano Ari, professor at Daito Bunka University in Japan, concluded "the Vietnamese government's predisposition to try and cover up inconvenient truths is fundamentally unchanged. The construction of the [NPPs], wrapped in a political climate with inadequate disclosure of information and the suppression of free speech, is already plagued with serious problems before they can even produce technological or economic results."[75]

A move toward greater transparency as Vietnam undertakes building nuclear power plants is impeded by endemic corruption. In 2012, *Transparency International* reported that "in spite of improvements over the past years, corruption is still considered widespread throughout the country and Vietnam still lags behind other Asian countries in terms of control of corruption and most governance indicators."[76] Opportunities for corruption in building NPPs include the use of substandard materials, cost overruns, and delays. Human resource development and training could also be subject to corrupt practices that lead to unqualified people gaining positions in the nuclear industry.[77] Faulty road construction is just one indicator of the kind of problem that may arise in building nuclear power plants.[78] On October 9, 2014, the IAEA, at the conclusion of its peer review of Vietnam's radiation and nuclear regulatory framework, emphasized that "the effective independence of the regulatory decision-making process needs to be urgently addressed."[79] As Vietnam seeks to enhance its incipient nuclear capability, the role of the IAEA is significant. Vietnamese officials are coming under increasing questioning from both domestic and international critics—something that is proving difficult to ignore.

Nuclear Security, Safeguards, and Nonproliferation

Presuming the problems of cost, safety, and transparency can be overcome, a critical issue remains: whether the international community can be assured of Vietnam's ability to track and secure all its stocks of radioactive materials. The threat of fissile materials falling into the hands of terrorists is serious. Likewise, so is the possibility of nuclear technology proliferating to states with inadequate controls or the intention to create weapons of mass destruction. In contrast to cost and safety matters, which have been more contentious in Vietnam's nuclear debate, there seems to be far greater domestic consensus on a fairly Globalist-oriented nonproliferation policy. Vietnam, in its public pronouncements and through its accession to relevant conventions and treaties, has affirmed its commitment to the peaceful use of nuclear energy and its staunch opposition to the proliferation of nuclear weapons.[80] A *Congressional Research Service* report noted, "Vietnam's Law on Atomic Energy passed in 2008 forbids the development of nuclear weapons and all forms of nuclear proliferation."[81] Furthermore, the deputy director of VARANS confirmed that Vietnam will not be conducting any reprocessing of spent fuel, and that France is ready to help with technology for storing spent fuel.

To clear the way for U.S. businesses to engage with Vietnam in the nuclear sphere, the U.S. Nuclear Regulatory Commission produced a statement in December 2013 and concluded that the proposed 123 Agreement "includes all of the provisions required by law and provides a sufficient framework for civilian nuclear cooperation between the United States and Vietnam."[82] The Nonproliferation Assessment Statement lists all of Vietnam's nuclear energy and nonproliferation policies as well as the extensive commitments the country has taken with respect to international regimes, conventions, and treaties on nuclear nonproliferation (see chapter 8 for details on Hanoi's role in crafting the Southeast Asian Nuclear-Weapon-Free Zone). Vietnam's stance did reflect one Realist view on sovereignty and future uncertainty by refusing to include a binding agreement to forego nuclear enrichment and instead adding a separate political statement outside the actual 123 Agreement, noting that Vietnam did not and would not seek to enrich its own nuclear fuel.[83]

The 123 Agreement did however open the door to increased commercial activity in the nuclear power sphere between Vietnamese agencies and United States firms. On August 14, 2014, the American firm Lightbridge Corporation and VARANS from MOST signed a memorandum of understanding on comprehensive cooperation in the field of radiation and nuclear safety for peaceful purposes. The cooperation between the two parties will focus on legal infrastructure development and regulatory law on radiation and nuclear safety, an integrated quality management system, regulatory activities on assessment and evaluation, licensing, inspection and sanctions, training, and other related issues.[84] On October 17, 2014, Lightbridge and VINATOM agreed to cooperate in promoting advisory capacity, research and development, training, and services in the field of atomic energy.[85]

In early 2015, the United States added Vietnam to the list of countries that are allowed to receive technology transfer and nuclear assistance, creating greater openness to strengthening the nuclear trade between the two countries.[86] In July 2015, EVN and the Westinghouse Corporation jointly held a two-day workshop in Hanoi to introduce the technology of the AP1000 nuclear power plant. Participants at the workshop shared experiences in building an NPP and employing advanced nuclear power technologies around the world.[87]

Aside from nuclear energy for power generation, it is important to examine Vietnam's record with respect to nuclear security and non-proliferation issues. In May 2014, Vietnam announced it would join the Proliferation Security Initiative (PSI), a U.S.-led effort to stop, through multilateral interdiction efforts, trafficking of weapons of mass destruction.[88] Vietnam's decision to join the PSI reflects the country's strong commitment to preventing the risk of nuclear proliferation, thereby contributing to global and regional security along Globalist lines. Joining the PSI can also be interpreted as a Realist move. Security analyst Carlyle Thayer suggests that the decision "could provide an opportunity for the United States to assist Vietnam in developing the capacity to conduct surveillance of its maritime zone of responsibility" and thereby deter China's designs on the South China Sea.[89]

These commitments as well as those undertaken at the 2010, 2012, and 2014 Nuclear Security Summits—President Barack Obama's global effort to prevent nuclear material from falling into the hands of terrorists—have demonstrated that Vietnam has held a positive attitude toward U.S. nonproliferation policies in particular, as well as broad strong support for nuclear security and nonproliferation in general. In 2013 at a meeting of the U.N. Conference on Disarmament (CD), Vietnamese Foreign Minister Phạm Bình Minh stated that "Vietnam's consistent policy is to uphold peace, oppose war, and support all efforts for disarmament, especially the disarmament of nuclear and other weapons of mass destruction, in accordance with the U.N. Charter and international law."[90] The Ministry of Defense further declares that Vietnam has no intention of pursuing nuclear weapons technology. These strongly articulated positions of the Vietnamese government from both political and defense quarters reflect a confluence of Globalist and Realist thinking. On the one hand, Globalists see the country's future best guaranteed by active and close participation with the international community through its organizations, norms, conventions, and laws. On the other hand, Realists see the country's security best managed by aligning its practices with those of the major powers on the critical issues of nuclear security and nonproliferation.

CONCLUSION

Vietnam's determination to become a nuclear power—in the realm of energy production, not armaments—is evident by its deliberate step-by-step approach taken in consultation with international experts, especially those from the IAEA, Russia, Japan, and the United States. Whereas the leading decision-makers in the Politburo do not publically divulge any differences of opinion with respect to the country's pursuit of nuclear power, the consistency of its pronouncements and actions over more than two decades lends credence to the assumption that the overall policy direction is uncontested within the leadership. These are core elements of the policy direction: (1) Vietnam must acquire nuclear energy for economic growth, and (2) Vietnam will cooperate with global efforts on nuclear security and the nonproliferation of nuclear weapons.

Nevertheless, discussions on the implementation of these core principles do offer scope for debate and a window into the schools of thought that influence the policy directions on domestic issues and foreign policy. Nationalist perspectives emerge in the quest to bring power to underserved populations through a nationwide electric grid and to don the enhanced status of a nuclear power, but the dominant threads are Realist and Globalist.

Realist thinking emanates from discussions about ensuring energy security, leveraging assistance from major nuclear powers, building capacity for domestic expertise in the nuclear field, and refusing to make a *binding* agreement (the so-called "Gold Standard") not to enrich its own nuclear fuel in the 123 Agreement with the United States. Meanwhile, Globalist thinking is evident in the country's attention to the issue of nuclear safety and security through its close cooperation with leading nuclear powers and the IAEA. The Vietnamese government's decision to delay construction of the first nuclear power plants to ensure its readiness to build and operate them safely demonstrates serious consideration of the issue.

Furthermore, from the beginning of its plans for nuclear power, the country has attached great importance to nuclear nonproliferation. In order to gain the benefits of the global economy, Globalists in Vietnam successfully argued for the need to reassure its neighbors and the international community that it is a benign and responsible power. Commitments taken with respect to various international regimes, conventions, and treaties on nuclear nonproliferation and the leadership's public pronouncements underscore the country's consistent policy against the proliferation of nuclear armaments. Even though Vietnam reserves the right to develop its own nuclear enrichment and reprocessing technologies for peaceful purposes as allowed under the Nuclear Nonproliferation Treaty,[91] Vietnamese leaders have demonstrated they do not have any intention to introduce nuclear weapons into the country's defense strategy.

Although debates simmer over whether or not Vietnam can overcome the challenges of insufficient human and financial resources, possible corrupt practices, and the lack of a fully independent regulatory regime to ensure safe and efficient development of nuclear power, the country has demonstrated its *intention* to engage in safe, secure, and peaceful uses of nuclear energy. As a U.S. State Department official observed, "Vietnam is acting like a responsible nuclear partner."[92]

NOTES

1. Linda J. Yarr thanks her coauthor, the Helsinki Collegium for Advanced Studies, and the Department of International Studies at the University of Social Sciences and Humanities

of the Vietnam National University for support of this work. She acknowledges the excellent research assistance provided by Timothy Westmyer. The chapter's authors translated Vietnamese language works cited in this section.

2. Reactor size was increased from 250 kW to 500 kW. "Nuclear Power in Vietnam," World Nuclear Association, December 2014, http://www.world-nuclear.org/info/Country-Profiles/Countries-T-Z/Vietnam.

3. "Vietnam GDP Growth Rate," *Trading Economics*, accessed January 30, 2015, http://www.tradingeconomics.com/Vietnam/gdp-growth.

4. "Aluminum and Sustainability," Aluminum Stewardship Initiative, accessed January 30, 2015, http://aluminium-stewardship.org/about-asi/aluminium-and-sustainability.

5. *The Economist* Intelligence Unit, "Vietnam: Forecast," *The Economist*, accessed January 21, 2015, http://country.eiu.com/article.aspx?articleid=122685996&Country=Vietnam&topic=Economy&subtopic=Forecast&subsubtopic=Forecast+summary.

6. "Vietnam Overview," *U.S. Energy Information Administration*, November 2014, http://www.eia.gov/beta/international/analysis.cfm?iso=VNM.

7. "Electricity in Vietnam: A Heavy Load," *The Economist*, August 31, 2013, http://www.economist.com/node/21584374/print.

8. Dao Nguyễn and Kevin B. Hawkins, "Vietnam: Vietnam Power Development Plan for the 2011–2020 Period," Mondaq: Mayer-Brown JSM, September 7, 2011, http://www.mondaq.com/x/144632.

9. "Cục trưởng Cục Năng lượng nguyên tử—Hoãng Anh Tuấn: Chỉ phát triển điện hạt nhân khi điều kiện chín muồi" [Director of VAEA—Hoảng Anh Tuấn: To develop nuclear power only when conditions are ripe], Vietnam Electricity (EVN), May 25, 2015, http://www.evn.com.vn/News/Tin-tuc-Hoat-dong/Dien-hat-nhan-NLTT/TS-Hoang-Anh-Tuan-Cuc-truong-Cuc-Nang-luong-nguyen-tu-Chi-phat-trien-dien-hat-nhan-khi-dieu-kien-chin-muoi-2015525.aspx.

10. Ibid.

11. Ibid.

12. Mark Fitzpatrick, ed., *Preventing Nuclear Dangers in Southeast Asia and Australasia* (London: International Institute for Strategic Studies, 2009), 161, http://www.iiss.org/en/publications/strategic%20dossiers/issues/preventing-nuclear-dangers-in-southease-asia-and-australasia-7e31.

13. Although written from the standpoint of making policy recommendations to the United States, for a brief summary of the South China Sea disputes, see Bonnie S. Glaser, "Armed Clash in the South China Sea," *Council on Foreign Relations—Contingency Planning Memorandum* 14, April 2012, http://www.cfr.org/world/armed-clash-south-china-sea/p27883.

14. Interview by L. Yarr with an official from the Ministry of Foreign Affairs in Hanoi, January 17, 2014.

15. Nguyễn Phạm Mười and Vũ Trọng Khanh, "Vietnam's Nuclear Energy Plan Likely Part of Russia Talks," *Wall Street Journal*, November 6, 2013, http://blogs.wsj.com/searealtime/2013/11/06/Vietnams-nuclear-energy-plan-likely-part-of-russia-talks.

16. Communications Department of the State Atomic Energy Corporation Rosatom, "Nuclear Power Deal Between Russia and Vietnam Signed," Rosatom, January 11, 2010, http://www.rosatom.ru/en/presscentre/highlights/dcd514804486eca5af4eaffdd2ede0d2; "Russia to Build Safest Modern NPP in Vietnam—Rosatom Head," *ITAR-TASS*, May 22, 2013, http://www.rosatom.ru/en/presscentre/nuclear_industry/5f6807804fb68be6855ccd798f00bb46.

17. "Bright Forecast for Trade Links between Viet Nam and Japan," *Viet Nam News*, October 10, 2013, http://Vietnamnews.vn/economy/246023/bright-forecast-for-trade-links-between-viet-nam-and-japan.html.

18. "Vietnam-South Korea Trade Surges," *Viet Nam News*, January 21, 2014, http://vietnamnews.vn/economy/250439/viet-nam-south-korea-trade-surges.html.

19. Law on Atomic Energy, Public Law No. 18/2008/QH12, (Hanoi: Socialist Republic of Vietnam, June 3, 2008), http://www.nti.org/media/pdfs/VNAtomicEnergyLAw2008.pdf?_=1333146100.

20. Dr. Trần Ngọc Ca, Head of the Office of Science and Technology, Embassy of the Socialist Republic of Vietnam in Washington, D.C., e-mail communication to Linda Yarr, December 31, 2014.

21. Resolution Approving Investment in Ninh Thuận Nuclear Power Project, National Assembly Resolution No. 41/2009/QH12 (Hanoi: Socialist Republic of Vietnam, November 25, 2009), http://www.varans.vn/vanbanphapluat/3_nghiquyetquochoi/Resolution_NA_2009_41_Investment_in_Ninh_Thuan_Nuclear_power_project.doc.

22. National Assembly of Vietnam, "Developing Atomic Energy: Top Priority Given to Safety" (Hanoi: Socialist Republic of Vietnam, May 14, 2008), http://www.na.gov.vn/htx/English/C1330/default.asp?Newid=18562#62U6eoLlgbMb.

23. Ibid.

24. "Hành lang Quốc hội, nóng chuyện điện hạt nhân" [Nuclear Power—A Hot Issue (Debated) in Congress's Lobby], *Tiền Phong*, March 22, 2011, http://www.tienphong.vn/Print.aspx?id=531942.

25. For examples of this, see "The Government on Detailing and Guiding a Number of Articles of the Law on Atomic Energy," Decree No. 07/2010/NĐ-CP (Hanoi: Socialist Republic of Vietnam, January 25, 2010); "Approving the Orientation for Planning Nuclear Power Development in Vietnam through 2030," Decision No. 906/QĐ-TTg (Hanoi: Socialist Republic of Vietnam, June 17, 2010); "Approving the Strategy for Peaceful Utilization of Atomic Energy up to 2020," Decision No. 01/2006/QĐ-TTg (Hanoi: Socialist Republic of Vietnam, January 2006).

26. "Thành lập Hội đồng phát triển, ứng dụng năng lượng nguyên tử quốc gia" [Establishment of the National Atomic Energy Council], *Vietnam Electricity*, May 11, 2013, http://www.evn.com.vn/News/Tin-tuc-Hoat-dong/Dien-hat-nhan-NLTT/Thanh-lap-Hoi-dong-phat-trien-ung-dung-nang-luong-nguyen-tu-quoc-gia.aspx.

27. According to the Director of the Vietnam Atomic Energy Institute, Dr. Trần Chí Thành, Vietnam Atomic Energy Commission (VAEC) is the former name of the institute and for some reason the name persists on the MOST website. Nevertheless, Vietnam Atomic Energy Institute (VINATOM) is the official name of the institute.

28. "Hội đồng an toàn hạt nhân quốc gia" [Officially launched Council of National Nuclear Security], ENERTEAM, accessed January 30, 2015, http://www.enerteam.org/tin-tuc-228/chinh-thuc-ra-mat-hoi-dong-an-toan-hat-nhan-quoc-gia-501.aspx.

29. "Vietnam Resort Town Concerned Nuclear Reactor Could Hit Tourism," *Thanh Niên News*, March 31, 2011, http://thanhniennews.com/special-report/Việt Nam-resort-town-concerned-nuclear-reactor-could-hit-tourism-24996.html.

30. For more information, see "Vietnam and IAEA Update an Integrated Work Plan (IWP) for the Development of Nuclear Power Infrastructure for 2011–2015," *International Atomic Energy Agency*, July 3, 2013, http://www.iaea.org/technicalcooperation/Home/Highlights-Archive/Archive-2013/07032013_Vietnam.html.

31. "Nuclear Safety Issues Should be Made Known to Public: IAEA," *Tuổi Tre'News*, January 11, 2014, http://tuoitrenews.vn/society/16783/nuclear-safety-issues-should-be-made-known-to-public-iaea.

32. Vũ Trọng Khanh, "Vietnam's Slowing Growth and Safety Concerns Delay Nuclear Plans," *Wall Street Journal*, January 23, 2015, http://blogs.wsj.com/frontiers/2015/01/23/vietnams-slowing-growth-and-safety-concerns-delay-nuclear-plans.

33. "Agreement Opens US-Vietnam Nuclear Trade," *World Nuclear News*, October 10, 2013, http://www.world-nuclear-news.org/NP-Agreement_opens_US_Vietnam_nuclear_trade-1010134.html.

34. "Japan to Support Vietnam's Nuclear Power Project," *Dang Cong San Viet Nam*, May 15, 2008, http://dangcongsan.vn/cpv/Modules/News/NewsDetail.aspx?co_id=30107&cn_id=227309; "Joint Statement between the United States of America and the Socialist Republic of Vietnam," Ministry of Foreign Affairs, accessed January 30, 2015, http://www.mofa.gov.vn/en/nr040807104143/nr040807105001/ns080626154459/view.

35. "Cooperation with Japan on Improvement of National Legal System on Compensation for Nuclear Damage," Vietnam Agency for Radiation and Nuclear Safety, August 15, 2012, http://www.varans.vn/default.asp?actType=2&TypeGrp=1&id_news=2365&menuup=102000&menuid=102000&menulink=100000.

36. Mike Ives, "Vietnam Presses Ahead with Ambitious Nuclear Plans," *Associated Press*, October 17, 2013, http://bigstory.ap.org/article/vietnam-presses-ahead-ambitious-nuclear-plans.

37. "VTB Offers $1Bln Loan for Vietnam Nuclear Plant," *RIA Novosti*, February 25, 2013, http://www.rosatom.ru/en/presscentre/nuclear_industry/e2aa04004eaec11dbd0ebd341cc28516.

38. "Nuclear Power in Vietnam," World Nuclear Association, December 2014, http://www.world-nuclear.org/info/Country-Profiles/Countries-T-Z/Vietnam/.

39. Asian Development Bank, *Asian Development Outlook 2014: Fiscal Policy for Inclusive Growth*" (Mandaluyong City, Philippines: Asia Development Bank, 2014), 232, http://www.adb.org/publications/asian-development-outlook-2014-fiscal-policy-inclusive-growth.

40. Asian Development Bank, *Asian Development Outlook 2015: Financing Asia's Future Growth* (Manila: Asian Development Bank, March 2015), 244, http://www.adb.org/publications/asian-development-outlook-2015-financing-asias-future-growth.

41. OECD Development Centre, *Economic Outlook for Southeast Asia, China, and India 2014: Beyond the Middle-Income Trap* (Paris: OECD, 2013), 2, http://www.oecd.org/site/seao/Pocket%20Edition%20SAEO2014.pdf.

42. Mycle Schneider and Antony Froggatt, "2012–2013 World Nuclear Industry Status Report," *Bulletin of the Atomic Scientists* 70 no. 1 (January 7, 2014), http://thebulletin.org/2014/january/2012-2013-world-nuclear-industry-status-report.

43. Interview with Dr. Lê Chí Dũng, Hanoi, January 2014.

44. Phạm Duy Hiển quoted in Thanh Phuong, "Professor Phạm Duy Hiển: Postponement of Nuclear Projects Would Be Wise Decision," *Radio France International*, February 3, 2014, http://vi.rfi.fr/viet-nam/20140203-gs-pham-duy-hien-%C2%AB-hoan-xay-nha-may-hat-nhan-se-la-quyet-dinh-sang-suot-%C2%BB/.

45. Ibid.

46. Index Mundi, "Historical Data Graphs: Electricity Consumption (billion kWh) for Vietnam," accessed October 1, 2015, http://www.indexmundi.com/g/g.aspx?c=vm&v=81.

47. "Vietnam Pushing Ahead with Nuclear Power Expansion," *Thanh Niên News*, November 11, 2013, http://www.thanhniennews.com/society/vietnam-pushing-ahead-with-nuclear-power-expansion-689.html.

48. Ibid.

49. "Tritium Levels at Fukushima No. 1 Top Pacific Ocean Dumping Limit, Tepco Reports," *Japan Times*, May 5, 2014, http://www.japantimes.co.jp/news/2014/05/28/national/tritium-in-fukushima-groundwater-tops-limit-for-discharge-into-sea/; Rianne Teule, "Fukushima's Hidden Impacts," *Greenpeace International: Nuclear Reaction Weblog*, October 10, 2013, http://www.greenpeace.org/international/en/news/Blogs/nuclear-reaction/Fukushima-Tamura-radiation/blog/46950.

50. For more, see Alissa de Charbonnel, "INTERVIEW-Russian Nuclear Exporter's Foreign Hires Battle Soviet-Style Secrecy," *Reuters*, November 14, 2013, http://in.reuters.com/article/2013/11/14/russia-nuclear-rosatom-idINL5N0IM0MW20131114.

51. Lubomir Mitev, "Safety Culture Protected Japan's Onagawa Nuclear Station, Researchers Say," *NucNet*, March 17, 2014, http://www.nucnet.org/all-the-news/2014/03/17/safety-culture-protected-japan-s-onagawa-nuclear-station-researchers-say.

52. Ibid.

53. Nick Carey, Margarita Antidze, and John Ruwitch, "Special Report: After Japan, Where's the Next Nuclear Weak Link?" *Reuters*, June 9, 2011, http://www.reuters.com/article/2011/06/09/us-nuclear-power-emerging-idUSTRE75828N20110609.

54. Changjian Wang, Qiang Wang, and Fei Wang, "Is Vietnam Ready for Nuclear Power?" *Environmental Science & Technology* 46 (2012): 5269–5270, http://pubs.acs.org/doi/pdf/10.1021/es301537t.

55. Nguyễn-Khắc Nhẫn, e-mail communication to Linda Yarr, January 10, 2014, translated from French by author.

56. Interview of Dr. Lê Chí Dũng with authors, Hanoi, January 17, 2014.

57. In a press release at the conclusion of its peer review of Vietnam's radiation and nuclear framework, the IAEA highly encouraged continued "efforts to increase the capacity of VARANS to regulate the developing nuclear power programme." For more, see "IAEA Mission Concludes Peer Review of Viet Nam's Radiation and Nuclear Regulatory Framework," *International Atomic Energy Agency* (Hanoi: IAEA, October 9, 2014), http://www.iaea.org/inis/collection/NCLCollectionStore/_Public/45/099/45099816.pdf.

58. "Vietnam: A Delay in Nuclear Power Raises Energy Security Concerns," *Manila Times*, January 23, 2014, http://www.manilatimes.net/Vietnam-a-delay-in-nuclear-power-raises-energy-security-concerns.

59. Vũ Trọng Khanh, "Vietnam's Slowing Growth and Safety Concerns Delay Nuclear Plans."

60. "Triển khai dự án điện hạt nhân Ninh Thuận: Nhiều chính sách hỗ trợ" [Implementation of NPP projects in Ninh Thuận: Many supportive of policy], Vietnam Electricity (EVN), May 20, 2015, http://www.evn.com.vn/News/Tin-tuc-Hoat-dong/Dien-hat-nhan-NLTT/Trien-khai-du-an-dien-hat-nhan-Ninh-Thuan-Nhieu-chinh-sach-ho-tro2015519.aspx.

61. On February 2, 2015, MET implemented its Joint Circular No. 208/2014/TTLT-BCT-BGDDT (dated October 14, 2013) to offer financial incentives and support for trainees in the field of atomic energy. For more, see "Nhiều ưu đãi đối với người học ngành năng lượng nguyên tử kể từ ngày 10/2/2015" [Incentives for people to study nuclear energy since February 02, 2015], Vietnam Electricity (EVN), February 3, 2015, http://www.evn.com.

vn/News/Tin-tuc-Hoat-dong/Dien-hat-nhan-NLTT/Tu-ngay-102-hang-loat-uu-dai-doi-voi-nguoi-hoc-nganh-nang-luong-nguyen-tu-duoc-ap-dung201522.aspx.

62. "Điều chỉnh quy hoạch hai Nhà máy điện hạt nhân Ninh Thuận" [Adjust the planning of the two NPPs in Ninh Thuận], *Datviet*, July 17, 2015, http://baodatviet.vn/khoa-hoc/khoa-hoc/dieu-chinh-quy-hoach-hai-nha-may-dien-hat-nhan-ninh-thuan-3277682.

63. Ibid.

64. "Bước tiến mới trong xây dựng nhà máy điện hạt nhân đầu tiên ở Việt Nam" [New step in building the first nuclear power plant in Vietnam], Project Management Committee of Ninh Thuận Nuclear Power, August 20, 2015, http://dienhatnhan.com.vn/index.aspx?u=nws&su=d&cid=499&id=23152.

65. "Hội thảo về Độ tin cậy lưới điện và giao diện giữa lưới điện đối với nhà máy điện hạt nhân" [Workshop on "reliability" of electric grid and grid interface for nuclear power plants], Vietnam Electricity (EVN), August 26, 2015, http://www.evn.com.vn/News/Tin-tuc-Hoat-dong/Dien-hat-nhan-NLTT/Lua-chon-cong-nghe-cho-du-an-nang-luong-chat-thai-ran-tai-Viet-Nam2015826.aspx.

66. Seven educational institutions would be upgraded so as to support the nuclear industry's staffing requirements with the goal of having 2,400 engineers and 350 staff with master's degrees or doctorates with nuclear plant experience. Another 900 trained personnel will be assigned to conduct research and ensure nuclear safety, and 500 managers and staff will be sent abroad for training. For more, see "Vietnam Seeks Experts to Aid Nuclear Power Plant," *Viet Nam News*, August 28, 2013,www.asianewsnet.net/Vietna-seeks-experts-to-aid-nuclear-power-plant-50886.html.

67. "Vietnam to Proceed with Nuclear Power Plants," *Thanh Niên News*, March 17, 2011, http://www.thanhniennews.com/society/vietnam-to-proceed-with-nuclear-power-plants-13008.html.

68. "VN Wastes Funds on Overseas Training for Nuclear Energy Workforce," *VietnamNet Bridge*, July 13, 2014, http://english.vietnamnet.vn/fms/science-it/107234/vn-wastes-funds-on-overseas-training-for-nuclear-energy-workforce.html.

69. "Những người đầu tiên được đào tạo cho dự án hạt nhân Ninh Thuận 2" [First trainees for Ninh Thuận NPP 2 project], *VN Express*, September 21, 2014, http://vnexpress.net/tin-tuc/giao-duc/nhung-nguoi-dau-tien-duo-c-dao-tao-cho-du-an-hat-nhan-ninh-thuan-2-3082613.html.

70. "Lễ ký kết Thỏa Thuận hợp tác giữa Trường Đại học điện lực và Cty GE Hitachi Nuclear Energy (Mỹ)" [MOU signing ceremony between Electric Power University and the GE Hitachi Nuclear Energy], Vietnam Electricity (EVN), October 8, 2014, http://www.epu.edu.vn/qlkh_htqt/Default.aspx?BT=13928.

71. Lê Đại Diễn, "Overview on Tsunami Risk Evaluation and NPP Project in Vietnam" (presentation, 1st Kashiwazaki International Symposium on Seismic Safety of Nuclear Installations, NIIT, Kashiwazaki, Niigata, Japan, November 24–26, 2010), http://www.nsr.go.jp/archive/jnes/seismic-symposium10/presentationdata/3_sessionB/B-09.pdf.

72. Dr. Nguyễn-Khắc Nhẫn e-mail communication to Linda Yarr, January 10, 2014.

73. "Guidance for National Regulatory Organisations Principles for Openness and Transparency," European Nuclear Safety Regulators Group, Draft 4 on January 24, 2011 for approval by ENSREG on February 1, 2011, European Nuclear Safety Regulators Group website, http://www.ensreg.eu/sites/default/files/HLG%20p(2011-14)%2057;Draft%204;%20Guidance%20for%20Natl.%20Reg.%20Org.;%20Principles%20for%20Openness+Transparency%20.pdf.

74. Ibid.

75. Ari Nakano, "Debate Suppressed in Vietnam, Questions on Japanese Cooperation," *Asahi Shimbun,* June 25, 2013, http://ajw.asahi.com/article/forum/politics_and_economy/southeast_asia/AJ201306250053.

76. Maira Martini, "Overview of Corruption and Anti-Corruption in Vietnam," *Transparency International—U4 Expert Answer 315*, January 27, 2012, http://www.u4.no/publications/overview-of-corruption-and-anti-corruption-in-Vietnam.

77. Dennis C. McCornac, "The Challenge of Corruption in Higher Education: The Case of Vietnam," *Asian Education and Development Studies* 1 no. 3 (2012): 262–275.

78. "Vietnam Contractors, Consultants Banned for 3 Years after Highway Sinks," *Thanh Niên News*, June 5, 2014, http://www.thanhniennews.com/society/Việt Nam-contractors-consultants-banned-for-3-years-after-highway-sinks-26940.html.

79. "IAEA Mission Concludes Peer Review," October 9, 2014.

80. Mary Beth D. Nikitin, Mark Holt, and Mark E. Manyin, "U.S.-Vietnam Nuclear Cooperation Agreement: Issues for Congress," Congressional Research Service, September 15, 2014, 8.

81. Ibid.

82. The 123 Agreement was initialed on October 10, 2013, officially signed on May 6, 2014, approved by President Barack Obama in February 2014, approved by the United States Senate on July 22, 2014, and came into force in August 2014. For more on the Nonproliferation Assessment, see *Letter of Allison M. Macfarlane, Nuclear Regulatory Commission, to President Barack Obama, December 3, 2013*, Attached to Proposed Agreement between the Government of the United States of America and the Government of the Socialist Republic of Vietnam Concerning Peaceful Uses of Nuclear Energy (Washington, D.C.: U.S. Government Printing Office, 2014), http://www.gpo.gov/fdsys/pkg/CDOC-113hdoc109/pdf/CDOC-113hdoc109.pdf.

83. "Vietnam: Overview," *Nuclear Threat Initiative—Country Profiles*, December 2014, http://www.nti.org/country-profiles/vietnam.

84. "Lễ ký Biên bản ghi nhớ giữa Cục An toàn bức xạ và hạt nhân và Công ty Lightbridge" [Signing ceremony of Memorandum of Understanding between the Vietnam Agency for Radiation and Nuclear Safety and the Lightbridge Corporation], Ministry of Science and Technology, August 15, 2015, http://www.most.gov.vn/Desktop.aspx/ND-hop-tac-QT/Tin-tuc-hop-tac-quoc-te/Le_ky_Bien_ban_ghi_nho_giua_Cuc_An_toan_buc_xa_va_hat_nhan_va_Cong_ty_Lightbridge/?SearchTerms=T%E1%BB%8Fng+c%C3%B4ng+ty+Lightbridge

85. "Lễ ký kết Bản Thỏa Thuận Hợp tác giữa Viện Năng lương nguyên tử Việt Nam và Công ty Lightbridge Hoa Kỳ" [Signing ceremony of cooperation agreement between the Vietnam Atomic Energy Institute and the U.S. Lightbridge Corporation], *Ministry of Science and Technology*, October 21, 2014, http://www.most.gov.vn/Desktop.aspx/Chi-tiet-bai-viet/Tin-tong-hop/Le_ky_ket_Ban_Thoa_thuan_Hop_tac_giua_Vien_Nang_luong_nguyen_tu_Viet_Nam_va_Cong_ty_Lightbridge_Hoa_Ky/?SearchTerms=T%E1%BB%8Fng+c%C3%B4ng+ty+Lightbridge.

86. "Mỹ đưa Việt Nam vào danh sách trợ giúp hạt nhân" [United States to add Vietnam in the list of nuclear assistance], *Vietnam Net*, March 6, 2015, http://vietnamnet.vn/vn/chinh-tri/224177/my-dua-vn-vao-danh-sach-tro-giup-hat-nhan.html.

87. "Hội hội thảo về công nghệ nhà máy điện hạt nhân AP1000" [Workshop on technology of the NPP of AP1000], Vietnam Electricity (EVN), July 28, 2015, http://www.evn.com.

vn/News/Tin-tuc-Hoat-dong/Dien-hat-nhan-NLTT/Cong-nghe-nha-may-dien-hat-nhan-AP10002015728.aspx.

88. U.S. Department of State, Office of the Spokesperson, "Vietnam Supports the Proliferation Security Initiative," (Washington, D.C.: Department of State, May 22, 2014), http://www.state.gov/r/pa/prs/ps/2014/05/226449.htm.

89. Carl Thayer, "Vietnam Mulling New Strategies to Deter China," *The Diplomat*, May 28, 2014, http://thediplomat.com/2014/05/vietnam-mulling-new-strategies-to-deter-china/.

90. Phạm Bình Minh, "Foreign Minister's Statement at the Conference on Disarmament," *Ministry of Foreign Affairs* (Hanoi: Socialist Republic of Vietnam, February 27, 2013), http://www.mofa.gov.vn/en/nr040807104143/nr111027144142/ns130227213538.

91. Tanya Ogilvie-White and Maria Rost Rublee, "The Nuclear Energy Aspirants: Egypt and Vietnam," in *Slaying the Nuclear Dragon: Disarmament Dynamics in the Twenty-First Century*, eds. Tanya Ogilvie-White and David Santoro (Athens: University of Georgia Press, 2012), 201.

92. Daniel Horner, "Arms Control Today: U.S., Vietnam Initial Civil Nuclear Pact," Arms Control Association, November 4, 2013, http://armscontrol.org/act/2013_11/US-/Vietnam-Initial-Civil-Nuclear-Pact.

8

Southeast Asia

A Measured Nuclear Policy

Catharin Dalpino and Timothy Westmyer

In the twenty-first century, Southeast Asia is a region at relative peace with itself, but one in which global economic and security dynamics increasingly intrude. This tension is seen in the region's nuclear affairs. Only one country—Vietnam—has nearly realized its nuclear energy ambitions while another—Thailand—began down that path before later diverging away. A handful of others—Indonesia, Malaysia, and the Philippines—are still considering nuclear plants. Throughout this experience, the specter of nuclear weapons proliferation loomed, with concerns about military programs in Indonesia and Myanmar. In the latter years of the Cold War, the Association of Southeast Asian Nations (ASEAN) declared itself to be a zone free of nuclear weapons, but it has never invoked the principles of this regional framework upon a member state or an external power. Today, member governments broadly view the threat of nuclear proliferation as one external to the region, although they acknowledge it could ultimately affect Southeast Asia.

Nationalist positions held by leaders in the Philippines and Indonesia during the Cold War have faded away in recent decades. Today, with the brief possible exception of Myanmar, ASEAN's members have been consistently Globalist in outlook on nuclear weapons and nuclear energy. This is despite the differing security conditions Southeast Asia has faced with the end of the Cold War, the heightened geopolitical tensions surrounding the rise of China and maritime disputes in the South China Sea, and the mixed alliance picture ranging from formal alliances with the United States (the Philippines, Thailand) to "quasi alliance" (Singapore, Vietnam) to non-alliance (Malaysia, Indonesia). The attempt to embark on a nuclear energy route, partly driven by a Realist aim to build up domestic capability and achieve a more diversified energy portfolio to meet rising demand, has been held up or even scuttled by domestic opposition in the wake of the Fukushima disaster. Evidence suggests that the Globalist nuclear consensus across ASEAN states is unlikely in the near term to translate into strong regional nuclear energy or nonproliferation architecture for both internal reasons and the pressures of great power politics.

This chapter examines three topics within the broader subject of nuclear policy in Southeast Asia. First, it focuses on the region's nuclear energy aspirations and why there has yet to be an operating nuclear reactor. Second, with case studies in Indonesia and the Philippines, it explores the significant role of civil society organizations in shaping domestic nuclear debates. Finally, the region's engagement with the nuclear nonproliferation and nuclear security regimes is discussed, including Southeast Asia's inherent proliferation challenges as well as its unique contributions to these efforts.

NUCLEAR ENERGY IN SOUTHEAST ASIA

Vietnam, despite the delays starting in 2013 on the construction of its first nuclear energy reactors (as explored in chapter 7), has perhaps the region's most advanced plans for nuclear power and is thus worthy of separate attention in this edited volume. The rest of Southeast Asia, however, has a rich history in the nuclear sphere—a history that lends a fuller understanding of domestic nuclear debates in Asia. President Eisenhower's Atoms for Peace program established the region's first small nuclear research reactors and promoted nuclear literacy among scientists and engineers.[1] Some countries like Indonesia, the Philippines, and Thailand were close to realizing their nuclear energy ambitions before hitting a litany of roadblocks: global economic crises; international incidents like Chernobyl, Three Mile Island, and Fukushima; and domestic political upheaval. Others like Singapore, Malaysia, and Myanmar have expressed at least a minimal level of interest in nuclear energy that could see resurgence in the future should conditions turn favorable.

Every nation that decides to gain membership to the nuclear energy club does so for its own set of reasons, but there are three basic impetuses seen in the quest for nuclear power in Southeast Asia: (1) meet energy demands, (2) reach climate change targets, and (3) achieve national prestige. First, energy security is a major challenge to Southeast Asia, a region that the Organisation for Economic Co-operation and Development estimates will achieve an average growth rate of over 5 percent.[2] Southeast Asia's primary fuels—oil, natural gas, coal, and hydropower—have come under increasing pressure to meet these economic targets. The share of natural gas in the region's energy portfolio is rising, but pipeline development and distribution concerns have thus far kept natural gas from being a total solution.[3] Even hydropower, one of the increasingly popular alternative energy sources in the region, does not come without risks.[4] In addition, the drive toward greater energy self-sufficiency is spurred on by disputes with China over the oil-rich Spratly Islands in the South China Sea as well as fears the freedom of navigation through the Straits of Malacca, through which most of Northeast and Southeast Asia's imported oil must pass, could be threatened.

After years of gasoline price hikes, ASEAN debated the idea of nuclear power at the 2007 and 2008 ASEAN Summits.[5] Since then, the nuclear energy question remained on the agenda to varying degrees. Host country Vietnam encouraged its fellow members at the 2010 ASEAN Summit to consider nuclear power as an option to meet energy needs.[6] After the 2011 nuclear accident at Fukushima in Japan, ASEAN leaders recognized "renewed concerns over the use of nuclear power" and reaffirmed efforts to promote nuclear safety as well as emergency preparedness and response.[7] The Energy Ministers Joint Statements in 2012 cited progress on engagement with stakeholders to study nuclear energy.[8] The 2014 statement noted support for collaboration in "the safe and efficient development of nuclear power among interested countries" to guarantee global energy security and sustainability.[9] Today, the Nuclear Energy Cooperation Sub-Sector Network serves as ASEAN's coordinating body to assist member states as they explore their nuclear options.

A second driver of nuclear power is that meeting climate change goals requires a reduction in the greenhouse gas emissions that accompany oil and coal use. References to the need for "clean energy" can be found in various ASEAN Summit statements in recent years. For most of Southeast Asia, however, energy security rationales outweigh climate change as a motive for national energy policy calculations.[10] It is important to note that since climate change and related natural disasters are expected to be acute in Southeast Asia, the risk of nuclear accidents may be high as a result.[11] Rising sea levels and other environmental impacts may threaten nuclear sites and put at risk considerable investments into nuclear power.

Finally, a distant but important historical driver for nuclear power aspiration is the pursuit of national prestige and a way to resist outside forces. The decline in foreign colonialism in Southeast Asia left many newly independent nations such as the Philippines and Indonesia in the 1950s and 1960s looking for ways to showcase their sovereignty. The prestige of being a player in the nuclear sphere was a sign of a modern and sophisticated country.[12] A related point is how the international nuclear marketplace targeted Southeast Asia as a priority for their marketing and sales pitches.[13] Both private and state-run vendors eyed the region as a client in the export of nuclear power technologies and services.

This section offers a look at individual countries in Southeast Asia (except for Vietnam, which deserves its own chapter in this volume), their history with nuclear energy, and their future outlooks. The subsequent section offers a more detailed look at how civil society organizations in the Philippines and Indonesia clashed to shape the direction of these nuclear debates.

Brunei, Laos, and Cambodia

For a variety of reasons, this trio of countries is unlikely to pursue a nuclear energy option in the near future. Their role as energy suppliers and consumers, however, is worth noting as they can have an influence on the direction of energy decisions by others in the region.

Brunei has not advanced beyond the information-gathering stage of a nuclear energy program. In November 2007, the government-established Brunei Energy

Association held a forum with South Korean officials to discuss nuclear energy. The association, however, did not expect Brunei to "move into the use of nuclear energy in the foreseeable future," but said it was important to start thinking about what happens "when the oil will run out."[14] This reality may be decades away given Brunei's considerable natural gas and oil reserves, but the BP World Energy Outlook forecasts enough oil for only twenty-two years at the current rate of extraction.[15] Since 2014, Brunei exchanges with the International Atomic Energy Agency (IAEA) have mostly centered on agriculture and health applications.[16] Therefore, if Brunei wanted to explore nuclear power, it would have to start work on the ground floor to build a nuclear infrastructure and regulatory system.

In a 2007 meeting with an Iranian ambassador, the vice prime minster of Laos defended every country's sovereign right "to take advantage of peaceful nuclear energy."[17] While he noted volatile oil prices leave developing nations with "no alternative but to make use of nuclear energy," Laos has not made a decision to move ahead with nuclear power.[18] The country has been an IAEA member since 2011, but it currently does not possess the nuclear infrastructure and regulatory systems necessary to support a nuclear power program. Still, Laos has regular exchanges with the IAEA on other nuclear sciences and is taking steps to advance its own capacities in order to coordinate nuclear plans with the region should they develop.[19] Instead, hydropower is Laos's energy priority as it tries to become the "battery" of Southeast Asia. The country's mountainous terrain and bountiful rainfall drive its hydropower exports to Thailand, Vietnam, and others in the region. As of 2014, Laos had twenty-three hydroelectric dams operating along the Mekong River, with plans for ninety-three by 2020 and another twenty-eight after that.[20]

Despite the lack of a nuclear infrastructure and being outside the IAEA, in 2008, Cambodian Prime Minister Hun Sen announced his country's plans to set up a nuclear plant.[21] The Ministry of Industry, Mines, and Energy (MIME) forecasted the country's energy demands to double by 2020.[22] The World Bank highlighted high electricity costs as a key constraint on Cambodia's manufacturing growth.[23] Believing its coal and hydroelectric power capacity would be insufficient to meet its energy needs, Phnom Penh wanted to see its first nuclear reactor by 2020-2025. "We cannot avoid nuclear power. It will be the only choice for producing cheap electricity," said Ngy San, then-deputy executive director at the NGO Forum on Cambodia.

Beginning in 2010, however, MIME Secretary Ith Praing scaled back these expectations. He said while nuclear power is a "possibility because our neighbors are doing it," Cambodia has "to study it and see how dangerous it is."[24] Praing pushed back on claims that Cambodia told ASEAN energy ministers in 2012 that it was planning on building a nuclear plant near the Thai border.[25] By 2014, Prime Minister Sen disputed his country had nuclear power ambitions and instead wanted to apply nuclear technologies to agriculture and health sciences.[26] Hydropower continues to be the nation's energy priority with its largest station opened in early 2015 and plans to build six plants by 2018. Even if Cambodia returns to its original nuclear energy goals, it is not expected to achieve them in the near to medium term due to the lack of an existing nuclear infrastructure.

Malaysia

In the 1970s when oil prices were skyrocketing, Malaysia issued plans to have two nuclear power plants operational by 1990 to diversify its energy mix.[27] This proposal was put on the back burner soon afterward, but the nuclear option has never entirely disappeared from Kuala Lumpur. Over the last decade, supply concerns about the country's once-extensive oil and natural gas provisions led officials to again reconsider nuclear energy. The Asian Development Bank projected Malaysia's annual energy demand to increase from 73 million tons of oil equivalent in 2010 to 109 million tons by 2035.[28]

While Malaysia does not have an extensive nuclear infrastructure in place today, the Malaysian Nuclear Agency has operated a research reactor (Reaktor Triga Puspati) under IAEA safeguards since 1982. Malaysia has worked closely with the IAEA since the mid-1970s with projects on applying nuclear science to medicine.[29] In July 2009, Fadillah Yusof, the deputy science, technology, and innovation minister, said Malaysia aimed to install nuclear power by 2025.[30]

Malaysia halted active plans to develop nuclear energy after the 2011 nuclear disaster at Fukushima in Japan, but the government continues to conduct feasibility studies with reports indicating the Atomic Energy Licensing Board had five undisclosed locations for potential nuclear plants.[31] The minister of energy, green technology, and water, Maximus Johnity Ongkili, said in 2013 "the sentiment at the moment is not quite in favor [of nuclear power], and we need to get a lot more buy-in." In early 2015, the Malaysia Nuclear Power Corporation (MNPC) said the country's legal framework was still being developed before any nuclear sites would be discussed.[32] While Malaysia has signed exploratory agreements with foreign nuclear vendors such as Westinghouse and the China National Nuclear Corporation, MNPC officials said a final decision remained years away.

The nuclear debate was met with fierce opposition from anti-nuclear groups such as Anak Malaysia Anti-Nuclear.[33] A National Energy Security Survey (NESS) by an NGO, the Association of Water and Energy Research Malaysia, found 98 percent of Malaysians as either not confident or unsure of the government's capacity to regulate nuclear power.[34] In response, MNPC announced they would conduct their own national survey in September 2015 to gauge the public's confidence.[35] Malaysia's nuclear debate is still evolving as the government and civil society clash over whether a nuclear program will be in the country's future.

Singapore

Similar to most of Southeast Asia in the late 2000s, Singapore articulated an interest in exploring nuclear energy. Though Singapore's nuclear infrastructure and experience with nuclear technologies was limited to medical and industrial applications, the country's tightly packed geography and population make smaller modular reactors or those kept on sea barges more attractive ideas if safety and economic hurdles can be overcome.[36] In December 2008, Prime Minister Lee Hsien Loong refused to rule out nuclear power due to rising oil prices and climate change goals.[37] Another option discussed would be to take advantage of a nuclear power station in Malaysia by interconnecting the two nations' power grids, but there has been little substantive progress to date on this proposal.

Without nuclear power, Singapore's economy and electrical production continues to be fueled by natural gas, which makes up 95 percent of its energy portfolio as of 2015.[38] Still, the appeal of nuclear power did not fade away from Singapore's domestic energy debates. In 2010, Prime Minister Lee Hsien Loong said the island state "cannot afford to dismiss the option of nuclear energy" since it was an "important part of the solution to mankind's energy problems" and a possibility for Singapore in his lifetime.[39] He ordered the Energy Studies Institute to conduct a two-year prefeasibility study on whether Singapore would be ready for a nuclear energy program. That study, however, concluded in 2013 that nuclear technology would not be appropriate for a small city-state and that the risks would outweigh the benefits.[40]

This report echoed the conclusions of another national energy policy report issued in 2007.[41] The primary reason noted by researchers in 2013 was that the city-state's unique geographic and population densities would not allow for the required emergency planning and response strategies in the event of an accident. Nevertheless, since Singapore's neighbors in the region were still actively considering nuclear energy, the report urged leaders to continue engaging in nuclear safety discussions and to boost capacities in other nuclear sciences to be in a position to take advantage of nuclear energy at a later date.[42]

Thailand

Thailand had once seriously considered adopting nuclear energy but later encountered obstacles and other dissuasions to that path. Thailand's Ministry of Energy (MOE) estimates that energy needs in the country will double in the next decade as the supplies of natural gas and other power sources decline.[43] Meeting this demand and advancing the cause of reducing energy vulnerability have been goals of nearly every political administration for the past two decades. The MOE calculated that nuclear energy could reduce the country's natural gas demands from 70 percent to 40 percent. Prior to the Fukushima disaster, Thailand planned to construct five nuclear power plants.[44] Although these plans had caught the attention of certain environmental groups in the country, some of which launched anti-nuclear protests, pre-2011 public opinion polls were nevertheless positive about the use of nuclear energy in Thailand.[45]

Thailand's nuclear development can be traced to 1954 with the creation of the Thai Atomic Energy Commission.[46] Under Atoms for Peace, Thailand received a nuclear research reactor in Bangkok that operated 1-MW of power starting in 1962 and later upgraded to a 2-MW model.[47] Thailand's nuclear establishment took advantage of the IAEA's technical and financial assistance programs to advance its nuclear science capabilities in medicine and agriculture.

Thailand's full-scale nuclear energy ambitions, however, have been less successful. In 1966, the country's public utility's first proposal for a nuclear power plant was approved by the government and the IAEA, but never advanced further due to financial reasons. Safety concerns, economic turmoil, and the discovery of natural gas reserves in the Thai Gulf derailed another attempt in the late 1970s.[48] Feasibility studies from 1996 to 1998 conducted by the Ministry of Science and Technology were undercut by the Asia financial crisis. It wasn't until the late 2000s that Thailand seriously resumed its pursuit of nuclear power as part of a national policy to diversity its energy portfolio away from an overreliance on natural gas. In 2007, Thailand established a Power Development Plan (PDP) that aimed for 2,000 MW of nuclear energy by 2020 and another 2,000 MW the following year. The government assigned the Nuclear Power Infrastructure Establishment Coordination Committee to convene various agencies with a role in energy policy to advance the nuclear program.[49] A revised 2010 PDP set more impressive goals for five nuclear plants operational in 2020, 2021, 2024, 2025, and 2028.[50]

Immediately following the Fukushima incident, then–Prime Minister Abhisit Vejjajiva ordered the Ministry of Energy to suspend plans on the nuclear plants and to examine emergency safeguards and the potential vulnerability of the locations that had been identified.[51] The 2011 PDP deferred the nuclear plant schedule by three years and later until 2023. In 2012, MOE resumed plans to complete site surveys for the plants. However, in the immediate aftermath of Fukushima, an ABAC (Assumption Business Administration College) public opinion poll showed a sharp reversal with more 83 percent opposing the construction of nuclear power plants.[52] Although the polls have become somewhat more moderate since then, they have yet to return to positive territory on the question of a move to nuclear energy.

Thailand's 2015 PDP included plans for nuclear power—two 1,000-MW plants by 2036 accounting for 5 percent of the country's energy mix—that were less ambitious than previous official proposals.[53] The deputy permanent secretary for the energy ministry, Kurujit Nakornthap, said, "We haven't discarded nuclear as an option for strengthening the country's energy security" and meeting climate change goals. However, six key factors have made the prospect of developing nuclear energy in Thailand even more remote. First, continued political instability in the country would make a sustainable political decision to move to nuclear power all but impossible. The overthrow of former Prime Minister Yingluck Shinawatra in 2014, the installation of an interim military government, and an uncertain constitutional reform process make it unlikely that the Ministry of Energy would be able to gain the long-term political and budgetary support needed to see its nuclear energy plans to completion. Moreover, although the military junta in power since May 2014 constricts public protests, Thailand's environmental movement has deep roots and would likely defy bans on public demonstrations to protest nuclear plans.

Second, Bangkok's relations with its neighbors have seen slight improvements that could give Thailand greater access to regional energy supplies. Tensions with Cambodia have eased somewhat, which may pave the way for plans for joint development of natural gas resources in an offshore area of overlapping claims. Thailand's relations with Myanmar have strengthened with the latter country's reform movement, making Thailand's supplies of natural gas from a Myanmar pipeline more secure.

Third, Laos's goal of becoming the "battery" of Southeast Asia has seen its primary customer for hydropower in Thailand. Fourth, the security of Thailand's imported energy and its international trade in general will likely strengthen with the planned joint development with Myanmar of a deep port in Dawei, with road and rail links to be added from Thailand. This would give Thailand an alternative to the Straits of Malacca since Thai imports and exports could travel overland from the Indian Ocean. Fifth, since 2014, the global drop in fuel prices has made Thailand's dependence upon natural gas less costly for the time being. In August 2014, the military government announced it would further explore coal-fired power plants instead of nuclear energy.[54]

Finally, Thailand's foreign relations with external powers have been affected by the May 2014 military coup. If the Ministry of Energy were able to go ahead with plans to build nuclear facilities, it would have a limited choice of partners. Immediately after the coup, the United States suspended its agreement with Thailand on civilian nuclear cooperation and let the deadline to renew it lapse. For similar reasons, it is unlikely that European companies would bid on contracts to build nuclear plants while the military rule is still in place. This would likely limit Thailand's choices to Japan and Russia. Although the Ministry of Energy has not formally abandoned its plans for nuclear energy, the collective weight of these factors may mean that Thailand's "nuclear moment" has passed.

Myanmar

For years, credible information on Myanmar's nuclear energy programs has been a limited resource. The country has shown varying degrees of interest in nuclear research since 1956 when it created the Atomic Energy Centre under the Union of Burma Applied Research Institute to foster training programs for its scientists and scholars. After joining the IAEA in 1957, the country engaged in several technical cooperation programs, including work in using nuclear science in agriculture and medicine.[55] A small neutron generator set up by the IAEA went into operation in the mid-1970s at Rangoon University.

For a while, this was the extent of the country's nuclear aspirations. However, the State Peace and Development Council (SPDC), Myanmar's military government in power since 1988, raised the priority of nuclear technology in the national debate in the late 1990s. Around 1996, the Ministry of Science and Technology started to explore the nuclear reactor technology necessary to establish an isotope production capability. The junta reached out to Russia about developing this facility; other unproven reports include China and Pakistan in this outreach.[56] The IAEA rejected a 2001 request by Myanmar for help to acquire the research reactor, stating it had "no confidence that Myanmar either needs a reactor or has the infrastructure and funding required to support such a project."[57] In early 2002, much to the displeasure of Myanmar's traditional benefactor in China, Russia stepped in to fill the gap with an agreement to build a 10-MW research reactor and two additional laboratories for related work. A groundbreaking ceremony was seemingly ready for a January 2003 project launch until SPDC suddenly announced the facility would be delayed for several years, evidently because of a dispute with Russia over financing.[58] In 2007, SPDC again undertook plans with Russia to build a 10-MW light-water research reactor along with a medical-isotope production site, a radioactive waste and burial facility, and other related work.[59] Both countries remain interested in nuclear cooperation, yet work on the research reactor has not begun.[60]

With regard to nuclear energy, state energy officials told their Japanese counterparts in 2003 that "nuclear power production is desirable for the long-term" with small-scale nuclear reactors by 2025.[61] This idea appears to have faded away by 2009 when the Ministry of Energy concluded nuclear energy was "not environmentally friendly" and should be considered "as a last resort, when other means cannot be implemented."[62]

The country's vast natural gas reserves and its reliance on hydropower, gas, and oil for electricity production puzzled experts who questioned Naypyitaw's surge of interest in nuclear energy technology, training, and equipment. While the Russian nuclear facilities would have had IAEA safeguards in place, the decision prompted international and regional anxieties about possible weapons uses for the nuclear site. Similar to the early nuclear motivations of Indonesia and the Philippines, some of the driving force behind Myanmar's nuclear decision-making seems rooted in Nationalist thinking. The bureaucratic roots of the energy program can be traced to U Thaung, the minister of science and technology from 1998 to 2011, who reasoned nuclear research was a prerequisite for "a modern nation."[63] The deputy foreign minister in 2002 declared it imperative for the country to reap the benefits of nuclear technology or else become marginalized relative to its neighbors.[64] Experts have attached meaning to this logic (e.g., mastering nuclear energy authenticates a country's status and international prestige) in light of the region's public nuclear energy goals in the mid-1990s and late 2000s.[65] By 2012, however, Lt. General Hla Min publicly declared his country's nuclear activities were limited to medical application and had "achieved nothing so far in our country" with "no practical activities so far."[66]

The Philippines

In the early 1970s, Manila seemed destined to join the nuclear energy club. In 2016, however, the Philippines is the only country in the region with a nuclear power plant that was built but never made operational. Construction on the Bataan Nuclear Power Plant (BNPP) began in 1976 under Ferdinand Marcos, the dictator who served as the country's president from 1965 to 1986.[67] However, the fall of Marcos from power and safety concerns forced the government to close the plant before it ever went online. The Philippine Department of Energy, however, has not taken nuclear energy off the table as a solution to its serious energy demands.

A later section of this chapter includes a detailed look at Manila's nuclear energy history, but it is worth mentioning here that the Philippines case shows how the domestic political environment of the country hosting a nuclear plant is a critically important factor in the global nuclear market. In addition to the stability of the political system itself, the public's anxieties about nuclear power or its confidence in national leadership to operate with competence and without corruption can result in sudden tremendous change in an industry that traditionally relies on long-term construction and operating schedules.

Indonesia

The roots of Indonesia's nuclear energy ambitions can be seen as far back as the 1950s. The country's early leader, President Sukarno, saw nuclear power as not only a solution to meet energy demands but also a means to demonstrate the national power, worldwide prestige, and technological prowess of his Indonesia.[68] The institute that eventually became the National Nuclear Energy Agency (BATAN) was set up in 1959 and began civilian nuclear work with the United States through Atoms for Peace.[69] Originating from that program, Indonesia has operated three research reactors for decades in Bangun (2-MW), Yogyakarta (100-kilowatt), and Serpong (30-MW).

The nuclear energy option was popular with political elites and technocrats.[70] Practical concerns include anxieties about climate change and Indonesia's transformation into an oil importer in an unpredictable market, which encouraged a look at alternative sources of energy.[71] Around 40 percent of the country's electricity is produced by oil and gas and the government would like to free more of those fossil fuels for export. Prior to Fukushima, Indonesia was an active candidate for nuclear energy and arguably had more expertise in nuclear sciences than any other ASEAN state.[72] Nuclear energy advocates pointed to encouraging IAEA visits in 2007 as evidence the country had no major barriers to a nuclear future.[73] In 2010, the IAEA signaled it considered Indonesia, meeting all the preconditions except for public acceptance and a political decision, ready to develop nuclear energy for power generation.[74] The government subsequently planned to have its first nuclear power plant operating by 2016 and additional ones by 2025.

The next section of this chapter includes a detailed look at Jakarta's nuclear energy history, but these plans were suspended in the wake of Fukushima although they have steadily shown signs of a possible revival since 2013. In December 2013, the Ministry for Research and Technology heralded public support for nuclear power based on a poll by an independent agency, PT Iconesia Solution, showing 60 percent of Indonesians comfortable with a nuclear plant in the country. This was followed by a 2014 public survey by BATAN that revealed that 70 percent of Indonesians believed nuclear energy should be developed.[75] In 2013, BATAN reviewed its prior plans and recommended the construction of ten nuclear power plants with site considerations in Sumatra, Java, and Bali.[76] As of August 2015, BATAN's official target is to have the first plant constructed by 2024.

Indonesia's domestic nuclear debate has been quite active in the past few years. The Energy and Mineral Resources Ministry launched a "Go Nuclear" campaign in 2015 to meet BATAN's target of a nuclear reactor by 2024. In April 2015, the Russian state-owned nuclear corporation, Rosatom, announced it won a bidding war to submit a preliminary design for the smaller 10-megawatt electrical (MWe), high-temperature, gas-cooled reactor at Serpong that would be "a flagship project in the future of Indonesia's nuclear program."[77] Other reports in 2014 indicated BATAN plans for two 1000-MWe light-water reactors on Java and two more in Sumatra, although those plans have apparently been delayed. Indonesia's experience with research reactors, its relatively mature nuclear infrastructure, its past engagement with the IAEA and international vendors, and its pressing energy needs make the country likely to continue chasing nuclear energy for the foreseeable future.

The Mosaic of Nuclear Debates in Southeast Asia

Despite not having an operating nuclear energy reactor today, the rich diversity and various twists and turns of Southeast Asia's nuclear path illustrate the complexity of Asia's domestic nuclear debates. While some countries such as Brunei, Laos, and Cambodia may have initially considered the nuclear option, their limited economic resources, more attractive energy alternatives, and lack of a nuclear infrastructure will likely keep them from moving down that road. In recent years, Myanmar has largely set aside its nuclear energy aspirations in order to reassure the international community that its political reforms and non-nuclear weapon pledges are truly genuine.

For other countries like Indonesia, the Philippines, Malaysia, Singapore, and Thailand, nuclear aspirations have been halted by financial calculations, political instabilities, external events, and public apprehensions about nuclear power. With this understanding of the region's nuclear history, the next section looks further at how rising democratic pluralism in the Philippines and Indonesia has fostered robust civil society organizations to challenge the direction of these countries' nuclear debates.

CIVIL SOCIETY'S CHALLENGE TO SOUTHEAST ASIA'S NUCLEAR DEBATES

As discussed throughout this book, public opinion can have a real impact on the policy debates in democratic countries and even in more unitary governments. After the 2011 accident at Fukushima in Japan, for example, grand nuclear energy programs in China, India, Japan, South Korea, Taiwan, and Vietnam were paused, and in some cases significantly delayed, due to an uptick in public opposition. Groups intensified opposition to Indian uranium mining in Jharkhand, Andhra Pradesh, and Meghalaya and the Kudankulam Nuclear Power Project in India (chapter 3), Taiwan's proposed Lungmen Nuclear Power Plant (chapter 6), or uranium processing facilities in China's Guangdong province (chapter 2).

Southeast Asia has been no exception to this trend of public engagement in nuclear issues. There has been a rise of civil society organizations (CSO)—traditionally focused on aiding local communities with food, health care, and housing needs—that have jumped into domestic nuclear debates.[78] Many of these groups challenged the dominant position of elites in these nuclear debates, presented a different set of options, and altered how the various schools of thought approach the issue. To illustrate this development further, the history of CSOs in the Philippines and Indonesia is worth exploring in this section. A detailed look at the domestic challenges within these two countries, which experienced a steady rise of democratic pluralism since the late 1970s, can help us understand how public opposition to nuclear power has led to the current standstill of Southeast Asian's nuclear plans.[79]

The Philippines

As noted in the previous section, the Philippines came incredibly close to having an operational nuclear reactor. If it were not for grassroots organizations that convened a broad coalition of educators, youth, peasants, women, religious leaders, human rights supporters, and other parties to rally against the Bataan Nuclear Power Plant (BNPP), the Philippines might have been the earliest adopter of nuclear energy in the region.[80]

Some civil society organizations, including Greenpeace South Asia and the Philippine Climate Watch Alliance, focused on the environmental risks of nuclear power (e.g., accidents, radioactive materials, and waste storage). Although BNPP's proximity to earthquake and volcanic activity areas was the primary talking point, environmental CSOs also tended to categorically argue against nuclear power and instead promote renewable sources like hydropower, biomass, wind, geothermal, and solar. Other groups argued nuclear power should be rejected on economic grounds due to the financial costs and debt requirements that siting, construction, operation, and cleanup place on the country's citizens. The Freedom from Debt Coalition was a leading force in this group and has pushed back on claims that the Philippines energy demand is as serious as reported by nuclear proponents, especially during the economic downturn around the 2007 debate on reviving the Bataan site.

Another set of civil society organizations, however, were pro-nuclear energy advocates who viewed nuclear power as an unavoidable and important part of the government's plan to meet economic targets by satisfying energy demands. For example, the Philippine Chamber of Commerce and Industry (PCCI)—the country's largest business association—saw "nuclear power as a major component of a strategic energy policy."[81] In 2010, PCCI urged Philippine leaders to reconsider the merits of going nuclear with "matured and proven technology that provides clean electricity at predictable and competitive costs."[82] As Manila's neighbors examine a nuclear path, the PCCI president said the Philippines needed to keep its options open.

After years of civilian nuclear cooperation since 1955, under the Atoms for Peace program, the U.S. government and the private nuclear industry saw the Philippines as an emerging marketplace for their reactors.[83] The U.S. nuclear vendor Westinghouse won the bidding war to build a 621-megawatt (MW) light-water reactor at the Bataan Nuclear Power Plant. The Export-Import Bank of the United States offered a record $272 million loan to help Marcos finance the project.[84]

The 1979 Three Mile Island nuclear accident in the United States, however, shifted global attention to nuclear plant safety concerns.[85] The Marcos government, reeling with corruption accusations from how the nuclear deal came into fruition, responded to the accident by setting up a commission to study BNPP's safety and whether his contract with Westinghouse could be broken.[86] The panel declared the plant was based on outdated designs,[87] but Westinghouse initially refused to make safety modifications unless Marcos was willing to pay for them. Eventually, the Philippine Atomic Energy Commission (PAEC) issued a construction permit for the plant against the protests of the head of the PAEC, which Professor William Beaver concluded was a sign the Marcos government was never "serious about cancelling the project."[88] Construction resumed in 1981 as the project costs ballooned to $1.8 billion after inflation and higher interest charges.

In January 1985, Westinghouse declared BNPP ready—at a final cost of $2.2 billion—to be loaded with nuclear fuel. However, the plant never saw operation. Conflicting reports by the IAEA about outstanding safety concerns at the site and an anti-nuclear platform held by the incoming Corazon Aquino administration that ousted Marcos led to the plant's being scrapped in 1986.[89] One Aquino cabinet official called the project "a monument to the pillage and plunder that the Marcos administration has perpetuated upon our helpless people."[90]

The Bataan controversy, however, refused to go away. Rising electricity costs and frequent power outages pressured Aquino into reconsidering the nuclear option. The remaining balance on the loan, its high interest rate, and the salaries of the almost 700 workers at the site were costing the Philippines nearly $1 million a day.[91] In March 1992, after years of legal battles, international arbitration, bribery charges, and countersuits, the Aquino administration and Westinghouse compromised on a plan for the U.S. firm to operate the plant for thirty years while paying the Philippines hundreds of thousands of dollars in safety upgrades and other services.

In the end, the legal wrangling was rendered moot by the election of Fidel Ramos and his resumption of legal action against Westinghouse. According to several scholars, the Bataan plant came to symbolize citizens' anger and frustration over the Marcos years and the plant's actual safety condition was secondary to the fury of Ramos's constituency.[92] Ramos briefly explored nuclear power again with the creation in 1995 of the Nuclear Power Steering Committee.[93]

The Philippines finally paid off the BNPP loan in 2007.[94] Subsequently, there were years of debate about whether to convert the facility to a natural gas or coal-fired power plant. Efforts were made to restart the plant in the late 2000s, but the 2011 nuclear disaster at Fukushima brought these plans to a halt.[95] Today, the site is a tourist attraction with visitors treated to a presentation by the plant's staff on the virtues of nuclear power. The 2011 Household Energy Consumption Survey, a study conducted by the National Statistics Office, said 33 percent of the public was willing to support nuclear power with 47 percent undecided and 20 percent reluctant to support.[96]

The 2016 president, Benigno S. Aquino III, the son of Corazon Aquino, has expressed his aversion to resurrecting the plant, which the Department of Energy in Manila estimates could cost $400 to $600 million to refurbish.[97] The Philippines Nuclear Research Institute (PNRI)—previously PAEC—is the primary government agency under the Department of Science and Technology charged with the promotion of nuclear energy, regulatory affairs, and licensing decisions. Should the Philippines resume their quest for a nuclear power plant, it will likely continue to heavily rely on foreign vendors for the reactors and the fuel cycle.

The path toward nuclear power in the Philippines has been a rocky one with several false starts and moments of potential breakthroughs. For some in the nuclear industry, this cycle of well-intentioned plans and inescapable setbacks is a frustrating existence. Mauro Marcelo, a nuclear engineer at a state-owned utility, thought that "there are several dates when we could have become a nuclear country, but every time a catastrophic event happened. We don't need to hire nuclear experts but feng shui masters to get rid of the bad luck."[98] While civil society organizations debated the merits of nuclear power, the Bataan site became inseparable from the public's negative views of the Marcos government. Civil society organizations will likely continue to play a role in shaping this debate in the Philippines.

Indonesia

Compared to the Philippines, civil society organizations are a more recent development in Indonesia, starting with the country's nuclear debates in the early 1990s.[99] As part of this budding nuclear debate, an anti-nuclear camp centered their arguments on safety concerns that some of the planned nuclear sites, such as the one in Muria, are situated near earthquake-prone areas as in the "Ring of Fire," where frequent volcanic activity and earthquakes are a constant risk. Since much of the region is heavily populated in central locations near where nuclear reactors are discussed, anti-nuclear groups hold geography as a reason to never go nuclear.[100] In a series of nuclear socialization efforts in the late 2000s to inform the public on the country's safety standards, BATAN raised its own experience with research reactors and the pristine safety record of the similarly positioned Japan—although this was before Fukushima.[101] Furthermore, the public's confidence in the central government to respond to a nuclear accident further eroded in the wake of natural disasters in the 2000s. After the 2004 tsunami devastation in Sumatera, a 2006 mud-volcano eruption in East Java, and a 6.2 scale earthquake in Yogyakarta in 2006, the public saw the official response as incompetent and slow, resulting in additional unnecessary human suffering.[102] Anti-nuclear groups argued that, if Indonesia could not adequately handle these types of episodes, the country was not ready for a nuclear accident.

Other opposition groups questioned whether the government was capable of using nuclear energy to meet the country's energy needs. Indonesia's people and economic growth is geographically located in Java. However, the electricity grid connecting the archipelago's island chains is limited and may be unable to effectively and efficiently transmit power throughout the country, leaving many regions without the benefits of nuclear energy.[103] In 2013, this led BATAN to explore the possibility of smaller reactor units to service those areas. Anti-nuclear activists countered these moves by calling for more hydropower, geothermal, solar, and other energy sources that can be set up on a small to medium scale locally where the demand exists.

An interesting wrinkle to this case is the employment of Nationalist views used by groups such as Wahana Lingkungan Hidup Indonesia,[104] the Indonesian Antinuclear Society, and Greenpeace Indonesia that argued nuclear power plants could lock the country into reliance on outside powers to provide fuel to burn in the reactors.[105] Leery of trading one form of energy dependence for another, some anti-nuclear groups prefer an energy cycle with clean alternative and renewable sources found at home.

Another unique element of Indonesia's anti-nuclear CSOs is the engagement of the religious sphere through Islamic scholars or *ulama*. The Nahdlatul Ulama (NU) branch in Jepara, which has constituents in the Muria Peninsula near one of the proposed nuclear sites, hosted a full-day event in September 2007 where pro-nuclear and anti-nuclear experts presented their arguments to the local community and religious leaders. After impassioned debate, the forum issued a consensus declaration that the Muria nuclear project was *haram*, or forbidden, in Islamic law as a detriment to the local population.[106] According to Sulfikar Amir from Nanyang Technical University, the "denouncement of nuclear power by the NU ulamas in Jepara seemed to have increased the leverage of antinuclear groups to halt the implementation of BATAN's nuclear power program in Jepara."[107]

However, there are also pro-nuclear energy groups within Indonesia that are less organized but nevertheless well-informed about nuclear issues.[108] In addition to energy security arguments, these coalitions stress Indonesia's long experience with nuclear science in medicine and agriculture. Along with organizations founded by former BATAN officials such as the Energy and Environment Awareness Society, business groups such as Masyarakat Energi Terbarukan (METI), and professional associations like Women in Nuclear Indonesia (WIN) produced the Declaration on Nuclear Energy 2010 pushing leaders to prioritize nuclear power in national energy plans. In response to anti-nuclear Nationalist worries about becoming dependent on foreign uranium suppliers, this group points to uranium deposits in Kalimantan. Moreover, they advance their own Nationalist perspective that outside forces want to slow Indonesia's economic growth and deny it access to advanced nuclear equipment and technology.

The energy demands of a rising Indonesian economy have created room for a middle-ground camp that can still influence the direction of the country's energy debates in the future.[109] A study by Sulfikar Amir identified the lack of trust in government as a strong influence on the rise of challenges to Jakarta's nuclear energy plans.[110] Groups such as the Institute for Essential Services Reform point to not only safety and environment concerns, but also whether the government is ready to integrate large base-load nuclear power plants into the electricity grid and distribute power to where it is needed throughout the archipelago.[111] Unlike in the Philippines, however, the presence of a significant "middle-ground camp" leaves open the possibly nuclear energy can make a resurgence in Indonesia if the technical requirements are satisfactorily met and pro-nuclear advocates can overcome this suspicion of incompetent governance.

Amir observed that Suharto prioritized a state-driven "technological supremacy" and thus elevated the views of elites with this expertise within the nation's decision-making.[112] Nuclear energy as a solution to Indonesia's energy needs and a ticket to national prestige emerged from an elite set of technology-fixated policymakers with little input from the population at large. Amir wrote, "public affairs became dependent on technical knowledge" with state technocrats "politically dominant in defining technology discourses" until the collapse of the Suharto government opened opportunities for a broader democratic process. Amir concluded that the resistance against nuclear power in Indonesia was made possible as individuals concerned with nuclear energy gained the freedom, in the wake of rising democratic pluralism and democratic reforms, to develop organized civil society groups to oppose government energy plans. To the extent Indonesia's energy plans are stalemated, the efficacy of the anti-nuclear and middle-ground camps has played a major part. At a minimum, the rise of civil society organizations complicates then Indonesian debate with a higher number of influential actors with an array of rival views (e.g., both pro-nuclear and anti-nuclear groups appropriate the Nationalist school of thought) competing to determine Indonesia's nuclear future.

CSOs as Referees in Domestic Nuclear Debates

As evident in other chapters throughout this volume, rising democratic pluralism in many Southeast Asian countries has empowered civil society organizations to gain influence in domestic nuclear debates and blunt the Realist and Nationalist drivers toward nuclear energy in the region—at least for the time being.[113] After the 2007 ASEAN Summit in Singapore identified nuclear power as a priority, for example, Greenpeace warned nuclear power would be a "disaster for the region" due to the area's unique geography and the lack of technical expertise. For these anti-nuclear energy CSOs, Fukushima was just the latest piece of evidence against nuclear power and for non-nuclear, renewable energy sources.

For pro-nuclear groups, on the other hand, stringent energy demands make nuclear power inevitable and the public just needs to be educated about the "facts" on nuclear energy. For example, Thai officials told the IAEA, "The lack of knowledge and understandings are observed to be the cause" of nuclear technology fears and "public awareness and participation are being promoted to increase the accurate understandings of nuclear technology for its safety and protection."[114] Sulfikar Amir, however, questioned this idea. He believed public opposition to nuclear power did not stem from a lack of understanding of the details, but from the enabling of CSOs with their own organizing and research capabilities to challenge the very credibility of the scientific elite consensus.[115] The history of civil society organizations in the Philippines and Indonesia demonstrates that wherever the transformation toward democracy and a more open policy-making process can be found, groups can now compete to shape the views of elites and the public within these domestic nuclear debates.

THE NUCLEAR WEAPONS DEBATE IN SOUTHEAST ASIA

Despite proliferation challenges inherent to the region, the overall likelihood of new nuclear weapons emerging from Southeast Asia is low. The Southeast Asian Nuclear-Weapon-Free Zone (SEANWFZ or the Bangkok) Treaty enjoys universal membership among ASEAN states and bans the development of nuclear weapons. The treaty also has provisions for the safe and peaceful use of civilian nuclear technologies and the management of nuclear waste. This section outlines the region's engagement with the broad nonproliferation regime, its specific proliferation challenges, and its unique contributions to global efforts to stop the spread of nuclear weapons.

The Nonproliferation Regime in Southeast Asia

No state in Southeast Asia currently has or has ever possessed nuclear weapons. As seen in table 8.1, all the states in the region are signatories to the Nonproliferation Treaty (NPT), most have some version of safeguards agreements with the IAEA, and several have ratified the IAEA Additional Protocol, which gives nuclear inspectors enhanced powers to verify that civilian nuclear technologies are not being used toward military applications. A number have also joined the U.S.-led Proliferation Security Initiative (PSI), a global counter-proliferation interdiction effort aimed to stop the potential trafficking of weapons of mass destruction. Indonesia's reluctance to sign the PSI stems largely from sovereignty issues in the Straits of Malacca. Laos has expressed concerns that the PSI could have a negative impact on trade. Myanmar has yet to sign. However, with Naypyidaw's opening to the West after the 2010 elections, the government has been eager to demonstrate a nonproliferation commitment

Table 8.1. ASEAN Membership in Nuclear Security and Nonproliferation Regimes

	NPT[1]	*IAEA Member*[1]	*CSA*[2]	*SQP*	*Additional Protocol*[2]	*CTBT*[1]	*CPPNM*[2]	*2005 CPPNM Amendment*[2]	*NSS*	*NTI Nuclear Materials Security Index Score*[5]
Brunei	Yes (1985)	Yes (2014)	Yes (1987)	Yes	No	Yes (2013)	No	No	No[3]	40
Cambodia	Yes (1972)	Yes (2009)	Yes (1999)	Yes	Yes (2015)	Yes (2000)	Yes (2006)	No	No[3]	45
Indonesia	Yes (1970)	Yes (1957)	Yes (1980)	No	Yes (1999)	Yes (2012)	Yes (1987)	Yes	Yes[4]	65
Laos	Yes (1968)	Yes (2011)	Yes (2001)	Yes	Signatory (2014)	Yes (2000)	Yes (2010)	No	No[3]	34
Malaysia	Yes (1968)	Yes (1969)	Yes (1972)	No[6]	Signatory (2005)	Yes (2008)	No	No	Yes[4]	46
Myanmar	Yes (1992)	Yes (1957)	Yes (1995)	Yes	Signed (2013)	Signatory (1996)	No	No	No[3]	23
Philippines	Yes (1968)	Yes (1958)	Yes (1974)	No	Yes (2010)	Yes (2001)	Yes (1987)	No	Yes[4]	55
Singapore	Yes (1970)	Yes (1967)	Yes (1977)	Yes	Yes (2008)	Yes (2001)	Yes (2014)	Yes	Yes[4]	75
Thailand	Yes (1972)	Yes (1957)	Yes (1974)	No	Signatory (2005)	Signatory (1996)	No	No	Yes[4]	41
Vietnam	Yes (1982)	Yes (1957)	Yes (1990)	No	Yes (2012)	Yes (2006)	Yes (2012)	Yes	Yes[4]	52

	SEANWFZ[1]	*PTBT*[2]	*GICNT Member*	*Nuclear Terrorism Convention*[1]	*PSI*	*Convention on Nuclear Safety*[2]	*Convention on Early Notification of a Nuclear Accident*[2]	*Convention on Safety of Spent Fuel*[2]
Brunei	Yes (1995)	Signatory (1997)	No	No	Yes	No	No	No
Cambodia	Yes (1995)	Yes (2000)	Yes	Signatory (2006)	Yes	Yes (2012)	Yes (2012)	No
Indonesia	Yes (1995)	Signatory (1996)	No	Yes (2014)	No	Yes (2002)	Yes (1993)	Yes (2011)
Laos	Yes (1995)	Yes (2000)	No	No	No	No	Yes (2013)	No
Malaysia	Yes (1995)	Yes (2008)	Yes	Signatory (2005)	Yes	No	Yes (1987)	No
Myanmar	Yes (1995)	Signatory (1996)	No	No	No	No	Yes (1987)	No
Philippines	Yes (1995)	Yes (2001)	Yes	Signatory (2005)	Yes	Signatory (1994)	Yes (1997)	Signatory (1998)
Singapore	Yes (1995)	Yes (2001)	Yes	Signatory (2006)	Yes	Yes (1998)	Yes (1998)	No
Thailand	Yes (1995)	Signatory (1996)	Yes	Signatory (2005)	Yes	No	Yes (1989)	No
Vietnam	Yes (1995)	Yes (2006)	Yes	No	Yes	Yes (2010)	Yes (1987)	Yes (2014)

Abbreviations: Nuclear Nonproliferation Treaty (NPT); Comprehensive Safeguards Agreement (CSA) with IAEA; Small Quantities Protocol (SQP) with IAEA; Comprehensive Nuclear Test Ban Treaty (CTBT); Convention on the Physical Protection of Nuclear Material (CPPNM); Nuclear Security Summits (NSS); Southeast Asia Nuclear-Weapon-Free Zone (SEANWFZ); Partial Test Ban Treaty (PTBT); Global Initiative to Combat Nuclear Terrorism (GICNT); International Convention on the Suppression of Acts of Nuclear Terrorism (ICSACT or Nuclear Terrorism Convention); Proliferation Security Initiative (PSI); and Joint Convention on Safety of Spent Fuel Management and on Safety of Radioactive Waste Management (Convention on Safety of Spent Fuel)

Sources: IAEA; UN; Comprehensive Nuclear-Test-Ban Treaty Organization, and Nuclear Threat Initiative

Notes: [1] The year country joined the accord is in parentheses. [2] The year accord went into force for the country is in parentheses. [3] Country attended NSS regional outreach meeting in Thailand in 2011. [4] Country attended NSS in 2010, 2012, and 2014. [5] NTI Nuclear Materials Security Index Score (January 2016 edition, "Theft Rankings") assesses the contribution of states on their nuclear materials security conditions with scores between 0 (least favorable) and 100 (most favorable). [6] Rescinded in 2006.

to the international community and the United States in particular; it is possible Myanmar could move forward on the PSI at some point in the 2010s.[116]

In early 2015, Indonesia represented ASEAN in a discussion at the Conference on Disarmament on how to move forward negotiations on a fissile material treaty that is multilateral, nondiscriminatory, and verifiable.[117] In 2013, the Philippines became the first ASEAN member to join the Global Partnership Against the Spread of Weapons and Materials of Mass Destruction, a group of now twenty-eight countries convened to coordinate their nuclear and radiological security, biosecurity, chemical security, and scientist engagement efforts.

The risk posed by the spread of nuclear material has prompted several Southeast Asian countries to join the U.N. Convention on the Physical Protection of Nuclear Material (CPPNM).[118] The 2005 Amendment to the CPPNM, which binds members to "protect nuclear facilities and material in peaceful domestic use, storage as well as transport," includes three members from the region.[119] Due to Southeast Asia's limited nuclear infrastructure, none of ASEAN are currently members of major multilateral export control groups such as the Nuclear Suppliers Group or the Zangger Committee.

ASEAN has largely supported United Nations Security Council Resolution 1540—requiring states to develop and implement legal measures against the proliferation of weapons of mass destruction—after it was unanimously adopted in 2004. The ASEAN Regional Forum (ARF) has hosted several workshops on the 1540 implementation. Laos, however, has expressed concern that the resolution could interfere with national sovereignty, technology exports, and "the noninterference in other country's internal affairs."[120] Several ASEAN members have been active participants in the U.S.-launched Nuclear Security Summit process with Indonesia, Malaysia, the Philippines, Singapore, Thailand, and Vietnam sending high-level representatives to the 2010, 2012, and 2014 meetings. The various gatherings, which seek to coordinate countries' efforts at securing loose and vulnerable nuclear material around the globe, has helped rally nuclear security efforts in Southeast Asia, such as the removal of Vietnam's entire stockpile of highly enriched uranium from the country in 2013. Some ASEAN countries have taken the lead on several of the agenda items, in particular Indonesian efforts in 2014 to create a mechanism to assist states in adopting nuclear security conventions and treaties.[121] Thailand hosted a regional outreach meeting in September 2011 for ASEAN states not invited to the summits themselves, such as Brunei, Cambodia, Laos, and Myanmar. Thailand also hosted the January 2014 and February 2015 summit planning meetings.

Anxieties about nuclear weapons proliferation register high on surveys of the Southeast Asian public. When asked by the PEW Research Center to identify the first or second "greatest threats to the world," the "spread of nuclear weapons" was ranked near the top: the Philippines (39 percent in 2014; 49 percent in 2002), Indonesia (33 percent in 2014; 23 percent in 2007), Malaysia (22 percent in 2014; 20 percent in 2002), Thailand (20 percent in 2014), and Vietnam (38 percent in 2014; 34 percent in 2002).[122] In terms of specific nonproliferation challenges, a great number in 2013 considered the North Korean nuclear program a "major threat": Philippines (61 percent); Malaysia (45 percent); Indonesia (39 percent). Furthermore, in spring 2015, 79 percent in the Philippines, 55 percent in Indonesia, 40 percent in Malaysia, and 50 percent in Vietnam were very or somewhat concerned about the Iranian nuclear program.[123] These apprehensions and the robust engagement in the nuclear nonproliferation regime signal Southeast Asia's broad alignment with Globalist norms on nuclear weapons. But what about the region's specific nuclear dilemmas?

Southeast Asia's Proliferation Challenges

Two examples suggest that Southeast Asia is nevertheless vulnerable to proliferation in some form. First, there is continued worry over whether Myanmar has abandoned its suspected plans to acquire nuclear weapons capability, which was assumed to be tied to its relationship to North Korea. Concrete data on this issue is difficult to obtain and verify given the opaqueness of Myanmar's government during the five decades of military rule that preceded the current era of reform.[124] Nevertheless, images of Burmese nuclear fuel cycle components in 2010 suggested Naypyidaw may have intended to build a reactor, enrich uranium, and develop weapons.[125] Analysts believe Russia and/or North Korea were assisting Myanmar in this development, with the preponderant view leaning toward Pyongyang. However, even those analysts who believed Myanmar was attempting to acquire nuclear weapons capacity calculated that the program was in its very early stages and that the realization of those ambitions would require a long and difficult process.[126] In late 2013, U.S. Assistant Secretary of Defense Vikram Singh testified before Congress that the United States believed Myanmar had ended its military alliance with North Korea, although "other ties remain."[127]

Since President Obama's trip to Myanmar in 2012, the lessening of international sanctions against the military government has created opportunities for progress on nuclear issues. In 2013, Myanmar signed the IAEA Additional Protocol. The U.S. State Department welcomed the move as a sign the country was "increasingly in line with international nonproliferation norms and standards."[128] The government in Naypyidaw proclaims it does not seek to develop or acquire nuclear arms. An IAEA member since 1992, Myanmar has had a safeguards system in place under the Small Quantities Protocol since 1995. The country joined its ASEAN neighbors as a party to SEANWFZ as well as a signatory to the Comprehensive Nuclear Test Ban Treaty.

The second concern is that Southeast Asia is rife with nontraditional security threats and therefore ripe for nuclear proliferation, even if only as a facilitator or transit point. It has the highest incidence of small arms smuggling, and Malaysia and Thailand were placed in the "Tier 3" (most severe) category for human trafficking in a 2014 State Department report. Many land borders in mainland Southeast Asia are porous, and the amount of unpatrolled coastline in the maritime subregion encourages the spread of these threats.[129]

This was amply illustrated in 2004 with the discovery of nuclear centrifuges manufactured in Malaysia and bound for Libya.[130] The seizure of the centrifuge parts highlighted the black market nuclear proliferation network run by Pakistani nuclear scientist A. Q. Khan. In the face of strong criticism from the United States—the components had been seized under the PSI—the Malaysian government denied any knowledge of the centrifuges, and the manufacturer, Scomi Oil and Gas, claimed it did not know the ultimate end use. However, Kuala Lumpur took the point and agreed to share information with the United States on the manufacture or transportation through Malaysia of this dual-use material.[131] In 2014, Malaysia announced that it would sign and participate in the Proliferation Security Initiative.[132] Furthermore, Cambodia's "flag of convenience" policy that allows non-nations to operate ships under its flag has some experts concerned about the proliferation risks. The creation of the PSI was inspired by the interdiction of a North Korea–owned ship, the *So San*, registered in Cambodia and carrying fifteen Scud missiles, conventional warheads, and military parts.[133] Reports have also identified Laos as a potential worry for proliferation routes due to weak strategic trade controls and a porous border.[134]

In the past two decades, the nontraditional security threat environment in Southeast Asia has increased exponentially with the acceleration of terrorism. Radical Islamist groups such as the Jemaah Islamiyah and Abu Sayyaf were reinvigorated after September 11, 2001, although they have receded somewhat with increased cooperation on counterterrorism between the United States and Australia on one hand and several Southeast Asian nations—particularly Indonesia, Malaysia, and the Philippines—on the other. However, the region also hosts some groups of concern, such as Hezbollah cells discovered in Thailand in 2014. As well, the region's leaders fear that when Southeast Asian radicals with ties to the Islamic State return from Syria and Iraq, they will bring new technology and means of violence with them. This underscores the threat of radical groups using Southeast Asia as a route for the transportation of nuclear material or, even worse, using small nuclear weapons on targets in the region itself.[135]

The Development of a Nuclear-Weapon-Free Zone

Of the many lessons that the Vietnam War offered the ASEAN states, chief among them was the region's experience as a fertile ground for great power competition, one Southeast Asian states feared could escalate into a nuclear exchange. In 1967, for example, U.S. President Lyndon Johnson briefly considered using tactical nuclear bombs against North Vietnam before being persuaded that it would be both ineffective and unpopular with the international community.[136] ASEAN was founded in 1967 as a diplomatic forum with a broad goal: to protect the security and sovereignty of the comparatively weak and vulnerable states of Southeast Asia by increasing cooperation with among them.[137] Former Malaysian Foreign Minister Tun Ismail Abdul Rahman called for Southeast Asia to be a neutral zone, off limits to Cold War games.[138] In 1971, the five ASEAN states issued a declaration that they intended to keep Southeast Asia "free from any form or manner of interference by outside Powers," proclaiming the region to be a Zone of Peace, Freedom and Neutrality (ZOPFAN).[139]

One primary instrument of ZOPFAN envisioned by the founding ASEAN states was a Southeast Asian nuclear-weapon-free zone treaty. However, Cold War tensions, Vietnam's invasion of Cambodia, and political rivalries delayed finalization of the treaty for more than two decades.[140] The dissolution of the Soviet Union and the cessation of conflict between Vietnam and Cambodia opened up new opportunities leading to the 1995 ASEAN Summit where all ten ASEAN members, including newcomers Vietnam and Laos signed SEANWFZ.[141] The treaty took effect in 1997 after all but one state had ratified it; this process was completed in 2001 when the Philippines, a security ally of the United States, became the last ASEAN member to ratify the treaty.

SEANWFZ members pledged to not develop, manufacture or otherwise acquire nuclear weapons.[142] If signatories acquired nuclear energy facilities, they agreed to maintain IAEA safeguards over them. By the time the treaty was finalized, the threat of nuclear war carried by the Cold War had all but disappeared. Instead, ASEAN replaced that concern with two new ones: (1) the potential for nuclear proliferation in the hands of non-state actors or rogue states; and (2) the potential for nuclear proliferation within Southeast Asia itself as ASEAN considered nuclear energy. In 2005, M. C. Abad Jr., then-head of the ASEAN Regional Forum Unit in the ASEAN Secretariat in Indonesia, argued SEANWFZ had "continued relevance, beyond the area denial sense for [nuclear weapon states], in contributing to the nonproliferation of nuclear weapons or fissile material" in the post–September 11 era.[143]

In addition to regional proliferation, SEANWFZ had a pointed externally focused purpose as well: to prevent the transportation of nuclear weapons on ships through Southeast Asian territorial waters. Not without reason, officials in Washington believed this prohibition was directed specifically at them, since U.S. submarines on patrol in the Pacific carry nuclear warheads.[144] U.S. opposition to SEANWFZ and to a similar prohibition in New Zealand created tensions between Washington and ASEAN and also between the United States and its treaty allies Thailand and the Philippines. However, the treaty allows each member state to decide individually if it will allow foreign ships or aircraft, including those that are nuclear-powered or nuclear-armed to visit or transit through their airspace or territorial waters.[145]

The United Nations recommended that the nations forming a nuclear-free zone endeavor to persuade the five permanent members of the U.N. Security Council or P-5 powers to sign on to their treaties.[146] ASEAN has accordingly lobbied the United States, the United Kingdom, France, China, and Russia to sign a protocol to the SEANWFZ Treaty. Thus far none has done so, although it appeared that the P-5 group was close to signing in 2012. The United States has no categorical objection to signing protocols to nuclear-free zones. In 2011, the Obama administration submitted protocols to the African Nuclear-Weapon-Free Zone (ANWFZ) and South Pacific Nuclear-Free Zone (SPNFZ) to the Senate for consent to ratification. Consultations among the P-5 and the Central Asia Nuclear-Weapon-Free Zone (CANWFZ) nations have concluded, though the P-5 countries have not yet signed the protocol. Although Washington continues to participate in consultations among the P-5 and SEANWFZ parties, no conclusion to this process appears to be in sight.[147]

At best, movement toward the accession of the P-5 is uneven and slow. China has indicated that in the past it would have been willing to sign, but in recent years Beijing has raised concerns that the nuclear-free zone could affect Chinese sovereignty over its territory, exclusive economic zone, and continental shelves (see chapter 2 for more on China's changing attitudes toward nuclear nonproliferation). Russia and the United States voiced reservations over whether foreign ships and aircraft would be allowed passage and flyovers. France and the United Kingdom have questioned the definition of the zone itself.[148]

Momentum within ASEAN on negotiating with the P-5 is also uneven despite the fact that this item remains part of the boilerplate of nearly every meeting of the ASEAN Foreign Ministers. Although all ten countries give lip service to pressing the P-5 to sign, it is the smaller and weaker states that make it a high priority. In 2014, when Myanmar made its debut as the chair of ASEAN, its Foreign Ministry declared that it would work to finalize negotiations on the P-5 protocols.[149] However, the chair was not able to interest the other ASEAN states in making it a priority. At the 2015 ASEAN Summit in Malaysia, the country's foreign minister signaled ASEAN states were going to "intensify efforts" to encourage the P-5 to sign on, but that the SEANWFZ Commission was still reviewing the reservations expressed by France, Russia, and the United Kingdom.[150] Until there is greater unanimity on this issue within the group, ASEAN is likely to find that bringing the P-5 into the SEANWFZ Treaty is an elusive goal.

CONCLUSION: REGIONAL APPROACHES TO NUCLEAR ISSUES

The host of challenges that nuclear energy entails—from the costs, nuclear fuel and waste management, and nuclear safety, to regulatory systems, human resource capacities, and nonproliferation safeguards regimes—is a difficult endeavor to master for a large advanced economy, let alone a developing nation. Therefore, regional approaches to nuclear energy and nuclear nonproliferation challenges are often embraced as a means to overcome these hurdles.

Since countries remain interested in the benefits of nuclear power, there has been discussion of a regional network of nuclear cooperation within ASEAN via the ASEAN Economic Community, the ASEAN Political-Security Community, or both. One suggestion is to base a regional ASEAN nuclear energy program on the European Atomic Energy Community (Euratom), which has managed the nuclear market throughout the European Commission since the 1957 Euratom Treaty. Euratom coordinates and implements nuclear plans in Europe by addressing nuclear fuel demands, safety concerns, nonproliferation safeguards, and the security of facilities and radioactive material.[151] The model, sometimes referred to as "Asiatom," was briefly proposed by the Philippines in the 1990s.[152] In November 2007, the ASEAN Summit produced the Singapore Declaration on Climate Change, Energy, and the Environment that urged "cooperating for the development and use of civilian nuclear power, in a manner ensuring nuclear safety, security, and nonproliferation."[153]

ASEAN members have distrusted the ability of their neighbors to provide adequate safeguards in the event of an accident or malfunction in a nuclear power plant. Nevertheless, there are presently no ASEAN frameworks that would regulate nuclear power plants in individual countries. Some member states have proposed that ASEAN build one or more regional nuclear energy plants, but this has been met with fears that such facilities would be even more insecure than national plants. As a result, ASEAN has focused instead on the possibilities of developing regional electrical power grids through the use of hydropower.

Another benefit raised by Asiatom supporters is that negotiating with foreign vendors as a bloc of countries would give an ASEAN nuclear body the best deal possible. Rather than competing with each other, Andrew Symon argues Asiatom could secure a contract that meets the financial, security, and safety interests of the region as a whole.[154] Asiatom could also manage the acquisition of nuclear fuel for individual countries with nuclear energy programs too small for a domestic fuel fabrication capacity. Regional approaches to sharing these nuclear labors appear to hold merit, especially for countries with limited experience in managing a complex nuclear infrastructure and the related safety, security, and nonproliferation minutia. However, for better or worse, control of nuclear energy development in Southeast Asia is likely to remain with national governments for the foreseeable future.

Nuclear proliferation concerns for Southeast Asian states are confined for the most part to threats from terrorist groups or nontraditional security risks, rather than from great power rivalry in the region. Globalist views on the value of the international nonproliferation regime endure despite challenges from North Korea and Iran. While the sustained implementation and strengthening of the SEANWFZ Treaty can help reassure members about the continued non-nuclear weapon status of their neighbors, Southeast Asian powers seem less than eager to provoke China by strongly pursuing a nuclear-free zone in the context of the South China Sea disputes.

Several ASEAN states have, however, signed on to the Proliferation Security Initiative and all have abjured the pursuit of nuclear weapons. ASEAN's leadership on nuclear materials security is another proven opportunity for governments to work together to address existing vulnerabilities and build a regional network of support and accountability as nuclear energy ambitions advance further. Should ASEAN pursue the Euratom model for nuclear energy, one added nonproliferation benefit often discussed is that Asiatom could negotiate for fuel with a single existing state supplier or a multilateral vendor such as the IAEA, thereby limiting the spread of enrichment and reprocessing technologies.[155] However, it appears momentum toward the Asiatom model has stalled in recent years as most Southeast Asia countries—with the possible exception of Vietnam—have placed their nuclear energy aspirations on hold for the near term.

Most importantly, there seems to be little, if any, internal drivers or strong constituencies in ASEAN states that can reverse the fairly enduring Globalist consensus on nuclear energy and weapons issues. Even the Realist logic of reducing energy vulnerability through a buildup of indigenous nuclear energy capabilities seems to have lost its steam in the aftermath of the Fukushima disaster, the most recent drop in global oil prices, and the challenge of civil society opposed to nuclear power. Nationalist outlooks appear even weaker, as regional economic integration continues to take precedence in policy priorities, and countries like Myanmar seek to dampen their Nationalist past and become more active players in the global system.

NOTES

1. For example, Indonesia, the Philippines, Thailand, and then South Vietnam received help through Atoms for Peace to establish nuclear research reactors. The Thailand government nuclear research center is known as the Office of Atoms for Peace. The program's legacy is still present with the lockdown of nuclear material from these reactors a major goal of President Obama's Nuclear Security Summits.

2. Sahara Piang Brahim, "Southeast Asia's Nuclear Path Post-Fukushima," *Asia Times Online*, February 26, 2013, http://www.atimes.com/atimes/Southeast_Asia/SEA-01-260213.html.

3. Andrew Symon, "Southeast Asia's Nuclear Power Thrust: Putting ASEAN's Effectiveness to the Test?," *Contemporary Southeast Asia* 30, no. 1 (April 2008): 121; Andrew Symon, "Nuclear Power in Southeast Asia: Implications for Australia and Non-Proliferation," Lowy Institute for International Policy, April 2008, 8, http://www.lowyinstitute.org/publications/nuclear-power-southeast-asia-implications-australia-and-non-proliferation.

4. Concerns range from the dislocation of local communities to environmental damages stemming from mismanaged, large-scale, hydropower programs. For more, see Catharine Beck, "The Push and Pull for Hydropower in Vietnam and Cambodia," *Wilson Center China Environment Forum*, January 15, 2014, https://www.wilsoncenter.org/publication/the-push-and-pull-for-hydropower-vietnam-and-cambodia; "Chinese Hydropower Electrifies Southeast Asia, but at a Cost," *TheConversation.com*, November 14, 2013, http://theconversation.com/chinese-hydropower-electrifies-southeast-asia-but-at-a-cost-20275; Symon, "Southeast Asia's Nuclear Power Thrust," 121.

5. "ASEAN to Promote Nuclear Energy, Solar Power," *Agence France-Presse*, November 13, 2007.

6. Rachel Harvey, "Vietnam Calls on ASEAN Nations to Embrace Nuclear Power," *BBC News*, July 22, 2010, http://www.bbc.com/news/world-asia-pacific-10728876.

7. ASEAN Ministers, "Joint Ministerial Statement of the 29th ASEAN Ministers on Energy Meeting (AMEM)," ASEAN, September 20, 2011, http://www.asean.org/joint-ministerial-statement-of-the-29th-asean-ministers-on-energy-meeting-amem.

8. ASEAN Ministers, "Joint Ministerial Statement of the Sixth East Asian Summit Energy Ministers Meeting (6th EAS EMM)," ASEAN, September 12, 2012, http://www.asean.org/joint-ministerial-statement-of-the-sixth-east-asian-summit-energy-ministers-meeting-6th-eas-

emm/; ASEAN Ministers, "Joint Ministerial Statement the 9th ASEAN+3 (China, Japan and Korea) Ministers on Energy Meeting," ASEAN, September 12, 2012, http://www.asean.org/joint-ministerial-statement-the-9th-asean3-china-japan-and-korea-ministers-on-energy-meeting/.

9. ASEAN Ministers, "Joint Ministerial Statement of the Eighth East Asia Summit Energy Ministers Meeting," ASEAN, September 24, 2014, http://www.eria.org/2014_0924_Joint%20Ministerial%20Statement%20of%208th%20EAS%20Energy%20Ministers%20Meeting.pdf.

10. Symon, "Nuclear Power in Southeast Asia," 7.

11. Intergovernmental Panel on Climate Change, *Climate Change 2013: The Physical Sciences Basis* (Cambridge: Cambridge University Press, 2013), http://www.ipcc.ch/pdf/assessment-report/ar5/wg1/WG1AR5_ALL_FINAL.pdf; Asian Development Bank, *Addressing Climate Change and Migration in Asia and the Pacific: Final Report* (Manila: Asian Development Bank, 2012); Mely Caballero-Anthony et al., *The Sustainability of Nuclear Energy in Southeast Asia: Opportunities and Challenges* (Singapore: Centre for Non-Traditional Studies at S. Rajaratnam School of International Studies, October 2014), http://www.rsis.edu.sg/wp-content/uploads/2014/10/NTS-Report-October-2014.pdf.

12. Allison Macfarlane, "Where, How, and Why Will Nuclear Happen?: Nuclear 'Renaissance' Discourses from Buyers and Suppliers," in *The Nuclear Renaissance and International Security*, eds. Adam N. Stulberg and Matthew Fuhrmann (Stanford, Calif.: Stanford University Press, 2013), 50–72.

13. Santoro, "ASEAN's Nuclear Landscape—Part 1," *Pacific Forum CSIS—Issues & Insights* 13, no. 10 (July 2013): 1.

14. Nurkhayrul Salam, "BEnA to Host Forum on Nuclear Energy," *Brunei Times*, September 11, 2007, http://m.bt.com.bn/news/2007/09/11/bena_to_host_forum_on_nuclear_energy.

15. Nathan Vanderkippe, "Brunei's Oil-Fuelled Economy Running on Empty," *Globe and Mail*, February 2, 2015, http://www.theglobeandmail.com/report-on-business/international-business/bruneis-oil-fuelled-economy-looks-to-diversify/article22750399/.

16. Miklos Gaspar, "Brunei, IAEA Discuss Cooperation in Cancer Care, Agriculture," IAEA, February 2015, https://www.iaea.org/newscenter/news/brunei-iaea-discuss-cooperation-cancer-care-agriculture.

17. "Laos's Vice President: Use of Nuclear Energy among Iran's Rights," *Tehran Times*, December 1, 2007, http://www.tehrantimes.com/index_View.asp?code=158179.

18. Ernest Z. Bower, "ASEAN Going for Nuclear Power," *Center for Strategic and International Studies—Southeast Asia from the Corner of 18th and K Streets* 1, no. 20 (July 2010), http://csis.org/files/publication/100707_seasia_newsletter.pdf; Prashanth Parameswaran, "Southeast Asia's Nuclear Energy Future: Promises and Perils," *Project 2049 Institute—Futuregram 09-006*, December 23, 2009, http://project2049.net/documents/southeast_asia_nuclear_energy_future.pdf; International Institute for Strategic Studies (IISS), *Preventing Nuclear Dangers in Southeast Asia and Australasia* (London: IISS, 2009), 58.

19. In a Country Programme Framework with the IAEA, Laos identified eight priority areas for 2014–2018: human health, food and agriculture, radiation and nuclear safety, environmental protection, sustainable energy development, industry, nuclear security, and water resources management. For more, see "Lao People's Democratic Republic Signs its First Country Programme Framework (CPF) for 2014–2018," IAEA, May 13, 2014, https://www.iaea.org/technicalcooperation/Home/Highlights-Archive/Archive-2014/05132014_cpf_Lao.html;

Khamso Kouphokham, "Lao PDR Country Report," in *Analysis on Energy Saving Potential in East Asia*, ed. Shigeru Kimura (ERIA Research Project Report 2012-19, June 2013), 193–194, http://www.eria.org/RPR_FY2012_No.19_chapter_10.pdf.

20. The Laotian government expects hydropower to be its major revenue stream by 2025. For more, see "Unquiet Grows the Don," *The Economist*, December 20, 2014, http://www.economist.com/news/asia/21636978-laoss-neighbours-may-not-its-plans-dam-mekong-they-can-do-little-about-it-unquiet-grows; Guy Delauney, "Laos Hydropower a 'Battery' for Power-Hungry Region," *BBC News*, December 10, 2010, http://www.bbc.com/news/11962210.

21. Tracey Shelton, "Cambodia Sets Sights on Nuclear Power," *Phnom Penh Post*, September 29, 2008, http://www.phnompenhpost.com/business/cambodia-sets-sights-nuclear-power; Ker Munthit, "Cambodia Hopes to Build a Nuclear Plant," *The Irrawaddy*, September 26, 2008, http://www2.irrawaddy.org/article.php?art_id=14322.

22. Aaron Larson, "Cambodia's Largest Hydropower Plant Begins Operation," *PowerMag*, March 1, 2015, http://www.powermag.com/cambodias-largest-hydropower-plant-begins-operation/.

23. "Cambodia—World Bank Expects Slower Growth in Developing East Asia Pacific in 2014," *World Bank—East Asia and Pacific Update*, October 6, 2014, http://www.worldbank.org/en/news/press-release/2014/10/06/cambodia-world-bank-expects-slower-growth-in-developing-east-asia-pacific-in-2014.

24. Prak Chan Thul, "Cambodia Eyes Nuclear Power to Meet Energy Demands," *Reuters*, August 20, 2010, http://www.reuters.com/article/2010/08/20/oukwd-uk-cambodia-nuclear-idAFTRE67J11H20100820.

25. Karon Snowdon, "Cambodia Denies Building Nuclear Plant," *ABC News*, October 16, 2012, http://www.abc.net.au/news/2012-10-17/an-cambodia-denies-building-nuclear-plant/4317770.

26. "PM Talks Nuclear Options," *Phnom Penh Post*, January 16, 2014, http://www.phnompenhpost.com/national/pm-talks-nuclear-options.

27. "27 Years into the Nuclear Age," *New Straits Times*, July 13, 2009.

28. Eric Yep, "Malaysia Keeps Nuclear Plans on Hold," *Wall Street Journal*, October 17, 2013, http://www.wsj.com/articles/SB10001424052702304410204579140994069930438.

29. IISS, *Preventing Nuclear Dangers*, 87–88.

30. "Malaysia Aims for Nuclear Energy by 2025," *United Press International*, July 22, 2009, http://www.upi.com/Business_News/Energy-Industry/2009/07/22/Malaysia-aims-for-nuclear-energy-by-2025/29381248299408/.

31. Yep, "Malaysia Keeps Nuclear Plans on Hold."

32. Sheridan Mahavera, "Malaysia's Nuclear Power Plant Not a Done Deal, Says Atomic Power Body," *Malaysian Insider*, May 19, 2015, http://www.themalaysianinsider.com/malaysia/article/malaysias-nuclear-power-plant-not-a-done-deal-says-atomic-power-body.

33. AMAN rejects nuclear power as a weapons proliferation risk, poor solution to climate change, an accident and safety concern, and an environmental threat throughout the nuclear fuel cycle. For more, see "Malaysia Must Immediately Stop Plans on Nuclear Power Plants and Spearhead ASEAN Free-Nuclear-Energy Zone—Abdulhadi Khalid," *Malaymail Online*, April 27, 2015, http://www.themalaymailonline.com/what-you-think/article/malaysia-must-immediately-stop-plans-on-nuclear-power-plants-and-spearhead.

34. Public concerns focused heavily on the government's ability to respond to nuclear accidents, train an expert workforce to operate a plant, and manage nuclear waste, and the health safety of communities near a nuclear site. For more, see "Only 2% Malaysians Trust Our

Government with Nuclear Technology," *Astro Awani*, July 1, 2015, http://english.astroawani.com/malaysia-news/only-2-malaysians-trust-our-govt-nuclear-technology-64399.

35. Kavithah Rakwan, "Govt Conducting Public Survey on Nuclear Energy," *Malaysian Reserve*, September 11, 2015, http://themalaysianreserve.com/new/story/govt-conducting-public-survey-nuclear-energy.

36. The Centre for Radiation Protection and Nuclear Sciences (CRPNS) is responsible for regulating the country's nuclear activities, which have mostly included IAEA projects on radiation protection and radioisotopes. The country does not have a nuclear research reactor. For more, see IISS, *Preventing Nuclear Dangers*, 131–132.

37. Li Xueyi, "Singapore: Nuclear Power Not Ruled Out," *Straits Times*, December 6, 2008.

38. Audrey Tan, "Singapore Opts for Cleaner Energy Sources," *Straits Times*, July 21, 2015, http://www.straitstimes.com/singapore/singapore-opts-for-cleaner-energy-sources.

39. Calvin Lee, "Singapore PM Says Nuclear Power Plant Possible 'During My Lifetime,'" *Platts*, November 1, 2010; Rachel Chang, "Govt Preparing for Nuclear Power Option," *Straits Times*, November 2, 2010.

40. "Nuclear Not Currently an Option for Singapore," *World Nuclear News*, April 11, 2013, http://www.world-nuclear-news.org/NP-Nuclear_not_currently_an_option_for_Singapore-1104131.html; Brahim, "Southeast Asia's Nuclear Path Post-Fukushima."

41. The 2007 report cited challenges relating to Singapore's small landmass and population density as well as the storage of nuclear waste. For more, see Singapore Ministry of Trade and Industry, *National Energy Policy Report: Energy for Growth* (Republic of Singapore: Ministry of Trade and Industry, November 2007).

42. Grace Chua, "Singapore Starts Nuclear Safety and Science Research Programme," *Straits Times*, April 23, 2014, http://www.straitstimes.com/singapore/singapore-starts-nuclear-safety-and-science-research-programme.

43. This section was informed by Catharin Dalpino's research in Bangkok in October 2013, at which time the prospects for developing nuclear energy in Thailand were considered strong, although the Ministry of Energy was waiting for the "Fukushima effect" to fade before going forward with plans for the first two nuclear energy plants. Since then, a number of factors that are detailed in the section have made the development of nuclear energy resources in Thailand a back-burner item.

44. World Nuclear Association, *The World Nuclear Industry Status Report 2012, World Nuclear Industry Status Report*, July 5, 2012, http://www.worldnuclearreport.org/The-World-Nuclear-Industry-Status.html.

45. Catharin Dalpino's interviews with officials from the Ministry of Energy, the Electricity Generating Authority of Thailand (EGAT), Thammasat University, the National Defense Studies Institute, the Ministry of Foreign Affairs, and the Foreign Service Institute, Bangkok, October 2013.

46. This agency was originally called the Committee on Atomic Energy until renamed in 1956. For more, see IISS, *Preventing Nuclear Dangers*, 137–150.

47. The research reactor was light water cooled and heavy water moderated with low-enriched uranium supplied by the United States. For more, see Santoro, "ASEAN's Nuclear Landscape—Part 1," 8–9.

48. Parinya Pongsoi and Somchai Wongwises, "A Review on Nuclear Power Plant Scenario in Thailand," *Renewable and Sustainable Energy Reviews* 24 (May 2013): 590–591.

49. The group was originally called the Nuclear Power Infrastructure Preparation Committee. For more, see Pongsoi and Wongwises, "Review on Nuclear Power Plant Scenario in Thailand," 586–592.

50. Karol Takabut, "Status of Nuclear Power Development in Thailand" (presentation, Thai-French Technical Association, Bangkok, Thailand, August 26, 2010), http://www.tfta.or.th/download/Status_of_NPP_Development_in_Thailand.pdf.

51. Oranian Paweewun, "Thailand to Delay Development of Nuclear Power," *Wall Street Journal*, April 27, 2011, http://www.wsj.com/articles/SB10001424052748704099704576288782682893862.

52. "80% of People Oppose Nuclear Power Plants," *The Nation (Bangkok)*, March 27, 2011, http://www.nationmultimedia.com/2011/03/27/national/80-of-people-oppose-nuclear-power-plants-30151871.html.

53. Chavalit Pichalai, "Thailand's Power Development Plan 2015 (PDP 2015) (presentation, Ministry of Energy, Thailand), accessed September 1, 2015, http://www.renewableenergy-asia.com/Portals/0/seminar/Presentation/02-Overview%20of%20Power%20Development%20Plan%20(PDP%202015).pdf; Nareerat Wiriyapong, "Nuclear Still in the Local Mix, but 21 Years Away," *Bangkok Post*, February 16, 2015, http://www.bangkokpost.com/business/news/476349/nuclear-still-in-the-local-mix-but-21-years-away.

54. "Thailand Rules Out Nuclear Power," *Press Trust of India*, August 11, 2014, http://www.firstpost.com/world/thailand-rules-nuclear-power-says-will-look-coal-meet-energy-needs-1658985.html.

55. IAEA, "Radioisotopes in Burmese Agricultural Research," *IAEA Bulletin* 3 no. 3 (1961), https://www.iaea.org/sites/default/files/publications/magazines/bulletin/bull3-3/03305401515.pdf; IAEA, "Atomic Assistance in 1961," *IAEA Bulletin* 3 no. 2 (1961): 16–18, https://www.iaea.org/sites/default/files/publications/magazines/bulletin/bull3-2/03205301618.pdf.

56. IISS, *Preventing Nuclear Dangers*, 101–102.

57. The IAEA was also concerned about Myanmar's economic, educational, and technological base. For more, see Andrew Selth, "Burma and Nuclear Proliferation" in *Over the Horizon Proliferation Threats*, eds. James Wirtz and Peter Lavoy (Stanford, Calif.: Stanford University Press, 2012), 116–117.

58. There were reports about North Korea possibility taking over from Russia as the vendor for the SPDC's research reactor. For more, see IISS, *Preventing Nuclear Dangers*, 109.

59. Ibid., 102–103.

60. Mratt Kyaw Thu, "Myanmar and Russia Boost Nuclear Cooperation," *Myanmar Times*, April 3, 2015, http://www.mmtimes.com/index.php/national-news/13885-myanmar-and-russia-boost-nuclear-cooperation.html; Symon, "Southeast Asia's Nuclear Power Thrust," 128.

61. Paul Kerr, "U.S. Accuses Burma of Seeking Weapons Technology," *Arms Control Today*, May 2004, https://www.armscontrol.org/act/2004_05/Burma.

62. IISS, *Preventing Nuclear Dangers*, 104–105.

63. Bertil Lintner, "Burma Joins the Nuclear Club," *Far Eastern Economic Review* 164, no. 51 (December 2001–January 2002): 26.

64. "Burma to Build Nuclear Reactors," *BBC News*, January 21, 2002, http://news.bbc.co.uk/2/hi/asia-pacific/1773464.stm.

65. Andrew Selth, "Burma and Nuclear Proliferation: Policies and Perceptions," *Griffith Asia Institute—Regional Outlook Paper* 12 (2007): 8; William Ashton, "Burma's Nuclear Program:

Dream or Nightmare," *The Irrawaddy* 12, no. 5 (May 2004), http://www2.irrawaddy.org/article.php?art_id=968; IISS, *Preventing Nuclear Dangers*, 105.

66. "Deterrence and Regional Stability: Q&A Session," (Third Plenary Session, IISS Shangri-La Dialogue: 14th Asia Security Summit, Singapore, June 2, 2012), http://www.iiss.org/en/events/shangri%20la%20dialogue/archive/sld12-43d9/third-plenary-session-eb1a/qa-db91; "Burma 'Has Given Up Nuclear Power Research'—Minister," *BBC News*, June 2, 2012, http://www.bbc.com/news/world-asia-18309879.

67. The 1973 OPEC oil embargo caused electricity prices to spike upward. Marcos saw nuclear power as an option to meet rising energy demands, reduce its overreliance on imported fuel that accounted for 93 percent of its oil use, and showcase his nation's highly educated population and economic prowess.

68. Sukarno himself studied engineering at Technische School in Bandung. For more, see Ewout Frankema, "Colonial Education and Post-Colonial Governance in the Congo and Indonesia," in *Colonial Exploitation and Economic Development*, eds. Ewout Frankema and Frans Buelens (New York: Routledge, 2013), 170; Sulfikar Amir, "Challenging Nuclear: Antinuclear Movements in Postauthoritarian Indonesia," *East Asian Science, Technology and Society: An International Journal* 3 (2009): 347.

69. Symon, "Southeast Asia's Nuclear Power Thrust," 124–125.

70. Sulfikar Amir, "Nuclear Revival in Post-Suharto Indonesia," *Asia Survey* 50, no. 2 (March 2010): 265–286.

71. Will Hickey, "Indonesia Must Consider Nuclear Power in Base Load to Responsibly Reach 35GW," *Jakarta Post*, June 4, 2015, http://www.thejakartapost.com/news/2015/06/04/indonesia-must-consider-nuclear-power-base-load-responsibly-reach-35gw.html.

72. "Nuclear Power in Indonesia," World Nuclear Association, June 2015, http://www.world-nuclear.org/info/Country-Profiles/Countries-G-N/Indonesia/.

73. Mely Caballero-Anthony, Lina Alexandra, and Kevin D. G. Punzalan, "Civil Society Organizations and the Politics of Nuclear Energy in Southeast Asia," in *Nuclear Power and Energy Security*, eds. Rajesh Basrur and Koh Swee Lean Collin (New York: Routledge, 2012), 184.

74. "IAEA Considers Indonesia Ready to Build Nuclear Power Plant," *Antara News*, December 16, 2010, http://www.eco-business.com/news/iaea-considers-ri-ready-build-nuclear-power-plant.

75. "BATAN: Most Indonesians Support Nuclear Energy," *Antara News*, October 30, 2014, http://en.tempo.co/read/news/2014/10/30/055618349/BATAN-Most-Indonesians-Support-Nuclear-Energy.

76. "BATAN May Begin Construction of Nuclear Power Plant in 2015," *Antara News*, December 17, 2013, http://www.eco-business.com/news/batan-may-begin-construction-nuclear-power-plant-2015.

77. "Nuclear Power in Indonesia."

78. In an extensive study of the region's CSO development, Mely Caballero-Anthony, Lina Alexandra, and Kevin D. G. Punzalan survey how these groups influence the national decision-making process and their "discursive impact: the framing of nuclear issues and the presentation of alternative views and ideas." In the lexicon of this edited volume, the role of CSOs in Southeast Asia's domestic nuclear debates is to inject competing perspectives into the national decision-making cycle of elites and shift the various schools of thought toward a different set of options. For more, see Caballero-Anthony, Alexandra, and Punzalan, "Civil Society Organizations," 173–174.

79. While CSO efforts on nuclear issues can be seen in other Southeast Asia nations, such as Thailand, Singapore, and Malaysia, this chapter focuses on the Philippines and Indonesia due to their long history of CSO engagement, the level of intense activity by CSOs, and the diversity of views expressed in these domestic debates.

80. Caballero-Anthony, Alexandra, and Punzalan, "Civil Society Organizations," 175. Furthermore, Professor William Beaver added that the Three Mile Island accident and reports about corrupt dealings between the Marcos government, Westinghouse, and the Philippine negotiator in charge of arranging the deal provided further evidence for CSOs to use to prevent the Bataan reactor from going on line. For more, see William Beaver, "Nuclear Nightmare in the Philippines," *Journal of Business Ethnics* 13, no. 4 (April 1994): 271–279; Fox Butterfield, "Marcos Facing Criticism, May End $1 Billion Westinghouse Contract," *New York Times*, January 14, 1978, A1; Ann Crittenden, "Behind the Philippine Loan," *New York Times*, February 12, 1978, F9; Gabriella Stern, "Action by Philippines Threatens Pact with Westinghouse on Nuclear Pact," *Wall Street Journal*, December 2, 1992, A8.

81. Caballero-Anthony, Alexandra, and Punzalan, "Civil Society Organizations," 177.

82. Business World, "PCCI Pushes Nuclear Power Development," *GMA News Online*, March 13, 2010, http://www.gmanetwork.com/news/story/186042/money/pcci-pushes-nuclear-power-development.

83. This cooperation included a small research reactor that was operational from 1963 until 1988. For more, see Santoro, "ASEAN's Nuclear Landscape—Part 1," 119–130.

84. The United States would end up guaranteeing 60 percent of the plant's financing needs. For more on the contract negotiations, including accusation of corrupt dealings between Marcos, Westinghouse, and a third-party lobbyist, see Beaver, "Nuclear Nightmare," 271–274.

85. For example, Philippine environmental groups opposing the reactor site made their views known to the U.S. Nuclear Regulatory Commission, which in turn denied Westinghouse an export license until May 1980 for the plant's necessary nuclear components. William Beaver argued the U.S. NRC eventually granted an export license to Westinghouse to avoid "embarrassment to the U.S. government, who had guaranteed much of the plant's financing, and a blow to the prestige of the American nuclear industry." Beaver, "Nuclear Nightmare," 275–276. For more, see R. Jeffrey Smith, "Westinghouse Sues NRC," *Science* 250, no. 4409 (August 31, 1979): 876.

86. Beaver, "Nuclear Nightmare," 274–275.

87. Joanne Omang, "Philippines Again Halts Building of Nuclear Plant," *Washington Post*, November 14, 1979.

88. Beaver, "Nuclear Nightmare," 275.

89. Ibid., 276.

90. Christopher S. Wren, "Manila Cabinet Considers Scrapping Atomic Plant," *New York Times*, April 10, 1986, A11.

91. Beaver, "Nuclear Nightmare," 276.

92. Ibid., 278; Stern, "Action by Philippines Threatens Pact," A8.

93. Santoro, "ASEAN's Nuclear Landscape—Part 1," 2.

94. "Nuclear Power Plant Loan Finally Paid," *Agence France-Presse*, June 13, 2007.

95. These efforts include a bill introduced in 2008 requiring the immediate recommissioning of the Bataan site, IAEA survey teams, and a deal with Korea Electric Power Corporation.

96. Jesus T. Tamang, "Philippines Nuclear Power Program," (presentation, Technical Meeting on Building a National Position on a New Nuclear Power Program, IAEA Headquarters,

Vienna, Austria, June 24–26, 2014), https://www.iaea.org/NuclearPower/Downloadable/Meetings/2014/2014-06-24-06-26-TM-INIG/Philippine_Country_Report.pdf.

97. "Regulatory Body Pushed to Study Bataan Nuclear Power Plant Revival," *Rappler*, April 20, 2015; Norimitsu Onishi, "A Nuclear Plant, and a Dream, Fizzles," *New York Times*, February 13, 2012.

98. Onishi, "A Nuclear Plant, and a Dream, Fizzles."

99. Caballero-Anthony, Alexandra, and Punzalan, "Civil Society Organizations," 182.

100. Symon, "Nuclear Power in Southeast Asia," 10–11.

101. BATAN's "nuclear socialization" programs included workshops, seminars, public discussions, and a film on nuclear safety. For more, see Amir, "Challenging Nuclear," 349–350.

102. Ibid., 352–353.

103. Ibid., 362–363.

104. Wahana Lingkungan Hidup Indonesia called nuclear energy "the destroyer of human future." For more, see Amir, "Nuclear Revival in Post-Suharto Indonesia," 265–266.

105. Caballero-Anthony, Alexandra, and Punzalan, "Civil Society Organizations," 183.

106. Richard Tanter, "Nuclear Fatwa: Islamic Jurisprudence and the Muria Nuclear Power Station Proposal," *APSNET Policy Forum*, December 13, 2007, http://nautilus.org/apsnet/nuclear-fatwa-islamic-jurisprudence-and-the-muria-nuclear-power-station-proposal/.

107. Amir, "Challenging Nuclear," 362.

108. Caballero-Anthony, Alexandra, and Punzalan, "Civil Society Organizations," 184–185.

109. Ibid., 183–188.

110. Amir, "Challenging Nuclear," 343–366.

111. Caballero-Anthony, Alexandra, and Punzalan, "Civil Society Organizations," 183.

112. Amir, "Challenging Nuclear," 344.

113. "ASEAN to Promote Nuclear Energy, Solar Power," *Agence France-Presse*, November 13, 2007.

114. Ratanachai Namwong, "Nuclear Power Project in Thailand," IAEA-CN-164-1P04, IAEA, October 29, 2009, http://www-pub.iaea.org/MTCD/publications/PDF/P1500_CD_Web/htm/pdf/poster/1P04_R.Namwong.pdf.

115. Amir, "Challenging Nuclear," 345.

116. David Santoro, "Myanmar on a Nonproliferation Roll," *PacNet Number 75* (October 16, 2014), http://csis.org/files/publication/Pac1475.pdf.

117. "Indonesia's Views on the Main Elements to Be Taken into Consideration in the Drafting of the Possible Fissile Material Treaty," United Nations Office at Geneva, accessed September 1, 2015, http://www.unog.ch/80256EDD006B8954/(httpAssets)/B64E4681C402EC4BC1257B7C0041667E/$file/Indonesia.pdf.

118. According to IAEA, CPPNM is the "only international legally binding undertaking in the area of physical protection of nuclear material. It establishes measures related to the prevention, detection and punishment of offenses relating to nuclear material." International Atomic Energy Agency, "Convention on the Physical Protection of Nuclear Material," United Nations, IAEA, entered into force February 8, 1987, https://www.iaea.org/publications/documents/conventions/convention-physical-protection-nuclear-material.

119. The amendment entered into force in early 2016 after receiving the required number of ratifications. For more, see "Amendment to the Convention on the Physical Protection of Nuclear Material," United Nations, IAEA, accessed online March 3, 2016, https://www.iaea.org/Publications/Documents/Conventions/cppnm_amend_status.pdf.

120. Lao People's Democratic Republic, "Note Verbale dated 3 May 2005 from the Permanent Mission of the Lao People's Democratic Republic to the United Nations addressed to the Chairman of the Committee," S/AC.44/2004/(02)/117, United Nations Security Council, May 4, 2015, http://daccess-dds-ny.un.org/doc/UNDOC/GEN/N05/332/68/PDF/N0533268.pdf?OpenElement.

121. Michelle Cann, Kelsey Davenport, and Sarah Williams, *The Nuclear Security Summit: Assessment of Joint Statements* (Washington, D.C.: Arms Control Association and Partnership for Global Security, March 2014), 5, http://www.armscontrol.org/files/ACA_NSS_Report_2014_FINAL.pdf.

122. The other options for "greatest threats to the world" listed in the survey were religious and ethnic hatred, AIDS and infectious diseases, pollution and other environmental problems, and a growing gap of rich and poor. For more, see PEW Research Center, *Global Attitudes and Trends* surveys in Spring 2014, Spring 2007, and Summer 2002, PEW Research Center, http://www.pewglobal.org/.

123. PEW Research Center, *Global Attitudes and Trends* survey, Spring 2015, PEW Research Center, http://www.pewglobal.org/question-search/?qid=2092&cntIDs=&stdIDs=.

124. One difficulty that added to the international community's ability to ascertain Myanmar's nuclear intentions was the fact that some of the data that suggested such intentions came from exiled Burmese opposition figures. Moreover, some Southeast Asian officials believed that Western officials and analysts read too much into the historic secrecy that enveloped all government affairs in Myanmar.

125. Michael S. Malley and Tanya Ogilvie-White, *Nuclear Challenges in Southeast Asia: Promoting Cooperation and Consensus*, Report of the Second Southeast Asia Strategic Dialogue 2011, PASCC Report No. 2012 010 (Monterey, Calif.: U.S. Naval Postgraduate School, Center on Contemporary Conflict, August 2012), http://www.dtic.mil/dtic/tr/fulltext/u2/a565766.pdf.

126. Richard Norton-Taylor, "Burma's Nuclear Weapons Intent 'Clear and Disturbing,'" *The Guardian*, June 4, 2010, http://www.theguardian.com/world/2010/jun/04/burma-nuclear-weapons.

127. J. Taylor Rushing, "State, DOD Officials Defend Warming Relations with Myanmar," *Stars and Stripes*, December 4, 2013, http://www.stripes.com/news/pacific/state-dod-officials-defend-warming-relations-with-myanmar-1.255916.

128. Associated Press, "US Welcomes Myanmar's Signing of UN Nuclear Agreement," *Washington Examiner, September* 21, 2013, http://www.washingtonexaminer.com/us-welcomes-myanmar-signing-nuclear-agreement/article/feed/2110543/.

129. For an in-depth discussion of these threats, see Mely Caballero-Anthony, *Non-Traditional Security Challenges, Regional Governance, and the ASEAN Political-Security Community (APSC)*, Asia Security Initiative Policy Series Working Paper No. 7 (Singapore: Center for Non-Traditional Security Studies, S. Rajaratnam School of International Studies, September 2010), http://www.rsis.edu/sg/wo-content/uploads/rsis-pubs/NTS/resources/research_papers/MacArthur_working_paper_Mely_Caballero-Anthony.pdf.

130. The Malaysian company, Scomi Oil and Gas, manufactured and shipped the centrifuges in October 2003, but this information did not come to light until 2004.

131. For a more extensive account, see Sheldon W. Simon, "U.S.-Southeast Asia Relations: A WMD Discovery in Malaysia and Counter-Terrorism Concerns in the Rest of Southeast Asia," *Comparative Connections* 6, no. 1 (April 1, 2004), http://csis.org/files/media/csis/pubs/0401qu_seasia.pdf.

132. U.S. Department of State, "Malaysia Endorses the Proliferation Security Initiative," (Washington, D.C.: State Department Office of Spokesperson, April 29, 2014), http://www.state.gov/r/pa/prs/ps/2014/04/225349.htm.

133. IISS, *Preventing Nuclear Dangers*, 56–57.

134. One example cited was a 2003 investigation into whether a Thai national smuggled a small amount of caesium-137, which could be used in a radioactive "dirty bomb," from Russia through Laos. For more, see Shawn Crispin and Gary Fields, "U.S.-Thai Seizure Triggers Fears of 'Dirty Bombs,'" *Wall Street Journal*, June 18, 2003, http://www.wsj.com/articles/SB105587230278383100; IISS, *Preventing Nuclear Dangers*, 58–59.

135. Santoro, "ASEAN's Nuclear Landscape—Part 1," 1.

136. The feasibility study for this option was commissioned by the Johnson administration and conducted by the Institute for Defense Analysis (Jason Division). For more, see F. J. Dyson et al., *Tactical Nuclear Weapons in Southeast Asia*, *Study S-266* (Alexandria, Va.: Institute for Defense Analyses, Jason Division, March 1967), http://nautilus.org/wp-content/uploads/2011/12/dyson67.pdf.

137. At the time of its establishment, ASEAN was comprised of Indonesia, Malaysia, the Philippines, Singapore, and Thailand. Brunei was incorporated after it became independent in 1984. In 1995, Vietnam and Laos were admitted; Myanmar entered in 1996; and Cambodia was admitted in 1998.

138. Noordin Sopiee, "The Neutralization of Southeast Asia," in *Asia and the Western Pacific: Toward a New International Order*, ed. Hedley Bull (Canberra: Nelson and Australian Institute of International Affairs, 1975), 137.

139. ASEAN, *1971 Zone of Peace, Freedom and Neutrality Declaration*, adopted by the Foreign Ministers at the Special ASEAN Foreign Ministers Meeting in Kuala Lumpur, Malaysia on 27 November 1971, http://www.icnl.org/research/library/files/Transnational/zone.pdf.

140. For more on the attempts at establishing a SEANWFZ through ASEAN working groups, intense negotiations, and summit meetings, see M. C. Abad Jr., "A Nuclear Weapon-Free Asia and Its Continuing Strategic Significance," *Contemporary Southeast Asia* 27, no. 2 (August 2005): 168.

141. Although Camobdia, Laos, and Myanmar were not yet ASEAN members in 1995, they were consulted in the lead-up to the treaty's opening up for signature. For more, see Ibid., 170.

142. The zone covers territories and exclusive economic zones of ten countries, namely Brunei, Cambodia, Indonesia, Laos, Malaysia, Myanmar, the Philippines, Singapore, Thailand, and Vietnam. The treaty covers land territory, internal waters, territorial sea, archipelagic waters, seabeds, and the airspace above these features. Dumping radioactive waste within the zone was prohibited as well. For more, see ASEAN, "Treaty on the Southeast Asian Nuclear-Weapon-Free-Zone," signed December 15, 1995, entered into force March 28, 1997, http://www.asean.org/news/item/treaty-on-the-southeast-asia-nuclear-weapon-free-zone; Catharin Dalpino, "Policy Brief: Can ASEAN Sell Its Nuclear Free Zone to the Nuclear Club?" Rising Powers Initiative, October 2014, http://www.risingpowersinitiative.org/wp-content/uploads/PolicyBrief_Oct2014_SEANFWZ.pdf.

143. Abad, "Nuclear Weapon-Free Asia," 166–167.

144. Walter Pincus, "Financing a Global Navy," *Washington Post*, May 1, 2013, http://www.washingtonpost.com/world/national-security/financing-a-global-navy/2013/05/01/d3fda038-b1b0-11e2-baf7-5bc2a9dc6f44_story.html.

145. Dalpino, "Can ASEAN Sell Its Nuclear Free Zone to the Nuclear Club?"

146. General Assembly Resolution 30/3472, Comprehensive Study of the Question of Nuclear-Weapon-Free Zones in All Its Aspects, A/RES/30/3472, December 11, 1975, http://www.un.org/en/ga/search/view_doc.asp?symbol=A/RES/3472%28XXX%29.

147. U.S. Department of State, "Report of the United States of America Pursuant to Actions 5, 20, 21 of the 2010 Nuclear Non-Proliferation Treaty Review Conference," April 29, 2014, http://www.state.gov/documents/organization/225580.pdf.

148. Supalak Ganjanakhundee, "Nuclear States Shun ASEAN Treaty," *The Nation (Bangkok)*, July 9, 2012, http://www.nationmultimedia.com/national/Nuclear-states-shun-ASEAN-treaty-30185739.html.

149. Catharin Dalpino's interviews with Foreign Ministry officials in Naypyidaw in October 2013 indicate that Burmese officials may have had another objective in seeking the P-5 signatures—convincing the United States that Myanmar did not harbor ambitions to acquire nuclear weapons.

150. This statement followed consultations with nuclear weapon states in October 2014. For more, see ASEAN 2015, "ASEAN to Intensify Efforts to Encourage Nuclear Weapon States to Accede to SEANWFZ Protocol," August 3, 2015, http://asean2015.bernama.com/newsdetail.php?id=1158086; "Report of the ASEAN Political-Security Community Council to the 26th ASEAN Summit," ASEAN2015, April 26, 2015, http://asean2015.bernama.com/psdetail.php?id=785.

151. "Nuclear Energy: Safe Nuclear Power," European Commission, accessed September 7, 2015, http://ec.europa.eu/energy/en/topics/nuclear-energy.

152. Symon, "Southeast Asia's Nuclear Power Thrust," 130–133.

153. East Asia Summit, "Singapore Declaration on Climate Change, Energy, and the Environment," Singapore, November 21, 2007, http://www.mofa.go.jp/region/asia-paci/eas/dec0711.pdf. In particular, the host country, Singapore, championed a proposal to address nuclear safety and potential radioactive fallout concerns. For more, see Angelo T. Reyes, "Energizing ASEAN to Power a Dynamic Asia," Joint Ministerial Statement for the 25th ASEAN Ministers on Energy Meeting, Singapore, August 23, 2007, http://www.asean.org/news/item/singapore-declaration-on-climate-change-energy-and-the-environment; Geoffrey Gunn, "Southeast Asia's Looming Nuclear Power Industry," *Asia-Pacific Journal: Japan Focus*, February 11, 2008, http://www.japanfocus.org/-Geoffrey-Gunn/2659/article.html; Symon, "Nuclear Power in Southeast Asia," 15.

154. Symon, "Southeast Asia's Nuclear Power Thrust," 133.

155. Symon, "Nuclear Power in Southeast Asia," 11–13.

9

Pakistan

The Nuclear Consensus

Christopher Clary

The nuclear debate in Pakistan is characterized by its near absence. There is almost universal public agreement on the necessity of Pakistan's nuclear weapons program. It is a rare, high-profile success story for Pakistan, viewed as such across the political spectrum. Pakistan features elements of the Realist and Nationalist paradigms so common in other states. Globalist voices, while not entirely absent, are peripheral in Pakistan, with little to no influence on national security or most other matters. The nuclear weapons program and the nuclear infrastructure necessary to produce it enjoy widespread support among Realists and Nationalists, while Globalists fret on the margins.

This consensus is to some extent manufactured. In Pakistan, the military and in part the bureaucratic, diplomatic, and scientific communities hold strong opinions about the role of nuclear weapons and nuclear energy. In general, their worldview just so happens to be one in which policies that increase the resources available to their home institutions are also the ones presented as good for Pakistan. They successfully argue that the Pakistani national interest can best be advanced in ways beneficial to their parochial interest. As a result, Realist and Nationalist worldviews are promulgated and accepted as part of legitimate discourse. Globalist voices are ostracized as naïve and harmful to Pakistani interests.

Those individuals and institutions that strongly favor Pakistan's nuclear programs uniquely have access to resources to propagate that view and curtail opposing narratives. Maintaining social consensus in favor of nuclear weapons is a priority for these individuals. For individuals and institutions that doubt the benefit from Pakistan's nuclear program, few are willing to expend significant energy in contesting a nuclear narrative when there are other problems that they perceive as more urgent (e.g., social justice, Islamic radicalization, corruption, civilian control over the military, and countless other problems that bedevil modern Pakistan), especially with the constraints that exist on nuclear discourse.

While there is no way to imagine a nuclear discourse in Pakistan absent from the power imbalances that characterize Pakistani society and politics, what does seem clear is that the elite consensus on nuclear matters is largely reflected in a corresponding mass consensus. This chapter will show that available surveys of Pakistani public opinion regarding nuclear energy and nuclear weapons correspond to the Realist and Nationalist worldviews that dominate elite discourse. As a consequence, it seems unlikely that the Pakistani nuclear debate will shift away from its current trends barring a substantial exogenous shock.

This chapter will proceed in three parts. First, it will describe this volume's analytical framework as manifest in the Pakistani context. Second, it will survey the nuclear landscape, identifying key actors and their role in the nuclear debate. Third, it will review the answers by these actors to three nuclear questions: (1) Are nuclear weapons necessary for Pakistani security? (2) Are global nonproliferation efforts beneficial? (3) How important is nuclear energy to the future of Pakistan?

PAKISTANI NUCLEAR WORLDVIEWS

As discussed in the introduction to this volume, individuals and entities may hold blended worldviews or even contradictory worldviews across specific nuclear issues. The boundary between Realists and Nationalists is particularly permeable. However, it is possible to identify dominant strands of thinking for individuals with a center of gravity in one of the worldviews.

Nuclear Realists

Pakistani Realists stress that Pakistan lives in a dangerous neighborhood and is locked in an intense security competition with India. Pakistani Realists may generally share with Pakistani Nationalists views about how the conflict with India began, but their narrative does not depend on India's ideologically striving to destroy Pakistan. Instead, in the international system, two powers next to one another will often find themselves in conflict over finite resources. For Pakistan and India, that conflict has manifested itself primarily in the Kashmir dispute, important not simply for ideological reasons, but also because its geographic location is so important for Pakistani access to China and as the origin for water critical to Pakistan's still agriculturally dominated economy.

Pakistani Realists emphasize that their country turned to nuclear weapons only after (1) alliances proved unreliable, (2) the threat to Pakistan's survival was proven real, and (3) their primary adversary pursued nuclear weapons. Pakistan's American allies and Chinese friends failed to come to Pakistan's aid in 1965, and their weak efforts were inconsequential in stopping India's dismemberment of Pakistan in 1971.[1]

After 1965, the United States also stopped arms from flowing to Pakistan at a crucial point. Realists will admit Pakistan mismanaged its internal politics leading to 1971, but India was waiting and willing to seize the opportunity to vivisect West and East Pakistan. After 1971 and Zulfikar Ali Bhutto's ascent to power, Realists admit that Pakistan began to put in place a foundation for nuclear weapons. Even then, they stress, the effort did not turn serious until after India's 1974 "Peaceful Nuclear Explosion" (see chapter 3 for more on this event), an appellation that causes them to roll their eyes.[2]

Those weapons reached some sort of existential capability in the early to mid-1980s, and according to Realists, since then the weapons have proven to be necessary for Pakistan's security against a much larger India. Realists point to the nuclear weapons as deterring several crises from becoming large wars in 1987, in 1990, in 1999, in 2001–2002, and in 2008.[3] Therefore, Realists have concluded that nuclear deterrence has worked for Pakistan—similar to other states who trusted nuclear weapons to ensure their security: the United States facing down a conventionally superior Soviet Union and contemporary Israel, the latter being an especially ironic comparison.[4]

Why then, Realists ask, should Pakistan be singled out for behavior common among similarly positioned states? They share this sense of grievance with Nationalists. Realists dismiss criticisms of Pakistani association with violent non-state actors since others also play such dirty games. Faced with a larger India, militants serve as an inexpensive way to force India to expend significant resources and manpower on a counterinsurgency. Yes, Pakistan is struggling with violent groups at home as well, but Realists stress that Pakistan's nuclear security record is no worse than other states situated in such an unpleasant environment. Realists retort that if U.S. and British air bases have been attacked and breached in Afghanistan without concern, then why do outsiders worry that strikes against Pakistani airbases are proof that Pakistan cannot secure its arsenal?[5] If the United States cannot protect its secret correspondence from WikiLeaks, why challenge Pakistan's ability to maintain nuclear secrets?[6]

In recent years, the Realists are quick to ascribe causation to Western courtship of India as a strategic partner to contain China and a trading partner for Western financial benefit: because it courts India, the West shuns and punishes Pakistan.[7] Surveys have found approximately half of Pakistanis believe U.S. policies "favor India too much," while less than 10 percent believe Pakistan is favored, reflecting Nationalist and Realists views.[8] Washington's contemporary ties with New Delhi are added to the long list of historical examples that Pakistanis cite in their indictment of U.S. reliability as a Pakistani ally.

Realists also tend to be moderate members of Pakistan's national security professions, especially the military, foreign service, and Pakistan's small community of think tank and university researchers. There are Realists within the mainstream political parties, though the nature of Pakistani parliamentary politics is such that parties do not contain extensive reservoirs of national security expertise. Realists are rare in Pakistan's religious parties or the new Pakistan Tehreek-e-Insaf (PTI) party.[9]

Nuclear Nationalists

While Pakistani Realists have a sense of grievance about perceived international punishment for what they view as normal behavior, Nationalists feel that grievance all the more acutely. These Nationalists believe other states are both hypocritical and motivated by anti-Pakistan sentiment. For Nationalists, this starts with an India that cannot abide Pakistan's existence as a permanent threat to its secular ideals. Episodes of anti-Muslim violence in India are reminders to Nationalists of the raison d'être of the Pakistan project.[10]

To Nationalists, fears about Pakistan's nuclear stewardship are not just overblown—they are part of an international effort to delegitimize Pakistan's nuclear weapons. In Shireen Mazari's words, "Of course there is nothing new in these assertions [of nuclear danger] which emanate from the U.S. periodically—especially in times of worsening relations with Pakistan. One knows only too well the link between the U.S. media and the U.S. state and it is not a point on which more proof is needed."[11] Nor is it just Western critics that carry ulterior motives. When Pakistani voices of a Globalist bent question the safety and security of, say, new nuclear reactors, they are shouted down as being part of "a conspiracy against Pakistan."[12]

Pakistani nationalism is inseparable from the fact that Pakistan was created to be the homeland for the subcontinent's Muslims. Pakistan's successes are Muslim successes, which should both make Muslims proud and also raise Pakistan's status in the Muslim world. Such a view is apparent in the writings and statements of three very different Pakistani Nationalists: Zulfikar Ali Bhutto, Zia ul-Haq, and A. Q. Khan.[13]

Bhutto, despite being a secular Nationalist himself, was masterful at infusing secular projects with an Islamic element. After the military, led by Zia ul-Haq, placed him under house arrest, Bhutto lashed out from his jail cell, accusing the military of participating in a plot to stop the Pakistani nuclear weapons program. This belief was false—Zia's military government continued the nuclear program—but it did lead to a series of statements by Bhutto about the program as he saw it: "We know Israel and South Africa have full nuclear capability. The Christian, Jewish, and Hindu civilizations have this capability. The communist powers also possess it. Only the Islamic civilization was without it, but that position is about to change."[14] Zia ul-Haq, though more circumspect, made a similar point in an interview he gave soon afterward while in Saudi Arabia.[15] In his survey of nuclear rhetoric in South Asia, Haider Nizamani summarized the motif of this period as "get rid of Bhutto but keep his Nukespeak."[16] The Islamist and Nationalist themes that Bhutto developed in the 1970s surrounding nuclear weapons have proven enduring under civilian, military, secular, and religiously minded governments.

After the May 1998 nuclear tests, Pakistani politicians were quick to cloak the accomplishment in Islamic imagery. The day of the tests became a holiday, Youm-e-Takbeer. A defensible translation would label May 28th as "the day of God's greatness." The phrase, inaugurated by the Nawaz Sharif government, was employed by subsequent, secular-minded governments and continues to be invoked.[17] When mainstream political parties or technocrats employ Islamic rhetoric, they may not sincerely believe it, but at a minimum they find it useful because of the symbolic resonance it holds.[18]

Discrimination against Pakistan is portrayed as anti-Islam. A. Q. Khan claimed his work on Pakistan's nuclear program was really work for the Muslim world. In a 1995 speech, Khan lamented Western "efforts to curtail the development of the Muslim world, which the Western powers unjustifiably see as a potential threat to their monopoly."[19] He called for greater pan-Islamic cooperation to block efforts to divide the Muslim world and suppress its advancement.[20]

Despite Islamic and Muslim motifs in the writings and speeches of "Nuclear Nationalists," there is one area where it might be possible to identify distinct "Nuclear Islamists." This worldview is still too marginal and inchoate to deserve extensive treatment. The principal difference from Nuclear Nationalists might be a greater willingness on the part of Nuclear Islamists for the bomb to be used for extended deterrence in the Muslim world. Khurshid Ahmad of the religious Jamaat-e-Islami party speaks for this group when he says, "Pakistan as an Islamic state has a responsibility to the broader Umma . . . . Pakistan's nuclear weapons will inevitably be seen as a threat by Israel, and therefore Pakistan must include Israel in its defense planning . . . . Under the circumstances, the future of the Muslim world depends on Pakistan."[21]

Similarly, it is still unclear what motivated retired Pakistani nuclear scientists Sultan Bashiruddin Mahmood and Chaudiri Abdul Majeed to talk to Osama bin Laden and others in Afghanistan about nuclear weapons and nuclear energy in August 2001, but given their earlier writings and broader charity work (their charity was called Ummah Tameer-e-Nau, or Reconstruction of the Muslim Ummah), it seems likely their motives were pan-Islamic in nature.[22] Nuclear Islamists differ primarily then on this obligation to share Pakistani technology with others in the Muslim world. Nonetheless, they may share beliefs with Realists or Nationalists on most other matters. Mahmood, for instance, was noted for his opposition to Pakistan's signing the Comprehensive Nuclear Test Ban Treaty (CTBT) prior to his subsequent contacts with bin Laden.[23]

Moreover, Nationalist and Islamist thought is so similar on nuclear matters because both agree that Pakistan's nuclear program has been singled out due to Pakistan's being a Muslim state (rather than genuine concerns about what many outsiders see as Islamabad's high-risk nuclear posture). Despite public statements that U.S. policy sought to bolster Pakistani security and the Pakistani state, many elites in the country preferred to believe U.S. policy had another goal entirely: "to trigger the collapse of Pakistan."[24] In a 2012 survey of 800 respondents in two districts in Punjab province, researcher Niloufer Siddiqui found that 69 percent of respondents sensed "the United States' policies in the country are motivated primarily by a desire to neutralize Pakistan's nuclear arsenal."[25] These fears give energy to Nuclear Nationalists and Islamists. Realists, while not discounting Western bias and nonproliferation efforts against the Pakistani program, are less likely to see threats behind every corner.

Nationalists and Islamists also share a fear that politicians or technocrats may sell out the nuclear program, for temporary gains. Applauding the decision to test nuclear weapons, retired General Mirza Aslam Beg argued, "Not testing now would also have been an utter mockery of the democratic process, if the popular will was to be sacrificed for the sake of paltry fringe benefits, and in turn would have been tantamount to bartering Pakistan's sovereignty."[26] A few months after Beg, who served as chief of army staff from 1988 to 1991, made these remarks, the Pakistani government considered relenting to U.S. pressure and signing the CTBT. There were pragmatic reasons for Pakistani signature: given validated and tested designs, Pakistan's accession to the treaty would earn U.S. appreciation while hampering India's ability to test again. Pakistani Islamists placed intense pressure on the Sharif government to not accede to the treaty. In a protest before Parliament in September 1998, for example, Jamaat-e-Islami leader Qazi Hussain Ahmed warned, "You will dig your graves by signing the CTBT."

In May 2014, Hafiz Saeed addressed a large rally in Karachi to protest Nawaz Sharif's attendance at Indian Prime Minister Narendra Modi's inauguration. Saeed, who leads the terrorist group Lashkar-e-Taiba and the Islamist charity organization Jamaat-ud-Dawa, is believed to have a close relationship with Pakistan's Inter-Services Intelligence (ISI).[27] As Saeed urged Sharif not to attend Modi's inauguration, he did so on a stage festooned with a backdrop showing Pakistani ballistic missiles and a banner that read, "An atomic Pakistan guarantees security to Pakistan and the rest of the Ummah."[28] A few days later, Saeed shared the stage with another terrorist-cum-charity group leader, Maulana Fazlur Rehman Khalil, at a rally in Islamabad where Khalil told the group, "By acquiring atomic technology, [the] Pakistani nation has followed an important command of Allah, who has ordered the Muslims to stay strong for their defence."[29] As Khalil is also reported to have a close relationship with the ISI,[30] it seems likely that Nationalists in Pakistan's security services encourage nuclear rhetoric by Islamists, at least in part to constrain any dovish inclinations of mainstream politicians.

Nationalists and Realists disagree somewhat on the value of external alliances and which states are more reliable than others. Realists are skeptical of foreign friends, believing they are unlikely to help during the periods of the greatest peril for Pakistan. Nationalists, especially those of an Islamist bent, believe that the United States is not just a fickle ally, but also antagonistic. For them, Pakistan's natural allies reside in the Islamic world. Realists privately scoff at that notion, wondering how much the Organization of the Islamic Conference has done to help Pakistan on the Kashmir dispute. Both Realists and Nationalists—even Islamists, at least in public—are more confident Pakistani and Chinese interests are congruent.[31]

The Nationalist that perhaps best exemplifies these views is retired General Beg. While still serving, he expounded a view that Iran, Iraq, and Pakistan should cooperate in "strategic defiance" of the United States, with the added assistance of China to buttress this Islamic triad.[32] The lack of Realist prudence in Beg's vision is all the more apparent in his decision to articulate it in the midst of the U.S.-Iraq crisis over Kuwait in 1991. His logic was primordial and Nationalist: "Iran and Afghanistan are two countries that have never been enslaved, they have remained independent, and

Pakistan being a neighbor carried the same traditions."[33] There is open speculation as to whether Beg's belief in this natural alliance led him to authorize nuclear technology cooperation with Iran through A. Q. Khan's nuclear supplier network during his time at the top of the Pakistan army.[34]

As an aside, it is worth emphasizing that the A. Q. Khan episode remains highly politicized in Pakistan, preventing introspection as to the role of the Pakistani state in authorizing Khan's nuclear enterprise. Realists typically will downplay Khan's role in the nuclear program—relegating him to a modest role as one of many Pakistani scientists and engineers who made the program a success. Khan's downfall within the Pakistani establishment and the comparative ascension of his bureaucratic rivals has facilitated this historiographic shift. Prior to his fall, Khan worked assiduously with journalists and senior officials to propagate an image that he was central to all elements of the Pakistani nuclear program. Now his opponents suggest he was marginal to most significant achievements in the multifaceted, multi-decade nuclear effort. Realists assert, then, that this modest player with a bent for public relations created an unauthorized proliferation network to sell old technology to international clients. When evidence as to his nuclear trade was presented to Pakistan, they argue, the state quickly shut down Khan's network and punished him to the extent possible given the scientist's widespread popularity. Now, they assert, Pakistan is being unfairly ostracized for one individual's past proliferation sins in which the state was not complicit.

The evidence of state involvement in Khan's enterprise is not as clear as most outsiders suggest. The scale of the enterprise, especially its international shipments, provides circumstantial evidence of state authorization, these outsiders allege. In order to operate his portion of the Pakistani nuclear program, Khan was given wide latitude to use military aircraft to transport goods into Pakistan. Additionally, the entrepreneurial Khan also developed and sold several types of conventional weaponry, and similarly used military aircraft for delivery as well as import. There were widespread suspicions that Khan was earning income beyond his salary, but most assumed that Khan was merely corrupt, skimming off the considerable resources made available to his laboratory. Since Khan's laboratory was more productive than most Pakistani state enterprises, such corruption was viewed as an acceptable cost for Khan's labor. The evidence is certainly not so damning that a motivated Pakistani interlocutor cannot argue that whatever problems occurred are now past, and those problems were overstated from the outset.

Nationalists have less concrete to say on the Khan episode. They are more inclined to view Khan as a national hero who played an irreplaceable role in advancing Pakistan's nuclear ambitions. When pressed, they might suggest that the charges against him were a conspiracy. They might hint that, even if they were true, how bad would it be if Pakistan worked to defy such a discriminatory international nonproliferation regime. Under no circumstances will either Realists or Nationalists admit that the Khan episode raises legitimate questions about Pakistan's nuclear stewardship, though informed Realists are more likely to reference the many changes Pakistan has made to improve its nuclear security and personnel reliability since 2001.

There are Nationalists in the Pakistan military, perhaps slightly more in the army than the other military services (that tend to be more Realist in disposition). Nationalist thinking is widespread in Pakistan's religious parties and the new PTI party. While Realists may outnumber Nationalists in the Pakistani foreign service, there are many Nationalists in the diplomatic corps as well. Nationalist thinking has also been rewarded in Pakistan's strategic, think tank, and university research communities, which rely on the military for funding. Even if the military ranks are split between Nationalists and Realists, instrumentally the military may encourage more beneficial Nationalist rhetoric than more moderate Realist arguments.

Nuclear Globalists

Nuclear Globalists critique from the margins, many having such difficulty in Pakistan's Realist and Nationalist–dominated institutional environment that they now live in the West and criticize developments from abroad. The most prominent voices are Pakistani scientists enmeshed in a series of overlapping international bodies focused on arms control and disarmament, namely the International Panel on Fissile Materials, Princeton University's program on Science and Global Security, and at international Pugwash Conferences on Science and World Affairs. When Pervez Hoodbhoy, perhaps the archetypical example of a Nuclear Globalist, edited a recent volume titled *Confronting the Bomb* about the concerns held by scientists regarding the Indian and Pakistani nuclear programs, he enlisted only two other Pakistani authors, Zia Mian and A. H. Nayyar, to take part in the initiative.[35] There is simply a thin Globalist bench.

If such anecdotal evidence is not persuasive, there are some indications from polling data as to the small size of this group. Prior to the nuclear tests in 1998, the Kroc Institute commissioned a survey of 910 members of the Pakistani business, scientific, academic, political, and military elite.[36] They found 61 percent of individuals supported the government's then position to maintain a nuclear weapons option, 32 percent sought outright weaponization (even prior to India's 1998 tests), and only 6 percent said Pakistan should renounce nuclear weapons.

A public poll of provincial capital cities after the Indian 1998 nuclear tests, but before the Pakistani tests, found two-thirds of respondents favored Pakistani testing in response with an additional one-sixth favoring nuclear testing at some future date instead.[37] In other words, over 80 percent of respondents favored immediate or eventual testing, despite the sanctions that would result and the benefits the international community was offering Pakistan for not testing. Consistent with the profile of well-known Globalists, the survey did find lower support for immediate nuclear testing among university graduates than those with less education.

Even three years after South Asia's tests, adults in eleven Pakistani cities were asked, "How do you feel about our country's continuing to develop its nuclear weapons capabilities?" Only 7 percent of respondents disapproved, while 89 percent approved—of which 76 percent strongly approved![38]

Nuclear Globalists employ a mix of arguments. They argue that the biggest threat to Pakistan comes from internal militancy and terrorism, not India. They fear that funds used for the army and nuclear weapons program take away from the important goals of development. They are concerned that expanding the nuclear infrastructure is inherently dangerous because of the potential for natural disasters, as well as threats of a human origin involving insiders and outsiders. Even so, Globalists have largely been unable to sway their compatriots.

Nuclear Globalists have a potential ally in a larger group, which I will call "Nuclear Apathists" or perhaps "Nuclear Pragmatists." This group does not particularly care about nuclear weapons one way or the other. To the extent they think about them, they believe if India has nuclear weapons then Pakistan ought to as well. They perhaps take some mild pride in the angst the atomic arsenal causes in the United States. In the 1998 Kroc Institute survey among elites, only 6 percent of respondents identified nuclear weapons as a major area of professional interest and an additional 24 percent considered themselves well-informed on the matter. Among the public at large, there are indications that a substantial minority are disinterested in nuclear matters. A survey of the Sindh province in 2000 found 29 percent of respondents agreed with the statement, "I don't care about Pakistan's nuclear policy."[39] Meanwhile, in the 2001 A. C. Nielsen survey, about 17 percent of individuals did not respond or did not know how to respond when asked about the Pakistani signature of the CTBT, suggesting feelings about nuclear weapons in general might be more emotive than informed for a sizeable chunk of the population.

This larger group—for which national security issues have lower salience—largely chooses to focus on topics other than the difficult terrain of nuclear weapons. They are not a silent majority, but they are a silent minority. Nuclear Globalists represent individuals informed and passionate about nuclear issues who have thus far been unable to rouse this passive portion of society. Yet, Globalist concerns about safety and the unproven economics of nuclear energy—as well as the expenses of continued military fissile material production—might find sympathetic ears among citizens who have no special interest in nuclear weapons. Even so, as long as they are shut out from the national security bureaucracy and military establishment, their views are unlikely to alter Pakistani policy. There is some limited survey evidence that a plurality of Pakistanis also believe the views of ordinary Pakistanis are not important to nuclear policy decisions.[40] What Samina Ahmed wrote regarding Pakistani nuclear test decision-making in 1998 still holds true today for nuclear issues more broadly: "The internal elite discourse on the pros and cons of nuclear testing did not influence the ultimate decision to test, since the closed circle of Pakistani decision-makers, with its traditional aversion to dissenting views, is still unfamiliar with democratic bargaining processes."[41]

PAKISTAN'S NUCLEAR LANDSCAPE

Having introduced the major worldviews, this chapter now turns to documenting in greater detail the key actors and their orientation on nuclear matters.

The Pakistan Military

In the beginning, there was the Pakistan army. It was the first actor to think seriously about nuclear weapons in Pakistan, adding nuclear warfare material onto its curriculum at the Command and Staff College as early as 1954.[42] This likely predates by a few months the creation of even the barest scientific organization within Pakistan to explore nuclear energy; the establishment of a civilian Nuclear Research Center would not be announced until January 1955.[43] In both cases, the United States was likely an active spur to Pakistani developments. Military-to-military ties through the Southeast Asia Treaty Organization beginning in 1954 and the Central Treaty Organization beginning in 1955 exposed the Pakistan military to U.S. thinking on nuclear weapons and nuclear war fighting.[44] Eisenhower's 1953 "Atoms for Peace" program similarly spurred civilian thinking in Pakistan and elsewhere.[45]

The military has not monolithically supported nuclear weapons since 1954. Ayub Khan, army chief beginning in 1951 and military dictator from 1958 to 1969, famously was skeptical of nuclear development efforts and believed nuclear weapons could be purchased "off the shelf" if Pakistan needed them at a future point. It certainly is possible that the field marshal believed diverting resources to nuclear weapons might endanger the institutional goals of the military of acquiring modern conventional weaponry and training professional soldiers. In his opposition to nuclear pursuit, he disagreed with a young civilian minister in his government, Zulfikar Ali Bhutto. It was Bhutto, returning as civilian leader of Pakistan in 1971, who pushed the nuclear weapons program, relegating the military largely to an organization capable of assisting the civilian-led effort with difficult engineering and public works projects at home and large-scale procurement of sensitive technology overseas.

Toward the end of Bhutto's tenure, to oversee the nuclear effort, he set up a committee of technocrats and scientists composed of the defense secretary, the foreign secretary, an economic advisor and deputy chairman of the planning commission, and the head of the Pakistan Atomic Energy Commission.[46] When Zia ul-Haq replaced Bhutto via a military coup, Zia became the dominant decision-maker in the 1980s and continued the program and Bhutto's committee.[47] The army was institutionally largely peripheral to the effort, but Zia was both president and army chief and ran it from the president's house or through the civilian scientific organization in the 1980s. What changed in the 1990s is the army built the institutional in-house capacity to consider such matters.

When Ghulam Ishaq Khan served as president from 1988 to 1993 following Zia's death, control and oversight of the nuclear program largely stayed in civilian—albeit military-friendly—hands. Based on interviews with Mirza Aslam Beg, G. I. Khan then was first among equals for a troika composed of G. I. Khan, Beg, and Benazir Bhutto during this early period of civilian rule.[48] Bhutto, despite being the elected prime minister, had the least influence over nuclear matters, though whether the prime minister was forced to cede control or chose to do so is a matter of opinion.[49]

When G. I. Khan left in 1993, he handed over the operation of the nuclear weapons program to the army chief, who in turn entrusted it to the director-general of the

Combat Development Directorate.[50] This does not mean that the military was not a supporter or active player in the nuclear program before 1993. Throughout this period, the military had played a major role in engineering, construction, and procurement efforts, as well as ensuring the physical security of the nuclear installations.[51] Distinguishing between Zia as president and Zia as army chief is a largely semantic distinction. Similarly, statements made by Beg demonstrate he had considerable input on nuclear decisions. This only means day-to-day and strategic management of the program was not in army hands.

From 1993, the military has been the primary decision-maker regarding the development and potential employment of an increasingly capable nuclear force. Responsibility for the program was handed over to a newly formed Strategic Plans Division (SPD) under the Joint Staff Headquarters following the nuclear tests, further enhancing the military's dominance of nuclear planning and policy. The SPD oversees a military Strategic Forces Command and serves as the secretariat to the National Command Authority (NCA), which is nominally run by the civilian leadership but also has a large roster of military members. Many outsiders accuse the military of making important nuclear decisions without civilian input, pointing specifically at potentially escalatory nuclear moves during the 1999 Kargil conflict.[52] For its part, the military denies any important nuclear steps were taken during the Kargil crisis.[53]

While it is difficult to tell with much precision, the army likely has a slightly higher ratio of Nationalists to Realists than the SPD, which has a higher mix of Realists to Nationalists. In part because of additional personnel screening to serve in the SPD, there are few to no Nuclear Islamists in the organization, though there obviously are individuals who are devout Muslims.

The Inter-Services Intelligence Directorate (ISI), so prominent in accounts of other national security issues in Pakistan, is largely absent from nuclear debates—aside from ensuring the arsenal is not penetrated by outsider threats nor endangered by insiders. The ISI is not directly involved in nuclear decision-making, although the director general of the ISI is frequently a close confidant of the army chief. The ISI is also suspected of playing an informal role in policing the national discourse and may nudge actors to talk about nuclear weapons in ways that conform to military preferences.

The Scientists

The Pakistan Atomic Energy Commission (PAEC) is a direct successor to the Atomic Energy Council created in 1956. The PAEC was established formally through legislative ordinance in 1965 and authority for the Commission was transferred directly to the president's secretariat in 1972.[54] The PAEC was responsible for all nuclear activities, including uranium mining, enrichment, nuclear power generation, spent fuel reprocessing, weapons design, and even delivery vehicle development. In the late 1970s, a separate uranium enrichment effort, initially called the Engineering Research Laboratory (ERL) and later Khan Research Laboratories

(KRL), was separated from PAEC.[55] Similarly, in 1961, a Space Sciences Research Wing was created under the PAEC, but in 1964 it became an independent organization, the Space and Upper Atmosphere Research Commission (SUPARCO).[56]

Two additional organizations were added to the nuclear effort in the mid-1980s—the National Development Complex (NDC) and the Air Weapons Complex (AWC)—to which the Project Management Organization (PMO) was added in 1994. All three organizations were focused on delivery systems and warhead design. They would be merged into the National Engineering and Scientific Commission (NESCOM) in 2001. Together PAEC, KRL, and NESCOM constitute the three major strategic organizations reporting to the National Command Authority, along with SUPARCO, which, despite its predominantly civilian emphasis, also reports to the NCA.[57]

During the 1980s and 1990s, relations between scientific organizations—especially Khan Research Laboratories and PAEC—were competitive and antagonistic. Both organizations and their larger-than-life leaders A. Q. Khan and Munir Ahmed Khan (no relation) sought to win the battle for public esteem and additional resources that underlay the bureaucratic rivalry.[58] The reorganization of strategic entities during the Musharraf era sought to tamp down this competition while simultaneously preventing the reemergence of a nuclear proliferation network like that headed by A. Q. Khan. President Musharraf was embarrassed by the revelations of Khan's nuclear commerce, and the reorganization was an attempt to show that such malfeasance would not recur.

In the context of the global nuclear suppliers' cartel, Pakistani scientists have had to solve scientific problems often in isolation from their global peers and without access to technologies and materials commonplace in the West. Each success in the nuclear arena was not only a personal accomplishment, but also viewed as a national triumph.

The Diplomats

The Ministry of Foreign Affairs (MFA) has had a consistent role in nuclear oversight as well as maintenance of the nuclear maneuvering space for Pakistan in the midst of international nonproliferation efforts. The foreign minister serves as deputy chairman of the NCA's Employment Control Committee, which is in theory responsible for nuclear use decisions. As is the case with many countries, responsibility for nuclear and nonproliferation matters is partly retained by the headquarters, but is also shared with diplomatic missions at the Conference on Disarmament, the United Nations's First Committee, and the International Atomic Energy Agency. Within MFA headquarters, there is an additional secretary for the United Nations and international cooperation, a disarmament directorate, and a Strategic Export Control Division.

What is somewhat unique in Pakistan is that the military SPD has an arms control and disarmament directorate, which has the principal responsibility of ensuring military views are represented in international negotiations affecting Pakistan's nuclear arsenal. Consistent with a wider civil-military capabilities gap in Pakistan, the SPD has more resources and personnel than its considerably talented civilian diplomat counterparts, giving it a dominant voice in internal nuclear debates. In a way, the foreign office and diplomatic corps has been relegated to the status of an implementing office for decisions arrived at in Joint Staff Headquarters. This was not always the case. In the early years, Pakistani diplomats worked to preserve options for Pakistan, nudging reluctant government and military officials to consider the gravity of global nuclear weapons developments.[59] Once the military built the in-house capacity to consider nuclear policy, the military gained dominance on nuclear diplomacy, too.

Nongovernmental Organizations and Analysts

The nongovernmental debate in Pakistan continues to echo, often quite consciously, official voices. The political economy of civil society in Pakistan is difficult to discern from the outside, but abundant anecdotal evidence suggests that the SPD and the ISI are directly and indirectly involved in funding and directing research tasks in Pakistan. This sounds hopelessly conspiratorial, but is readily apparent when outside researchers attempt to arrange meetings in vain until a chat with the ISI resolves concerns, after which point meetings materialize rapidly. Similarly, researchers associated with the official position might find considerable resources at their disposal. Researchers advancing arguments contrary to the official position might find life difficult at their host institution. They are told that dissensus within Pakistan's strategic community regarding nuclear weapons will lead outside observers into thinking that more pressure on Pakistan can alter its nuclear trajectory and jeopardize its security. This obviously distorts the debate in many ways large and small.

This gives the most autonomy and influence to individuals with military backgrounds, who are insulated somewhat from the system's machinations. Prominent voices since the mid-2000s have included Feroz Hassan Khan and Naeem Salik, both retired brigadiers with experience in the SPD; Mahmud Durrani, a retired major general who subsequently served in official capacities in the Musharraf government; and Talat Masood, a retired lieutenant general. All of them could accurately be labeled Realists, with Salik perhaps being comparatively more hawkish and Masood being comparatively more dovish. Khan and Salik remain prominent commentators on the Pakistan strategic program, while Durrani and Masood have a diminished public profile since the late 2000s. Similarly, diplomats with nuclear expertise have enjoyed prominence as commentators on nuclear issues—for example, Munir Ahmed Khan and Tariq Osman Hyder have regularly penned columns for Pakistani newspapers.

On the margins of the debate, some voices of concern can be faintly heard. Globalists criticize the pace and direction of nuclear weapons and nuclear energy developments, though generally from a more permissive research environment outside of Pakistan.[60] They are joined occasionally by environmental or local activists wary of the risks associated with nuclear energy infrastructure in Pakistan.[61] Opportunistically, opposition politicians might latch onto environmental concerns to embarrass the sitting government,[62] but establishment nuclear voices have ultimately tended to prevail without great difficulty.[63]

DEBATING PAKISTAN'S NUCLEAR FUTURE

The above sections have outlined the paradigmatic landscape of the Pakistani nuclear debate, and the extent to which these viewpoints are represented in governmental organizations and civil society debates. Given this mix of attitudes, individuals, and entities, how should we expect the Pakistani debate among elites and the public to confront a series of policy controversies surrounding nuclear weapons in the future? On most issues, it seems safe to predict Pakistan's love affair with nuclear weapons—and, as a consequence, nuclear energy—to continue.

Are Nuclear Weapons Necessary for Pakistani Security?

There remains overwhelming support for Pakistani possession of nuclear weapons. Surveys conducted in the late 1980s and early 1990s suggested approximately half of Pakistani respondents believed India would be willing to use nuclear weapons against Pakistan in the event of a war, while only 13 percent thought India would not.[64] In 2011, A. Q. Khan wrote, "Don't overlook the fact that no nuclear-capable country has been subjected to aggression or occupied, or had its borders redrawn. Had Iraq and Libya been nuclear powers, they wouldn't have been destroyed in the way we have seen recently. If we had had nuclear capability before 1971, we would not have lost half of our country."[65] Khan may be incorrect historically—nuclear weapons did not prevent attacks against Israel or India, nor did nuclear weapons prevent the dissolution of the Soviet Union—but Pakistanis by a wide margin share his fundamental faith in nuclear deterrence. There is no evidence that Globalist skepticism will do anything other than very modestly erode this bedrock belief.

Are Global Nonproliferation Efforts Beneficial?

Having been the target of global nonproliferation efforts for decades, Pakistanis show pride at having been able to evade international export controls and develop indigenous solutions to vexing scientific and engineering problems. This pride is widely held despite considerable evidence that Pakistan illicitly acquired technological know-how and components from Western countries as well as obtained Chinese and North Korean assistance for its nuclear weapons and missile programs. If pressed, Pakistani interlocutors will downplay the role of outside assistance and emphasize the necessity of Pakistani scientists and engineers to devise local solutions to integrate the various elements acquired from abroad.[66] Despite this understating of outside help, elite Pakistanis of Nationalist or Realist bents are likely grateful for Chinese assistance in the nuclear realm, one of a dozen examples that demonstrate Beijing's dependability as an ally compared to Washington. They openly express their gratitude for civil nuclear cooperation with Beijing, which is still expanding.[67] These elite will hint at confidence that, even as China is pressured to accede to global nonproliferation norms, Beijing will continue to safeguard Pakistani interests. China's willingness so far to argue that existing Sino-Pakistani nuclear technological cooperation is "grandfathered"—and hence not subject to China's newer nonproliferation obligations as noted in chapter 2—suggests such confidence on the part of elite Pakistanis is not misplaced.[68]

Having so frequently been the target of global nonproliferation efforts, it should be no surprise that most Pakistanis are skeptical about the nonproliferation regime, even when its enforcement might be in Pakistan's interests. The Pew Global Attitudes Project has periodically asked Pakistanis about nuclear proliferation generally and nuclear programs in North Korea and Iran in particular. When asked broadly whether states without nuclear weapons should be stopped from developing them, a majority of Pakistani respondents say no.[69] Most Pakistanis have no opinion about North Korea, but those who do view the North Korean nuclear program as no threat or a minor threat.[70] With Iran, Pew has found an outright majority of Pakistanis *in favor* of Iranian nuclear acquisition in eight separate surveys conducted since 2006.[71] Even among the minority who oppose Iranian nuclearization, Pakistani respondents disapprove of tougher economic sanctions[72] and oppose military action[73] as means to achieve that end. For many ordinary Pakistanis, Iran's shared experience as a Muslim state targeted by the global nonproliferation regime outweighs the pragmatic concerns that would accompany a nuclear-armed Iran. In this case, the views of Pakistan's Realist national security elite, who would have to manage the security consequences of another nuclear neighbor, may be more in accord with the international community, although Pakistani Nationalists and Islamists might be attracted to another Muslim bulwark against Western domination, even if it happens to be Shi'a-majority Iran.

Across the spectrum, mainstream Pakistani elites believe Western-led nonproliferation efforts are rife with hypocrisy. Even so, having been burned by the discovery of the A. Q. Khan nuclear supplier network, Pakistan is unlikely to actively flout global nonproliferation norms to help other states, even if it adheres to those norms without conviction and does its best to prevent new rules that might "discriminate" against Pakistan.

Since 2009, Pakistani diplomats at the United Nations Conference on Disarmament have stopped international nonproliferation negotiations on a fissile material cut-off treaty (FMCT) that would prohibit fissile material production for weapons purposes. These negotiations proceed on a consensus basis, so Pakistani opposition alone is sufficient to halt progress. While most open-source estimates suggest Pakistan has parity or slightly more nuclear weapons than India today,[74] Pakistani diplomats are wary that India can draw upon substantial existing stockpiles of reactor-grade plutonium to manufacture weapons and hence evade an FMCT that does not cover existing stocks of fissile material (see chapter 3 for the perspective on an FMCT held by India's schools of thought). Additionally, Pakistani planners have increasingly relied on nuclear weapons to compensate for Pakistan's conventional military inferiority compared to India.[75] Since conventional military asymmetry is only likely to worsen, Pakistan may fear that an FMCT might leave Pakistan without sufficient nuclear capability to deter India in future decades even if Islamabad is comfortable with its nuclear stockpile today.

Opposition to an FMCT is also one of the few ways that Pakistani diplomats can express their continuing grievance about the U.S.-India civil nuclear agreement, which permitted civilian technology transfers to India despite its outlier status in the Nuclear Nonproliferation Treaty. Overwhelmingly, Pakistani elites opposed the U.S.-India nuclear deal, and ordinary Pakistanis that were aware of the deal held similar views.[76] To grant Pakistan access to technology controlled under the Nuclear Suppliers Group guidelines, Pakistani officials continue to seek a similar exemption to that granted to India despite lack of support in the international community.

Some Pakistani Realists may ultimately be persuaded that the additional deterrence benefits of an ever-expanding arsenal do not merit the opprobrium that comes from halting a major initiative like the FMCT on the global nonproliferation agenda. Until that point, however, Realists and Nationalists will support "more time" for Pakistan to modernize its arsenal, even as Globalists are again disappointed in this course.

There is lip service paid across the worldviews in favor of global nuclear disarmament, but this is mostly disingenuous for all except the Globalists. It postpones the need for any additional constraints on Pakistan until the established nuclear powers—principally the United States and Russia—achieve substantial drawdowns in their nuclear capabilities. As a negotiating stance, it also calls attention to the deep unfairness at the heart of the nonproliferation regime. Pakistani elites might believe that global disarmament is fairer than the currently unequal nuclear order, but in practice Realists and Nationalists would likely be unwilling to lose the benefits of nuclear deterrence in a truly nuclear-disarmed world.

Realists and Nationalists might disagree over the issue of arms control and confidence-building measures with India. Some Pakistani Realists may be willing to embrace arms control as a means to slow unnecessary arms expenditures if reasonable accommodations could be found. Pakistani officials proposed a raft of measures designed to create a "strategic restraint regime" in South Asia shortly after the 1998 tests.[77] Given what they see as likely Indian malfeasance in the future, Pakistani Nationalists, on the other hand, fear that such measures might constrain Pakistani freedom of action while creating a false sense of security. This split within the elite has not been tested, largely because India faces structural difficulties in agreeing to arms control measures with Pakistan that do not involve China.[78]

How Important Is Nuclear Energy to the Future of Pakistan?

The final area where there is widespread agreement among elite and ordinary Pakistanis is the conviction that nuclear energy should play an important role in meeting Pakistan's energy needs. Pakistan is a net importer of energy and faces considerable difficulty in paying for foreign oil and natural gas, as well as generating and distributing energy throughout Pakistan. Elites worry about the cost of such imports in hard currency and the loss in economic production that accompanies power shortages. Ordinary Pakistanis have less abstract worries: the immediate discomfort that comes from insufficient heat or erratic air-conditioning during cold winters and warm summers that overwhelm the domestic energy infrastructure. This is true even though nuclear energy will never constitute a substantial source of production in Pakistan. The Pakistan Atomic Energy Commission optimistically projects nuclear energy to provide no more than 8 percent of installed capacity in 2022, up from around 4 percent in 2012.[79] Despite this modest contribution, nuclear energy is attractive given the variable capacity rates of the main renewable alternatives (wind and solar), the strong seasonal variation in hydroelectric generation, and the currency requirements for fossil fuel imports.

Even immediately after the Fukushima nuclear disaster discussed in chapter 4, a majority of Pakistanis reported favorable views about nuclear energy, with only about a quarter of Pakistanis recording opposition to nuclear energy for electricity production. This gap between those in favor and those opposed to nuclear energy is substantial in Pakistan, which had the sixth most net favorable views on nuclear energy among forty-seven countries polled in 2011.[80] Nuclear Globalists correctly question the economics and safety of nuclear energy in Pakistan. They worry about unproven designs and note the comparative success of non-nuclear renewable energy production elsewhere in the developing world.[81] Their arguments are detailed and well-argued, but there is little evidence to date they have been successful in persuading Pakistani elites or the majority of the public.

Unless public sentiment shifts, the PAEC intends to more than double the number of nuclear power reactors operating in Pakistan by 2020. As of autumn 2015, Pakistan operates the Karachi Nuclear Power Plant-1 (KANUPP-1) pressurized heavy water reactor in Sindh province, which has a modest 137 megawatt capacity, and two pressurized water reactors at the Chashma Nuclear Power Complex in Punjab with 325 and 340 megawatt capacities. KANUPP began operations in 1972, while Chashma-1 and -2 were ready for operation in 2000 and 2011 respectively. That slow pace of building is no longer evident in PAEC's plans. To these existing reactors, Pakistan is already adding two new reactors at Chashma, currently under construction and scheduled to become operational in 2016 and 2017, with similar capacities as the earlier Chashma units. Moreover, two new 1100-megawatt reactors are planned near the KANUPP-1 site. Construction of the first reactor began in August 2015, and both new reactors are scheduled for completion as early as 2019.[82]

Pakistan's current and planned nuclear energy production depends on Chinese assistance. Chinese firms built the existing Chashma reactors and are constructing the new ones. Previously, China was unwilling to build reactors in Pakistan that had more generative capacity but for which China did not have outright ownership of the intellectual property rights (IPR) associated with the design. The newer units being planned for Karachi, however, involve a new design that China asserts does not contain foreign IPR.[83] Thus the newer units have nearly three times the production capacity as the older Chashma designs. Except for KANUPP-1, which is fueled from indigenous sources, China also provides all of the nuclear fuel for the Chashma units and has agreed to do so for the lifetime of the new units.[84] The importance of Chinese assistance is self-evident, and Pakistani long-term energy planning documents repeatedly reference the fact that "China is presently the single supplier of nuclear technology and equipment for Pakistan."[85]

CONCLUSION

There is a consensus in Pakistan in favor of nuclear weapons as a means to protect Pakistan from its foes and in favor of nuclear energy as one component of a broader solution to its energy needs. That consensus is to some extent a deliberate construct. Pakistan's uniformed nuclear custodians follow the debate closely and nudge participants in preferred directions. Despite that, there is widespread Pakistani pride in their nuclear accomplishment and conviction that nuclear weapons possession is necessary so long as India also possesses them. Nuclear Globalists worry that Pakistan's ever-increasing arsenal is expensive and dangerous and that Pakistan's nuclear energy plans are uneconomical, unsafe, and unwise. Outliers in this volume's set of case studies, Pakistan's Globalists remain a vocal, well-informed minority. Dominated by Realist and Nationalist voices and bolstered by the military's resources, the nuclear consensus in Pakistan is likely to prove enduring.

NOTES

1. Feroz Hassan Khan, "Comparative Strategic Culture: The Case of Pakistan," *Strategic Insights* 4, no. 10 (October 2005); and Naeem Salik, *The Genesis of South Asian Nuclear Deterrence* (New York: Oxford University Press, 2009), 5–6.

2. Dennis Kux, *The United States and Pakistan, 1947–2000: Disenchanted Allies* (Washington, D.C.: Woodrow Wilson Center Press, 2001), 212.

3. Zafar Khan, *Pakistan's Nuclear Policy: A Minimum Credible Deterrence* (New York: Routledge, 2014), 132, 139; Munir Akram, "Gambling against Armageddon," *Dawn*, October 26, 2014.

4. For comparisons with Israel, see Feroz Hassan Khan, "Comparative Strategic Culture"; and Salik, *The Genesis of South Asian Nuclear Deterrence*, 5–6.

5. Naeem Salik and Kenneth N. Luongo, "Challenges for Pakistan's Nuclear Security," Arms Control Association: *Arms Control Today*, March 2013, http://www.armscontrol.org/act/2013_03/Challenges-for-Pakistans-Nuclear-Security.

6. Salmaan Taseer, governor of Punjab, Pakistan, Twitter post, December 9, 2010, 7:10 a.m., https://twitter.com/SalmaanTaseer/.

7. Zahir Kazmi, "Security and the FMCT," *Dawn*, September 20, 2011; Asif Ezdi, "Nuclear Bluff," *The News International*, February 27, 2012.

8. "On Eve of Elections, a Dismal Public Mood in Pakistan: Rising Concerns about the Taliban," Pew Global, May 7, 2013, 25.

9. There are Realists in the Pakistan Tehreek-e-Insaf political party, but their influence is hard to see. Ashraf Jehangir Qazi, for instance, appears to be a Realist and at points has been part of PTI's leadership.

10. "The Gujarat riots . . . had the perverse effect of strengthening Pakistan's resolve to resist what it views as Hindu chauvinism." Stephen Cohen, *India: Emerging Power* (Washington, D.C.: Brookings Institution, 2001), xiv.

11. Shireen Mazari, "Our 'Ally from Hell': A Rejoinder from U.S. Hawks," *The News International*, November 10, 2011. For years, Mazari researched and led government-funded think tanks in Islamabad. More recently, she has served as spokesperson for the quintessentially Nationalist Pakistan Tehreek-e-Insaf political party.

12. Ikram Junaidi, "Voices Raised against Government's Policy Shift toward Nuclear Energy," *Dawn*, February 3, 2014.

13. Though A. Q. Khan relied on resources from politicians and the military, he was regularly chastised by the military for saying statements contrary to Pakistani declaratory policy.

14. Zulfikar Ali Bhutto, *If I Am Assassinated*, 151, http://bhutto.org/Acrobat/If-I-am-assassinated-by-Shaheed-Bhutto.pdf.

15. Quoted in "Amembassy Islamabad to Secretary of State Washington, D.C.," Cable Islamabad 7624, "Nuclear Reprocessing," August 6, 1978, http://www.gwu.edu/~nsarchiv/nukevault/ebb333/doc11.pdf.

16. Haider Nizamani, "Rewriting Third World Security: A Comparative Study of Nuclear Discourse in Pakistan and India" (PhD diss., University of British Columbia, Vancouver, Canada, December 1997), 136.

17. In May 2014, I learned of the phrase when I passed a large gathering in Islamabad and asked my driver what was occurring. He told me it was for Youm-e-Takbeer.

18. For an example of how a technocrat invokes nuclear imagery, see Jeffrey Goldberg, "The Education of a Holy Warrior," *New York Times Magazine*, June 25, 2000.

19. A. Q. Khan, "Restricted Areas of Science and Technology and Ways to Develop Such Technologies in the Muslim World" (speech, International Conference on Science in Islamic Polity in the Twenty-First Century, Islamabad, Pakistan, March 26–30, 1995), reprinted in *Dr. A. Q. Khan on Science and Education*, eds. S. Shabbir Hussain and Mujahid Kamran (Lahore: Sang-e-Meel, 1997), 169–182.

20. To what extent such Nationalist sentiment was also self-serving, given the later revelations about A. Q. Khan's illicit nuclear commerce with both Muslim countries and North Korea, is an open question.

21. Quoted in Feroz Hassan Khan, *Eating Grass: The Making of the Pakistani Bomb* (Stanford, Calif.: Stanford University Press, 2012), 383.

22. David Albright and Holly Higgins, "A Bomb for the Ummah," *The Bulletin of the Atomic Scientists* 59, no. 2 (March-April 2003): 49–55.

23. "Sattar Wants Pakistan to Join CTBT before India," *Kyodo News International*, January 4, 2000.

24. Ijaz Hussain, "Is the U.S. after Our Nukes?" *Daily Times*, June 3, 2009.

25. Niloufer Siddiqui, "Who Do You Believe? Political Parties and Conspiracy Theories in Pakistan" (Midwest Political Science Association annual conference, Chicago, Ill., April 3–6, 2014).

26. Mirza Aslam Beg, "Dynamics of Counter-Blasts," *The News International*, May 30, 1998.

27. C. Christine Fair, "Lashkar-e-Tayiba and the Pakistani State," *Survival* 53, no. 4 (2011): 1–23.

28. Hasan Mansoor, "JuD Terms PM Attending Modi's Oath-Taking 'Betrayal' of Kashmir Cause," *Dawn*, May 26, 2014.

29. "Relations with India: Govt Urged to Get Tough on Kashmir Issue," *Tribune Express*, May 29, 2014.

30. Carlotta Gall, Pir Zubair Shah, and Eric Schmitt, "Seized Phone Offers Clues to Bin Laden's Pakistani Links," *New York Times*, June 23, 2011.

31. In 2014, 3 percent of Pakistani respondents identified the United States as the country that Pakistan could most rely upon as a dependable ally in the future; 57 percent said China. In contrast, 38 percent identified the United States as the greatest threat to Pakistan, while no respondents identified China. Pew Global Attitudes Project, Spring 2014 survey, http://www.pewglobal.org/question-search/?qid=1865&cntIDs=@36-@36.5-&stdIDs.

32. Husain Haqqani, *Pakistan: Between Mosque and Military* (Washington, D.C.: Carnegie Endowment for International Peace, 2005), 280–281; Brian Cloughley, *A History of the Pakistan Army* (New York: Oxford University Press, 2000), 309–313; and Dennis Kux, *The United States and Pakistan, 1947–2000: Disenchanted Allies* (Washington, D.C.: Woodrow Wilson Center Press, 2001), 312–313.

33. "Interview with Gen. Mirza Aslam Beg," *NBC News*, February 9, 2004, http://www.nbcnews.com/id/4223687/ns/nbc_nightly_news_with_brian_williams/t/interview-gen-mirza-aslam-beg/#.U7BWV41dVqk.

34. David Rohde, "Nuclear Inquiry Skips Pakistani Army," *New York Times*, January 30, 2004.

35. Pervez Hoodbhoy, ed., *Confronting the Bomb: Pakistani and Indian Scientists Speak Out* (New York: Oxford University Press, 2013).

36. Samina Ahmed and David Cortright, eds., *Pakistan and the Bomb: Public Opinion and Nuclear Options* (Notre Dame, Ind.: Notre Dame University Press, 1998), esp. Appendix A.

37. Mehtab S. Karim, "Figuring it Out," *Newsline*, June 1998, 34A; see also Gallup Pakistan, "Gallup Pakistan Poll Findings on Strong Support for Nuclear Capability," Gallup Pakistan, January 25, 1986, http://www.gallup.com.pk/Polls/Press%20release%2025-1-1986.pdf.

38. A. C. Nielsen for the U.S. Information Agency, "USIA Poll #2001-I200131: Domestic Issues/Institutions/Government/Security/International Issues/Relations with U.S.," September 13–October 8, 2001, available via the Roper Center for Public Opinion Research archives.

39. Haider K. Nizamani, "Whose Bomb Is It Anyway? Public Opinion and Perceptions about Nuclear Weapons and Policy in the Post-Explosions Phase in Pakistan" (Washington, D.C.: Social Science Research Council, n.d. [survey conducted June-August 2000]), accessed March 4, 2003, http://www.ssrc.org/programs/gsc/gsc_activities/nizamani.page.

40. Ibid..

41. Samina Ahmed, "The Nuclear Debate," *Newsline*, June 1998, 30–32.

42. Zia Mian, "The Coming of the Atomic Age to Pakistan," in *Confronting the Bomb*, ed. Hoodbhoy, 48.

43. Ibid., 58.

44. The Pakistan Army's Command and Staff College's official history records that a visit by "a special United States Nuclear Warfare Team" in 1957 led to syllabus changes. For more, see "History Command and Staff College, Quetta," Command and Staff College, Quetta, Pakistan, accessed October 1, 2015, http://armystaffcollege.gov.pk/Hist_college.php.

45. Mian, "The Coming of the Atomic Age to Pakistan," 56–60; and Matthew Fuhrmann, *Atomic Assistance: How "Atoms for Peace" Programs Cause Nuclear Insecurity* (Ithaca, N.Y.: Cornell University Press, 2012).

46. Interview with Agha Shahi, Islamabad, Pakistan, June 19, 2005; A. Q. Khan, "Bhutto, GIK, and Kahuta," *The News International*, July 29, 2009. See also Feroz Hassan Khan, *Eating Grass*, 150.

47. Interview with K. M. Arif, Rawalpindi, Pakistan, June 20, 2005; A. Q. Khan, "Bhutto, GIK, and Kahuta."

48. Interview by Feroz Hassan Khan and Peter Lavoy with Mirza Aslam Beg, Pakistan, 2006.

49. According to Saeed Shafqat, Bhutto largely ceded national security decision-making to the Army. See *Civil-Military Relations in Pakistan: From Zulfikar Ali Bhutto to Benazir Bhutto* (Boulder, Colo.: Westview Press, 1997), 227.

50. Feroz Hassan Khan, *Eating Grass*, 325.

51. A. Q. Khan says that Zia made Lt. Gen. Ali Zamin Naqvi the security advisor to the PAEC and Khan Research Laboratories and also recalls then-Brig. Zahid Ali Akbar Khan overseeing construction for the laboratories. A. Q. Khan, "Bhutto, GIK, and Kahuta."

52. The best evidence for the military's sidelining of the civilian Pakistan Prime Minister Nawaz Sharif in the conflict and nuclear signaling comes from an insider account of Sharif's meeting with President Bill Clinton in Washington, D.C., as the conflict was unfolding. See Bruce Reidel, *American Diplomacy and the 1999 Kargil Summit at Blair House Policy Paper Series* (Philadelphia, Pa.: Center for the Advanced Study of India, 2002). Three months later, Sharif was deposed by a military coup.

53. Feroz Hassan Khan, *Eating Grass*, 314.

54. The Pakistan Atomic Energy Commission Ordinance, no. 17, May 29, 1965, http://www.iaea.org/inis/collection/NCLCollectionStore/_Public/21/072/21072764.pdf; Pakistan Atomic Energy Commission, "History of PAEC," PAEC, http://www.paec.gov.pk/paec-hist.htm (accessed January 1, 2009).

55. A. Q. Khan refers to ERL as a separate organization as early as 1976, though Feroz Hassan Khan's history says the separation from PAEC occurred after Zia's takeover in 1977. A. Q. Khan, "Bhutto, GIK, and Kahuta" and Feroz Hassan Khan, *Eating Grass*, 151.

56. Pakistan Space and Upper Atmosphere Research Commission (SUPARCO), "History," SUPARCO, accessed June 14, 2013, http://www.suparco.gov.pk/pages/history.asp.

57. Feroz Hassan Khan, *Eating Grass*, 185–186, 239; also SUPARCO, "History."

58. Christopher Clary, "The A. Q. Khan Network: Causes and Implications" (MA thesis, Naval Postgraduate School, Monterey, Calif., December 2005), 30–35.

59. Christopher Clary, "Policy Brief: How the Pakistan Military Learned to Love the Bomb," Rising Powers Initiative, September 2013, http://www.risingpowersinitiative.org/wp-content/uploads/PolicyBrief_Sept2013_Pakistan1.pdf.

60. Waqas Naeem, "Power Crisis: 'Untested' Proposed N-Plants Worry Top Nuclear Scientists," *Tribune Express*, February 3, 2014.

61. Pakistan Press International, "Keamari Town's Nuclear Power Projects Irk Fisherman," *Tribune Express*, November 29, 2013; Faiza Ilyas, "Call for EIA of Nuclear Plants Project," *Dawn*, January 30, 2014.

62. "PPP Wants Debate in Parliament on Karachi N-Reactors," *Dawn*, February 6, 2014.

63. Faiza Ilyas, "Nuclear Plants Seen as Answer to City's Water, Power Needs," *Dawn*, February 21, 2014; Faiza Ilyas, "Safety Measures Incorporate Lessons Learnt from Fukushima Disaster," *Dawn*, February 17, 2014; Faiza Ilyas, "'Nuclear Plant Project Okayed after Secret EIA Meeting,'" *Dawn*, February 3, 2014.

64. Gallup Pakistan, "Gallup Pakistan Poll Findings on: Pakistan Is Concerned about India's Nuclear Threat," Gallup Pakistan, March 14, 1987, http://gallup.com.pk/bb_old_site/Polls/14-3-1987.pdf.

65. A. Q. Khan, "My Nuclear Manifesto," *Newsweek*, May 16, 2011. This article appeared two weeks after the U.S. raid on the bin Laden compound in Abbottabad, Pakistan.

66. C. Christine Fair, review of *Eating Grass: The Making of the Pakistani Bomb*, by Feroz Hassan Khan, *Journal of Strategic Studies* 36, no. 4 (2013): 625–627.

67. "PM Sharif in Karachi, Inaugurates KANUPP-2 Power Project," *GEO TV*, August 20, 2015, http://www.geo.tv/article-194757-PM-Sharif-in-Karachi-inaugurates-KANUPP-2-power-project.

68. Saeed Shah, "China Agrees to Sell Pakistan Two More Nuclear Reactors," *Wall Street Journal*, October 15, 2013.

69. Pew Global Attitudes Project, Spring 2006 survey, http://www.pewglobal.org/question-search/?qid=929&cntIDs=@36-@36.5-&stdIDs.

70. Pew Global Attitudes Project, Spring 2013 survey, http://www.pewglobal.org/question-search/?qid=1666&cntIDs=@36-@36.5-&stdIDs.

71. Pew Global Attitudes Project, surveys from Spring 2006 to Spring 2012,, http://www.pewglobal.org/question-search/?qid=747&cntIDs=@36-@36.5-&stdIDs.

72. Pew Global Attitudes Project, Spring 2010 and Spring 2012 surveys.

73. Pew Global Attitudes Project Spring 2010 and Spring 2012 surveys.

74. For example, Hans M. Kristensen and Robert S. Norris, "Worldwide Deployments of Nuclear Weapons," *Bulletin of the Atomic Scientists* 70, no. 5 (2014): 97.

75. Christopher Clary, "The Future of Pakistan's Nuclear Weapons Program," in *Strategic Asia, 2013–2014*, eds. Ashley Tellis, Abraham Denmark, and Travis Tanner (Seattle, Wash.: National Bureau of Asian Research, 2013).

76. Pew Global Attitudes Project, Spring 2006 survey.

77. Feroz Hassan Khan, *Eating Grass*, 296–300.

78. In 1988, Benazir Bhutto and Rajiv Gandhi did reach a landmark agreement to not attack each other's nuclear installations and it has stood the test of time.

79. Syed Yusuf Raza, "Drivers and Impediments: Regional Cooperation on the Way to Sustainable Nuclear Energy Systems," IAEA, July 31, 2012, https://www.iaea.org/INPRO/4th_Dialogue_Forum/DAY_3_01_August-ready/2._-_DG-C3-4-31-07-2012.pdf. This is the most recent (as of early 2016) forecast available that includes both nuclear capacity and an estimate of total energy production capacity in Pakistan.

80. WIN-Gallup International, "Impact of Japan Earthquake on Views about Nuclear Energy: Findings from a Global Snap Poll in 47 Countries by WIN-Gallup International," March 21–April 10, 2011, Gallup Pakistan, April 10, 2011, http://www.gallup.com.pk/JapanSurvey2011/PressReleaseJapan.pdf.

81. Pervez Hoodbhoy, Zia Mian, and A. H. Nayyar, "The Nuclear Shadow over Karachi," *Newsweek Pakistan*, March 17, 2014.

82. The capacity of Chashma-2 is frequently stated as 325 MW. Above, I use 340 MW, consistent with Mohsin's 2014 presentation. The World Nuclear Association suggests that

upgrades to the plant have increased its capacity to 330 MW. Groundbreaking began at the KANUPP-2 site in November 2013, but the first "concrete pour" for the plant did not take place until August 2015. For more, see M. Mohsin, Pakistan Atomic Energy Commission, "Knowledge Management (KM) Risk Assessment of Critical Knowledge Loss in an Organization with Expanding Nuclear Power Program" (presentation, International Conference on Human Resource Development for Nuclear Power Programmes, Vienna, Austria, May 12–16, 2014), http://www-pub.iaea.org/iaeameetings/cn215p/Thursday/IAPs/Morning/Moshin.pdf; World Nuclear Association, "Nuclear Power in Pakistan," World Nuclear Association, last updated August 2015, http://www.world-nuclear.org/info/Country-Profiles/Countries-O-S/Pakistan/; "China Commits $6.5 Billion for Nuclear Power Project in Karachi," *The News International* [Pakistan], December 24, 2013; Leader of the Pakistan Delegation, "Statement by the Leader of the Pakistan Delegation" (statement, 59th IAEA General Conference, Vienna, Austria, September 14–18, 2015), https://www.iaea.org/sites/default/files/pakistan2015.pdf.

83. Mark Hibbs, "Chinese Chashma Poker Chip?" *Armscontrolwonk.com*, March 23, 2013, http://hibbs.armscontrolwonk.com/archive/1526/chinese-chashma-poker-chip.

84. World Nuclear Association, "Nuclear Power in Pakistan."

85. Economic Advisory Council, Ministry of Finance, Government of Pakistan, *Integrated Energy Plan, 2009–2022* (Islamabad: Government of Pakistan, March 2009): 81–82.

10

Conclusion

Policy Implications of Changing Domestic Debates

Mike Mochizuki

The Asian countries examined in the preceding chapters represent a diversity of cases. Some have democratic political systems (India, Indonesia, Japan, the Philippines, South Korea, and Taiwan), while others are governed by authoritarian regimes (China and Vietnam). Pakistan and Thailand have oscillated between authoritarian and quasi-democratic politics. The country cases also vary in terms of their level of economic development. Japan, South Korea, and Taiwan are advanced economies, while India, Indonesia, Pakistan, the Philippines, and Vietnam are relatively low-income countries. China and Thailand fall in the middle in terms of per capita income. The countries also differ regarding their relationship to the U.S. security alliance network. Japan, the Philippines, South Korea, and Thailand are formal treaty allies of the United States. Both Pakistan and Taiwan had formal defense pacts with the United States, but now their security relations with Washington are less formalized. Although the United States and India are not formally allies, Washington and New Delhi have promoted security cooperation since 2004 on an ad hoc basis. Security interactions between China and the United States have become increasingly competitive; Beijing is increasingly suspicious that the Washington-led alliance network is being directed against China, and Washington sees China as being more assertive against neighboring countries in East Asia.

The states included in this volume vary in terms of their nuclear policies as well. China, India, Japan, and South Korea have relatively well-developed nuclear energy programs, while the Pakistan and Taiwan programs are more modest in scale. Vietnam's nuclear energy program is just about to get started, while Indonesia, the Philippines, and Thailand have put their consideration of nuclear energy on hold. India and Pakistan have joined China in the Asian club of nuclear weapon states. South Korea and Taiwan initiated clandestine nuclear weapons programs but abandoned them under U.S. pressure. Notwithstanding its formidable technological capabilities in the nuclear field, Japan has chosen to forgo nuclear weapons.

Despite their diversity, the nations included in this study manifest some common features. In all of the cases, decision-making on nuclear issues was on the whole an elite process; to the extent there was pluralism, it usually involved competition among agencies within the governmental system. This elitist characteristic applied to both democratic and authoritarian states. Although policymaking in the nuclear sector tended to be top down, civil society and local communities in some states did resist the construction of nuclear plants. Japan was exceptional given the critical role of civil society and opposition parties in compelling a conservative government to commit to a "no nuclear weapons" policy.

After the Fukushima disaster, the techno-nationalist elite enclaves for nuclear policymaking found in many Asian states have weakened. In Japan, the mobilization of anti-nuclear sentiment has challenged the so-called "nuclear village," the collusive network of interests that had heretofore managed the nation's extensive nuclear energy programs. In Taiwan, growing public opposition has compelled previous proponents of nuclear energy to change their positions. Similarly, civil society groups have contributed to a pullback from nuclear energy aspirations in Indonesia, the Philippines, and Thailand. Nuclear policymaking in India and South Korea has not been immune from this trend. Regarding this tendency toward greater openness and pluralization, Pakistan is the main outlier among the countries examined in this volume.

Globalist, Realist, and Nationalist worldviews and discourses found representation in virtually all of the countries examined in this volume. Despite widespread concerns about both horizontal and vertical nuclear proliferation in Asia, one conclusion of this volume is that in every country except for Pakistan, the center of gravity of domestic debates regarding nuclear issues has shifted toward a Realist or a Realist-Globalist worldview. Only in Pakistan is Nationalist discourse a primary driver of nuclear policy. This finding is indeed remarkable given the geopolitical suspicion that has accompanied the regional power transition (especially China's ascendancy) and the rise of so-called "resource nationalism" in the context of increasing demand for energy. This suggests the relative insulation of nuclear policy from the whims of populist Nationalism. These overall discourse trends notwithstanding, however, there were significant cross-national variations in how different worldviews and discourses mattered for cleavage patterns on specific dimensions of nuclear policy: nuclear energy, nuclear weapons, and nonproliferation. The United States has played an important role in shaping both the discourse and policies in all of the Asian countries, which in turn have implications for future U.S. foreign policy toward the region.

NUCLEAR ENERGY

As seen in chapter 2, there appears to be a strong consensus in China in favor of developing nuclear power. Rather than a debate among Globalists, Realists, and

Nationalists, each of these schools of thought provide different rationales for the pursuit of nuclear energy. Globalists in China favor nuclear power as a way to reduce greenhouse gas emissions and thereby help the global effort to address the challenge of climate change. Realists pragmatically see nuclear energy as a way to meet the country's growing energy demand, diversify energy sources, and enhance energy security. Nationalists argue for nuclear energy as a means of achieving energy self-reliance. Although nuclear energy policy in China has been determined through an elite-centered process that is largely insulated from public discussion, there has been a modest change after the Fukushima accident. Before Fukushima, local communities tended to welcome the construction of nuclear facilities because of potential economic benefits; after Fukushima, residents near the sites of planned nuclear facilities have voiced concerns about safety. This has challenged the Chinese government to enhance safety measures and transparency, improve communication, and foster public understanding about nuclear issues.

In Japan as seen in chapter 4, a combination of techno-Nationalism and a Realist interest in energy security drove the country's nuclear energy ambitions. Globalist discourse was also a factor in two ways. First, Japan served as a critical case for the U.S. global vision of the peaceful use of atomic power. Second, with concerns about climate change, some environmentalists began to support nuclear energy as a way to reduce Japan's greenhouse gas emissions. The Fukushima disaster, however, transformed the Japanese debate by weakening the position of environmental Globalists who favored nuclear energy, and anti-nuclear Globalists gained traction by emphasizing the merits of renewable energy sources as well as the dangers of nuclear power. In addition to invoking energy security and economic necessity arguments, Realists in Japan have sought to incorporate Globalist discourse on behalf of nuclear energy by highlighting Japan's continuing contribution to nuclear technology and safety.

As in the case of Japan, chapter 5 shows how a Realist calculation to enhance energy security motivated South Korea to pursue nuclear energy. And again similar to the Japanese case, local community concerns about nuclear safety have compelled the South Korean government to engage in public education campaigns and to monitor public opinion in order to get residents to acquiesce to the construction of nuclear facilities. The Fukushima disaster has weakened public support for an expanded nuclear energy program. Although there has not been an across-the-board shutdown and review of existing nuclear plants as in Japan, the South Korean government has had to curtail significantly its ambitious program to have nuclear energy generate over 40 percent of the country's electricity.

The debate about nuclear energy in Taiwan explored in chapter 6 has been contentious with large-scale public demonstrations, extensive media coverage, and partisan manipulation. This was the case before Fukushima and has become even more so after Fukushima. Moreover, this debate cannot be neatly defined according to Globalist, Realist, and Nationalist modes of thinking. There are proponents and opponents of nuclear energy in each of these three perspectives. Some Globalists align with anti-nuclear movements in other countries, while other Globalists support nuclear power as a means to address the challenge of global climate change and to compete more effectively in the global economy. Some Realists believe that energy conservation and renewable energy sources can meet Taiwan's energy needs without nuclear power, while other Realists insist that without nuclear power, Taiwan's electricity costs would increase. Finally, while many Nationalists see nuclear energy and even reprocessing as a way to achieve greater energy security, some Nationalists oppose nuclear energy to protect the "purity and integrity" of the homeland from nuclear contamination and waste. The effect of this messy and highly politicized debate has been to block the expansion of nuclear power—and perhaps even to have Taiwan abandon nuclear energy.

Although Vietnam is economically smaller and less developed than China, chapter 7 demonstrates that elite discourse about nuclear energy exhibits similarities with that of China. As in the Chinese case, Globalists, Realists, and Nationalists in Vietnam do not take different sides regarding the pursuit of nuclear energy. The three perspectives instead offer complementary rationales for nuclear power: Nationalists want to enhance their country's standing in the nuclear domain and to increase energy self-sufficiency; Realists aim to strengthen ties with the major nuclear states; and Globalists seek to show the world its will to follow international rules and norms and to help address the problem of global climate change. And as in China, Vietnamese policymakers have had to address issues of safety and transparency after Fukushima, but the motivation to develop nuclear energy remains strong.

In most of the other Southeast Asian countries discussed in chapter 8, however, the initial consideration of nuclear power to address long-term energy needs subsided in the wake of Fukushima. For example, before Fukushima, Thailand had plans to construct five nuclear power plants, and the Thai public was generally supportive of this policy despite anti-nuclear protests by environmental groups. However, the Fukushima accident triggered a sharp reversal of public opinion. Moreover, a reenergized environmental movement in the context of political uncertainties compelled the Thai Ministry of Energy, the leading proponent of nuclear energy, to put their plans on hold in favor of alternative non-nuclear energy sources. In the Philippines, nuclear power advocates tried to overcome myriad financial and safety issues to put into operation the Bataan Nuclear Power Plant, but the Fukushima disaster tipped the political balance enough to halt this plan. Democratization in Indonesia allowed anti-nuclear groups to mobilize and restrain the nuclear energy ambitions of technocrats.

In chapter 3, one can see Indian discourse about nuclear energy initially exhibited features that differed substantially from the above cases. Before the 1990s, the Nationalist worldview prevailed in discussions about nuclear energy with Globalist and Realist rationales virtually nonexistent. The international technology sanctions imposed on India after the 1974 atomic test explain in part the predominance of Nationalist discourse in the nuclear energy field. But after India emerged from its nuclear commercial isolation with the improvement of U.S.-India relations in 2002, Realist and Globalist arguments moved to the forefront with the former highlighting the technological benefits of cooperating with the United States on nuclear energy and the latter embracing some international controls as part of India's becoming more integrated into global regimes. An unlikely coalition of pro-nuclear Nationalists and anti-nuclear Leftists and Rejectionists, however, mobilized to oppose the U.S.-India nuclear deal. Although debate about nuclear energy cooperation with the United States has been contentious, at least at the elite level, discussion about the development of nuclear power itself as an energy source has not been acrimonious. Realists, Nationalists, and even Globalists all seem to favor nuclear energy, and the main pockets of opposition tend to be the local populations who feel threats to their livelihood because of nuclear plant construction.

As in China, there is a robust consensus in Pakistan for pursuing nuclear energy. Beyond the role that civilian atomic energy policy can play in Pakistan's nuclear weapons program, chapter 9 observes that power shortages and concerns about the costs of importing energy have engendered strong support for nuclear power at both the elite and popular levels. According to a cross-national survey of views about nuclear energy taken soon after the Fukushima disaster, Pakistan ranked sixth out of forty-seven countries polled regarding positive views of nuclear energy. Although Globalists raise questions about safety and the costs of nuclear plants, their views have little influence.

NUCLEAR WEAPONS

In China, a robust elite consensus informed by Realist thinking exists in favor of a minimum deterrent capability and the maintenance of a No First Use (NFU) doctrine. Although some Chinese policy analysts question the NFU doctrine, chapter 2 notes this continues to be a minority view. The greater priority is on modernizing nuclear forces and insuring the survival of a reliable second-strike capability. Nationalist voices might argue for achieving numerical strategic parity with the nuclear superpowers (the United States and Russia), but the Chinese strategic community tends to be more pragmatic. Pursuit of numerical parity is unthinkable given the staggering expense of such a path and the Chinese leadership's focus on economic development. Although there is little or no public debate about nuclear strategy, the general public appears to embrace the existing policy.

In contrast to the absence of open debate about nuclear weapons in China, the discussions in India described throughout chapter 3 have become increasingly contentious. China's atomic tests in 1964 triggered a debate in India about nuclear weapons, but Nationalist and Globalist perspectives initially dominated the discourse. Although some insisted that India needed to develop nuclear weapons after China's nuclearization, more influential were those who counseled restraint. While acquiring a "recessed deterrence" capability by testing "peaceful nuclear explosives," India used both Nationalist and Globalist arguments to insist that the existing nuclear weapon states should be true to the principles embedded in the Nuclear

Nonproliferation Treaty (NPT). By the mid-1990s, however, Realist critiques of the predominant thinking emerged by pointing out that India's "recessed deterrence" was losing credibility. Nationalists joined the fray by criticizing the hypocrisy of the nuclear weapon states. With the breakdown of the Nationalist-Globalist consensus, Realism and Nationalism worked in tandem to support the 1998 nuclear tests and India's explicit declaration as a nuclear power. Nevertheless, significant differences between Nationalists and Realists now exist regarding nuclear strategy and the desirable size of India's nuclear arsenal.

Whereas India's nuclear weapons discourse was driven by a consensus between Nationalism and Globalism, Nationalism has dominated Pakistan's discourse as examined in chapter 9. For Nationalists, nuclear weapons are the only way to guarantee Pakistan's sovereignty as well as security in the face of what they portray as Indian and U.S. machinations. Although there are pragmatic Realists in the military, they tend to echo the Nationalist rhetoric in order to protect military interests and secure budget resources. Realist politicians in the mainstream civilian parties who might be open to greater nuclear restraint are in turn put on the defensive by Pakistan's powerful military that has spurned an NFU policy and adopted risky nuclear strategies. Since the early 2000s, the emergence of nuclear Islamism has reinforced Pakistan's commitment to nuclear weapons by arguing that Pakistan's natural allies are countries in the Islamic world and by even suggesting that Pakistan should share its nuclear technology with some Islamic states. The Globalist voice that might restrain Pakistan's nuclear weapons program is feeble, and many who use Globalism to advocate nuclear restraint tend to be Pakistanis who live in the West.

Although Japan has had the technological and economic capability to become a nuclear weapon state, chapter 4 shows a tacit coalition between anti-nuclear Globalists and security Realists have kept Tokyo committed to the three non-nuclear principles, even though these two schools of thought have diverged in terms of how vigorously to promote global nuclear disarmament and U.S.-Japan defense cooperation. While embracing the non-nuclear principles, some Realists support maintaining a nuclear hedge through an advanced nuclear energy program. Although Nationalist voices in favor of a nuclear weapons option have been a perennial feature of the Japanese debate, this school of thought continues to remain the minority position despite North Korea's nuclearization and the rise of Chinese military power and assertiveness.

During the 1960s and 1970s, a combination of Nationalist and Realist calculations motivated both Taiwan (chapter 6) and South Korea (chapter 5) to clandestinely pursue a nuclear weapons capability. Strong U.S. pressure ultimately forced both Taipei and Seoul to abandon the nuclear option. The Nationalist motivation for nuclearization has weakened considerably in both Taiwan and South Korea. In Taiwan in the early twenty-first century, Nationalist voices focus more on cruise missile and ballistic missile programs. In South Korea, a blended Globalist-Realist faction of defense hawks advocate the reintroduction of U.S. tactical nuclear weapons to the Korean peninsula, but hardly anyone supports reviving a nuclear weapons option to counter North Korea's nuclear weapons program. As in the Japanese case, Realists and Globalists now dominate the debate in Taiwan and South Korea, thereby blocking a path toward nuclear weapons.

Although Vietnam's nuclear energy program is still in its infancy, there are no signs that a confluence of Nationalist and Realist calculations might steer Vietnam toward a clandestine nuclear weapons program as it did in South Korea and Taiwan during the Cold War era. To be sure, chapter 7 concludes that any hint of this would prevent Vietnam from acquiring the technology and materials needed for the civilian use of nuclear power. But thus far, Globalist and Realist thinking have joined together to enact an Atomic Energy Law forbidding the development of nuclear weapons and to eschew the pursuit of nuclear weapons technology.

The ten member states of the Association of Southeast Asian Nations (ASEAN), including Vietnam, have signed the Southeast Asian Nuclear-Weapon-Free Zone (SEANWFZ) Treaty, which declares that each of the signatories will not develop, manufacture, or possess nuclear weapons. SEANWFZ echoes Japan's three non-nuclear principles. And as in the case of Japan's non-nuclear principles, there have been Realist concerns about how the treaty might restrict U.S. naval operations in Southeast Asian waters. The nuclear-weapon-free zone indeed created tensions between the United States on the one hand and its treaty allies Thailand and the Philippines on the other. To finesse this issue, chapter 8 discusses how SEANWFZ permits each signatory country to determine whether or not it wants to restrict the visit or transit of naval vessels that are nuclear-powered or may be carrying nuclear weapons. Globalist arguments have motivated ASEAN to try to persuade the five nuclear weapon states that are permanent members of the U.N. Security Council to sign SEANWFZ, but various Realist calculations have thus far blocked this effort to use the concept of nuclear-weapon-free zones to promote a global nuclear disarmament agenda.

NONPROLIFERATION AND REPROCESSING

As table 10.1 shows, Asian countries, with some notable exceptions, have generally supported international agreements regarding nuclear security and nonproliferation. The issue of reprocessing, however, continues to be problematic in efforts to reduce further the risks of nuclear proliferation; and its saliency is likely to increase as more Asian states confront the difficult task of nuclear waste disposal.

Chinese policy toward nuclear nonproliferation is driven by a coalition of Globalists and Realists. In chapter 2, it is clear that Globalists argue that supporting the international nonproliferation regime gives China an opportunity to demonstrate that it is a responsible stakeholder and favors greater integration in the global economic system. Realists in China are skeptical of U.S. and Russian intentions regarding nuclear disarmament, but they believe that nuclear modernization might enhance China's leverage vis-à-vis the nuclear superpowers regarding the global nonproliferation

Table 10.1. Membership in Nuclear Security and Nonproliferation Regimes

	NPT[1]	*IAEA Member*[1]	*CSA*[2]	*SQP*	*Additional Protocol*[2]	*CTBT*[1]	*CPPNM*[2]	*2005 CPPNM Amendment*	*NSS*	*NTI Nuclear Materials Security Index Score3*
Brunei	Yes	Yes	Yes	Yes	No	Yes	No	No	No[4]	40
Cambodia	Yes	Yes	Yes	Yes	Yes	Yes	Yes	No	No[4]	45
China	Yes	Yes	Yes	No	Yes	Yes	Yes	Yes	Yes[5]	60
India	No	Yes	No[6]	No	Yes	No	Yes	Yes	Yes[5]	46
Indonesia	Yes	Yes	Yes	No	Yes	Yes	Yes	Yes	Yes[5]	65
Japan	Yes	Yes	Yes	No	Yes	Yes	Yes	Yes	Yes[5]	78
Laos	Yes	Yes	Yes	Yes	Signatory	Yes	Yes	No	No[4]	34
Malaysia	Yes	Yes	Yes	No[7]	Signatory	Yes	No	No	Yes[5]	46
Myanmar	Yes	Yes	Yes	Yes	Signed	Yes	No	No	No[4]	23
Pakistan	No	Yes	No[8]	No	No	No	Yes	Yes	Yes[5]	42
Philippines	Yes	Yes	Yes	No	Yes	Yes	Yes	No	Yes[5]	55
Singapore	Yes	Yes	Yes	Yes	Yes	Yes	Yes	Yes	Yes[5]	75
South Korea	Yes	Yes	Yes	No	Yes	Yes	Yes	Yes	Yes[5]	90
Taiwan	N/A[9]	N/A[10]	Yes[11]	No	Yes[11]	N/A[10]	N/A[10]	N/A[10]	No[12]	73
Thailand	Yes	Yes	Yes	No	Signatory	Yes	No	No	Yes[5]	41
Vietnam	Yes	Yes	Yes	No	Yes	Yes	Yes	Yes	Yes[5]	52

Sources: IAEA; UN; Comprehensive Nuclear-Test-Ban Treaty Organization, and Nuclear Threat Initiative

Abbreviations: Nuclear Nonproliferation Treaty (NPT); Comprehensive Safeguards Agreement (CSA) with IAEA; Small Quantities Protocol (SQP) with IAEA; Comprehensive Nuclear Test Ban Treaty (CTBT); Convention on the Physical Protection of Nuclear Material (CPPNM); Nuclear Security Summits (NSS); Southeast Asia Nuclear-Weapon-Free Zone (SEANWFZ); Partial Test Ban Treaty (PTBT); Global Initiative to Combat Nuclear Terrorism (GICNT); International Convention on the Suppression of Acts of Nuclear Terrorism (ICSACT or Nuclear Terrorism Convention); Proliferation Security Initiative (PSI); and Joint Convention on Safety of Spent Fuel Management and on Safety of Radioactive Waste Management (Convention on Safety of Spent Fuel)

	SEANWFZ	*PTBT*[1]	*GICNT*	*Nuclear Terrorism Convention*[1]	*PSI*	*Convention on Nuclear Safety*[2]	*Convention on Early Notification of a Nuclear Accident*[2]	*Convention on Safety of Spent Fuel*
Brunei	Yes	Yes	No	No	Yes	No	No	No
Cambodia	Yes	Yes	Yes	Yes	Yes	Yes	Yes	No
Indonesia	Yes	Yes	No	Yes	No	Yes	Yes	Yes
Laos	Yes	Yes	No	No	No	No	Yes	No
Malaysia	Yes	Yes	Yes	Yes	Yes	No	Yes	No
Myanmar	Yes	Yes	No	No	No	No	Yes	No
Philippines	Yes	Yes	Yes	Yes	Yes	Yes	Yes	Yes
Singapore	Yes	Yes	Yes	Yes	Yes	Yes	Yes	No
Thailand	Yes	Yes	Yes	Yes	Yes	No	Yes	No
Vietnam	Yes	Yes	Yes	No	Yes	Yes	Yes	Yes
Japan	No	Yes	Yes	Yes	Yes	Yes	Yes	Yes
China	No	Yes	Yes	Yes	No	Yes	Yes	Yes
South Korea	No	Yes	Yes	Yes	Yes	Yes	Yes	Yes
Pakistan	No	Yes	Yes	No	No	Yes	Yes	No
Taiwan	N/A[10]	N/A[13]	N/A[8]	N/A[8]	No[12]	N/A[10]	N/A[10]	N/A[10]
India	No	Yes	Yes	Yes	No	Yes	Yes	No

Notes: 1 Membership in the accord via signing or ratification; 2 Accord has entered into force in the country; 3 NTI Nuclear Materials Security Index Score (January 2016 edition, "Theft Rankings") assesses contribution of states on their nuclear materials security with scores between 0 and 100 (most favorable); 4 Attended NSS regional outreach meeting in Thailand in 2011; 5 Attended NSS in 2010, 2012, and 2014; South Korea hosted 2012 summit; 6 India has facilities-specific safeguards agreement on facilities New Delhi classifies as civilian; 7 Rescinded in 2006; 8 Pakistan has safeguards on some but not all of its nuclear facilities; 9 Although Taiwan signed in 1968, the NPT and IAEA consider Beijing since 1971 as the center of power in China. Nevertheless, Taiwan considers itself bound to these obligations; 10 Taiwan is not considered a legal sovereign state by these bodies and is unable to officially participate; 11 Safeguards under trilateral agreement with United States, Taiwan, and IAEA (INFCIRC/133 and INFCIRC/158) and Additional Protocol is applied in Taiwan on a nongovernmental basis; 12 Taiwan not invited to participate but has since removed all HEU from its territory and engaged in PSI efforts through interdiction of ships suspected of carrying WMD-related material; 13 Although Taiwan signed in 1963, the PTBT and United Nations consider Beijing as the center of power in China. Nevertheless, Taiwan considers itself bound to these obligations.

and arms control agenda. Since the mid-1980s, China has pursued a closed fuel cycle policy that includes reprocessing. Although this policy has been embraced with hardly any debate, China's plans for reprocessing facilities and fast-breeder reactors have been modest. Government planners have focused more on building reactors than on addressing the spent fuel issue. Unlike other countries in East Asia where the nuclear waste problem is becoming acute, China's vast geography has permitted the shipment of spent fuel to facilities in western China for interim storage and possible future reprocessing.

Because India used its civilian nuclear energy program to ultimately become a nuclear weapon state, its stance toward the existing nonproliferation regime and reprocessing has been ambivalent, and chapter 3 suggests this ambivalence has been shaped by a complex internal debate. For example, Nationalists favor India's entry into various aspects of the technology control regime, but they oppose signing a possible fissile material cut-off treaty (FMCT) because an FMCT could inhibit India's ability to expand its nuclear arsenal. By contrast, both Realists and Globalists are more open to the FMCT because they contemplate India having only a small nuclear arsenal. This same divide is reflected in India's approach to reprocessing and the nuclear fuel cycle. Nationalists oppose the U.S.-India nuclear deal for its equivocal stance regarding Indian reprocessing, while Realists and Globalists are more willing to accept possible constraints on reprocessing as well as IAEA inspections of civilian nuclear facilities.

As seen in chapter 9, Pakistan's stance toward global nonproliferation is similar to the Indian Nationalist critique of the NPT regime. Pakistani elites across the board see the NPT as a hypocritical regime to preserve what some have called "nuclear apartheid." Like their Nationalist counterparts in India, they oppose an FMCT because it would block fissile material production for weapons development. Moreover, strong opposition to an FMCT is one of the ways that they can express their defiance against the U.S.-India nuclear deal as being unfair and detrimental to Pakistan's interests. Pakistani elites also take pride in their ability to evade international export controls, and they tend not be alarmed by the prospects of a nuclear-armed Iran as well as a nuclear-armed North Korea. Realists in Pakistan might raise concerns about the negative security consequences of an Iran with nuclear weapons, but the dominant Nationalist discourse discounts that security risk by considering such a development to be another Muslim bulwark against Western domination. Therefore, in sharp contrast to India, the Realists in Pakistan are unable to steer their country toward a more cooperative policy regarding global nonproliferation.

Japan has been a strong supporter of nonproliferation by accepting stringent IAEA inspections, by embracing additional international safeguards, by being a core member of the Proliferation Security Initiative, and by sticking to its three non-nuclear principles. A combination of Realism and Globalism has provided robust backing for this policy line. But at the same time, chapter 4 stresses that Japan has been the target of international criticism for its continuing support of reprocessing and a closed fuel cycle. Initially, techno-Nationalism as well as Realist interest in energy security motivated reprocessing in Japan. Even after the Fukushima nuclear accident, Nationalist and Realist voices continue to back the existing policy. But increasingly, the complicated politics of nuclear waste disposal, more than the quest for energy security or autonomy, has kept Japan committed to reprocessing. At the same time, anti-nuclear Globalism, with its emphasis on renewable energy sources and an eventual abandonment of nuclear energy, has broadened its base into the conservative political camp because of concerns about nuclear waste disposal as well as the safety of nuclear power plants.

In South Korea (chapter 5), in addition to the acute problem of nuclear waste disposal, Realist interest in greater energy self-sufficiency and Globalist calculations of expanded exports of nuclear technology have motivated South Korean interest in pyroprocessing and even the development of fast-breeder reactors. But thus far, the United States has been reluctant to provide "advance consent" to South Korea to reprocess spent fuel as it did to Japan because of the proliferation risks. Globalist and Realist arguments, however, have steered both Taiwan (chapter 6) and Vietnam (chapter 7) in a different direction than South Korea. The Globalist perspective that animates the DPP's policy has emphasized Taiwan's cooperation with and commitment to global nonproliferation norms. Realist calculations among Taiwan's government officials have reinforced this policy orientation because of their pragmatic interest in enhancing Taiwan's international image. As a result, Taiwan has pledged to forego plutonium reprocessing and nuclear enrichment. Similarly, a marriage of Globalist and Realist thinking was behind Vietnam's statement that it would not seek to enrich nuclear fuel.

Because the Southeast Asian states do not have any nuclear power plants in operation, the challenge of nuclear waste and the option of reprocessing and a closed fuel cycle have yet to become salient policy questions in Southeast Asia, in contrast to the Northeast Asian countries. And the risk of nuclear proliferation in Southeast Asia is relatively low. All of the ASEAN countries have signed the NPT, and many of them have decided to participate in the Proliferation Security Initiative. The countries (outlined in chapter 8) that are reluctant to join PSI assume this posture not because they are lukewarm about nonproliferation, but because they are concerned about potential negative implications for trade or the management of sovereignty issues. Of the Southeast Asian countries, Myanmar provoked the greatest international suspicion that it might be seeking to develop nuclear weapons via clandestine ties with North Korea. After embarking on domestic political reform, however, Myanmar's leadership has become more forthcoming about confirming its non-nuclear intentions. Therefore, for the time being, the key proliferation challenge that Southeast Asia faces is the problem of clandestine transit and proliferation of nuclear technology by non-state actors.

ROLE OF THE UNITED STATES AND REGIONAL COOPERATION

Although this volume has focused on the domestic debates about nuclear policy questions, the United States has played a critical role in shaping both the discourse and policies in all of the Asian countries. In most cases, Washington has dampened the influence of Nationalist discourse and steered countries toward nuclear restraint. By reaffirming its regional security commitments, the United States has prevented pro-nuclear Nationalist discourse from gaining ground in allied states like Japan and South Korea. Moreover, the American ability to regulate access to nuclear technologies has compelled states like South Korea and Taiwan to abandon clandestine nuclear weapons programs or encouraged countries like India and Vietnam to accept constraints on their civilian nuclear programs. But at the same time, as a provider of nuclear technology, the United States, through both government policy and private sector links with foreign nuclear providers, has supported the expansion of civilian nuclear energy programs in Asia. And in the case of Japan, Washington has helped to impede the Japanese movement to abandon nuclear energy altogether.

The major caveats to the above American role in discouraging nuclear proliferation relate to Pakistan, North Korea, and China. In the case of Pakistan, U.S. nuclear rapprochement with India has intensified Nationalist sentiment and even encouraged arguments, especially by "nuclear Islamists," in favor of nuclear cooperation with Islamic states in defiance against the West. Regarding North Korea, the absence of robust domestic political support in the United States for a sustained policy of comprehensive engagement, as well as North Korea's own sabotage of delicate provisional agreements between Washington and Pyongyang have complicated efforts to stop North Korea's nuclear weapons program. Finally, although China continues to adhere to minimal deterrence and NFU doctrines, Beijing perceives itself to be strategically vulnerable to U.S. nuclear and conventional forces—especially Conventional Prompt Global Strike weapons—and has thus proceeded to modernize and enlarge its nuclear arsenal.

Due in part to the predominant role the United States has played in shaping the nuclear environment in Asia, the level of multilateral regional cooperation has been minimal. Washington has indeed preferred to negotiate bilateral nuclear deals with Asian countries as evident in agreements with China (1985 and 2015), India (2008), Japan (1988), South Korea (1974 and 2015), and Vietnam (2014). Intensifying economic competition among Japanese, South Korean, and Chinese vendors in Asia's growing nuclear market has also inhibited regional cooperation. Common challenges and interests nevertheless exist that should provide strong incentives to collaborate. The Fukushima disaster demonstrated that a nuclear plant mishap taking place in any part of the world can derail nuclear energy programs in one's own country. In short, the issue of nuclear safety affects the business prospects of the nuclear industry as well as the health and welfare of citizens. Fukushima revealed that all of the Asian countries lacked strong independent regulatory regimes because nuclear oversight tended to be in the hands of government agencies who are the promoters of nuclear energy. In the wake of Fukushima, both Tokyo and Seoul have moved to rectify this regulatory deficiency. The Indian government initiated steps in this direction as well. But as other Asian countries begin to develop and expand their civilian nuclear energy programs, multilateral cooperation is urgently needed to share best safety and regulatory practices, to develop peer review mechanisms, to train technical and regulatory personnel, to promote a strong and effective safety culture, and to plan responses to potential nuclear accidents.

During the 1990s, a number of experts in Asia advocated the creation of a regional nuclear organization modeled after Europe's Euratom. In addition to promoting Asian cooperation and transparency regarding the peaceful use of nuclear power, what some called "Asiatom" or "Pacificatom" would address the problem of regional enrichment and reprocessing.[1] These proposals, however, did not garner much support in the U.S. government because of concerns that such a regional regime might legitimize and encourage reprocessing in Asian countries beyond Japan and thereby undermine U.S. nonproliferation policy. In light of the expansion of Asian nuclear energy programs since the 1990s, the time may now be ripe to revisit the concept of a regional nuclear regime.

As discussed in chapter 8, the ASEAN Summit in 2007 supported regional cooperation "for the development and use of civilian nuclear power, in a manner ensuring nuclear safety, security, and nonproliferation," but the halt of nuclear energy programs in almost all Southeast Asian countries makes ASEAN a less pivotal voice in regional regime building. More relevant is the South Korean proposal for a Northeast Asia Peace and Cooperation Initiative (NAPCI). South Korean President Park Geun-hye has advocated a Northeast Asian consultative body for nuclear safety that builds on the trilateral China, Japan, and South Korea "Top Regulators Meeting," which was launched after Fukushima.[2] Although the NAPCI concept for nuclear safety now encompasses Canada, France, Russia, and the United States, the participation of other Asian nuclear energy states like India, Taiwan, and even Vietnam would expand the reach of this initiative. Because reprocessing and a closed fuel cycle entail proliferation risks, there should be more regional cooperation to address the challenge of disposing of spent fuel and of maintaining a steady supply of uranium. Geographic constraints in Japan, South Korea, and Taiwan make national solutions to the problem of nuclear waste management and disposal difficult, if not impossible, to find. The United States could become more proactive in developing international solutions by considering and supporting multinational approaches to the nuclear fuel cycle, including storage and disposal of spent nuclear fuel.[3] In short, the interests of nuclear safety and nonproliferation point to the need to transition from an Asian nuclear energy network dominated by the United States through bilateral relationships to a more multilateral structure that promotes cooperation among Asian countries as well as between the United States and individual Asian states.

NOTES

1. Richard H. Speier and Brian G. Chow, *Asiatom: Proposals, Alternatives, and Next Steps* (Santa Monica, Calif.: RAND, July 1996).

2. Republic of Korea Ministry of Foreign Affairs, *Northeast Asia Peace and Cooperation Initiative 2015* (Seoul: MOFA, 2015), http://www.mofa.go.kr/ENG/North_Asia/res/eng_2015_0310.pdf.

3. Miles Pomper et al., "Nuclear Power and Spent Fuel in East Asia: Balancing Energy, Politics and Nonproliferation," *The Asia-Pacific Journal* 8, issue 25, no. 2 (June 2010), http://japanfocus.org/-Stephanie-Lieggi/3376/article.html; and Charles McCombie and Thomas Isaacs (eds.), *Multinational Approaches to the Nuclear Fuel Cycle*, (Cambridge, Mass.: American Academy of Arts and Sciences, 2010), https://www.amacad.org/multimedia/pdfs/publications/researchpapersmonographs/isaacsinside.pdf.

Index

Page locators in italics refer to figures and tables

About the Contributors

Christopher Clary is a postdoctoral fellow at the Watson Institute for International and Public Affairs at Brown University. Previously, he was a predoctoral fellow at the Belfer Center for Science and International Affairs at Harvard University (2014–2015), a Stanton Nuclear Security Predoctoral Fellow at the RAND Corporation in Washington, D.C. (2013–2014), and a Council on Foreign Relations International Affairs Fellow in India (2009). He also served as country director for South Asian affairs in the Office of the Secretary of Defense (2006–2009), a research associate at the Naval Postgraduate School in Monterey, California (2003–2005), and a research assistant at the Henry L. Stimson Center in Washington, D.C. (2001–2003).

Catharin Dalpino teaches in the Washington Program of Seton Hall University as an adjunct professor. She has previously taught Southeast Asian Studies at Georgetown University, The George Washington University, and the Johns Hopkins School of Advanced International Studies. She has also taught International Relations at Simmons College. Dalpino has been a Deputy Assistant Secretary of State (1993–1997), and has held fellow positions at the Brookings Institution, Carnegie Endowment for International Peace, and the Atlantic Council of the United States.

Mike M. Mochizuki holds the Japan-U.S. Relations Chair in Memory of Gaston Sigur at the Elliott School of International Affairs at The George Washington University. He was also the director of the Sigur Center for Asian Studies from 2001 to 2005. He codirects the "Memory and Reconciliation in the Asia-Pacific" research and policy project of the Sigur Center. Previously, he was a senior fellow at the Brookings Institution, codirector of the Center for Asia-Pacific Policy at RAND, and a professor at the University of Southern California and Yale University. Mochizuki's books include *The Japan-U.S. Alliance* and *China-Taiwan Relations: Implications for*

Okinawa (coeditor and author, 2008), *Japan in International Politics: The Foreign Policies of an Adaptive State* (coeditor and author, 2007), *The Okinawa Question and the U.S.-Japan Alliance* (coeditor and author, 2005), and *Crisis on the Korean Peninsula: How to Deal with a Nuclear North Korea* (coauthor, 2003). He has published articles in such journals as *The American Interest, Asia Pacific Review, Foreign Affairs, International Security, Japan Quarterly, Journal of Strategic Studies, Nonproliferation Review, Survival,* and *The Washington Quarterly*. Mike M. Mochizuki has a PhD from Harvard University.

Deepa M. Ollapally is a research professor of International Affairs and the associate director of the Sigur Center for Asian Studies at the Elliott School of International Affairs at The George Washington University. She is also the director of the Rising Powers Initiative launched by the Elliott School in 2009. Her areas of research and teaching include Asian regional security, comparative politics of South Asia, identity politics in Asia, nuclear nonproliferation, extremism, and gender in international security. Ollapally's current research focuses on domestic foreign policy debates in India and their implications for regional security and global leadership of the United States. Her previous books include *Worldviews of Aspiring Powers: Domestic Foreign Policy Debates in China, India, Iran, Japan and Russia* (coeditor and contributor), *The Politics of Extremism in South Asia* (author), and *Confronting Conflict: Domestic Factors and U.S. Policymaking in the Third World* (coauthor). She has published extensively in journals such as *Foreign Affairs, Asian Survey, The National Interest,* and *Political Science Quarterly*.

Before joining The George Washington University in 2003, she directed the South Asia program at the U.S. Institute of Peace, taught at Swarthmore College, and headed the Strategic Studies Unit at the National Institute of Advanced Studies in Bangalore, India. Ollapally has a PhD from Columbia University.

Rajesh Rajagopalan is a professor of International Politics at the Centre for International Politics, Organization and Disarmament, School of International Studies at Jawaharlal Nehru University. Previously, he was a senior fellow at the Observer Research Foundation, New Delhi, and a research fellow at the Institute for Defence Studies and Analyses, New Delhi. He also served as deputy secretary in the National Security Council Secretariat, Government of India (2000–2001). He has taught at Hunter College, Brooklyn College, and Queens College of the City University of New York. Rajagopalan's areas of research interest are international relations theory, military doctrines, and nuclear weapons and disarmament. His books include *Fighting like a Guerrilla: The Indian Army* (2008) and *Counterinsurgency and Second Strike: Arguments about Nuclear War in South Asia* (2005). He has a PhD from the City University of New York.

Scott Snyder is Senior Fellow for Korea Studies and director of the Program on U.S.-Korea Policy at the Council on Foreign Relations (CFR), where he had served as an adjunct fellow from 2008 to 2011. Snyder's program examines South Korea's efforts to contribute on the international stage; its potential influence and contributions as a middle power in East Asia; and the peninsular, regional, and global implications of North Korean instability. Snyder is also the coeditor of *North Korea in Transition: Politics, Economy, and Society* (2013) and the editor of *Global Korea: South Korea's Contributions to International Security* (2012) and *The U.S.-South Korea Alliance: Meeting New Security Challenges* (2012). Prior to joining CFR, Snyder was a senior associate in the international relations program of The Asia Foundation, where he founded and directed the Center for U.S.-Korea Policy and served as The Asia Foundation's representative in Korea. He was also a senior associate at Pacific Forum CSIS (Center for Strategic and International Studies). Snyder has worked as an Asia specialist in the Research and Studies Program of the U.S. Institute of Peace and as acting director of Asia Society's contemporary affairs program. He was a Pantech visiting fellow at Stanford University's Walter H. Shorenstein Asia-Pacific Research Center during 2005–2006, and in 1998–1999 received an Abe Fellowship, administered by the Social Sciences Research Council. Snyder received a BA from Rice University and an MA from the Regional Studies East Asia Program at Harvard University and was a Thomas G. Watson Fellow at Yonsei University in South Korea.

Robert Sutter has been a Professor of Practice of International Affairs at the Elliott School of International Affairs at The George Washington University since 2011. His earlier full-time position was visiting professor of Asian Studies at the School of Foreign Service, Georgetown University (2001–2011). Sutter taught part-time for over thirty years at Georgetown, The George Washington University, Johns Hopkins Universities, or the University of Virginia. Sutter has published nineteen books, over 200 articles, and several hundred government reports dealing with contemporary East Asian and Pacific countries and their relations with the United States. His most recent book is the second edition of *U.S.-Chinese Relations: Perilous Past, Pragmatic Present* (2012). Sutter's government career (1968–2001) involved work on Asian and Pacific affairs and U.S. foreign policy for the Congressional Research Service, the Central Intelligence Agency, the Department of State, and the Senate Foreign Relations Committee. He was for many years the senior specialist and director of the Foreign Affairs and National Defense Division of the Congressional Research Service. He also was the National Intelligence Officer for East Asia and the Pacific at the U.S. government's National Intelligence Council, and the China Division Director at the Department of State's Bureau of Intelligence and Research. Sutter earned his PhD in History and East Asian Languages from Harvard University.

Nguyễn Thị Thanh Thủy is an associate professor and head of the Division of Americas Studies, Faculty of International Studies, University of Social Sciences and Humanities, Vietnam National University—Hanoi. She is the author of the 2012 book *United States and the Nuclear Weapons Issues over More than Half a Century, 1945–2010*.

Timothy Westmyer is a research and program associate at the Rising Powers Initiative located at The George Washington University's Sigur Center for Asian Studies. In addition to this study of domestic debates in Asia over the risks and rewards of nuclear energy and nuclear weapons, his projects include energy security and maritime strategies in Asia. Previously, he worked as a foreign affairs assistant at the Department of State's Office of the Coordinator for Threat Reduction, where he aided U.S. efforts at the 2012 Nuclear Security Summit and Ambassador Bonnie Jenkins as she chaired the Global Partnership Against the Spread of Weapons and Materials of Mass Destruction. He has also spent time working at the Council on Foreign Relations, the U.S. Senate, the Arms Control Association, and on a U.S. presidential campaign. Westmyer was a member of the 2013 Nuclear Scholars Initiative at the Center for Strategic and International Studies. His writing has been published by several nuclear and national security outlets, including *Democracy Arsenal*, the *China Institute of International Studies*, *The Bulletin of the Atomic Scientists*, *Arms Control Today*, *The Diplomat*, and the *Center for Strategic and International Studies*. Timothy holds a Georgetown University MA with honors in Security Studies with a concentration on Unconventional Weapons.

Yuan-Ming Alvin Yao is the director of research at the Prospect Foundation. He was previously an adjunct research associate at the MacArthur Center for Security Studies and the deputy director of research at the Foundation on Asia-Pacific Peace Studies. He has also spent time as a visiting fellow at the Center for Northeast Asia Policy Studies at the Brookings Institution and an associate research fellow at the Elliott School of International Affairs. Yao earned his PhD of East Asian Studies at the National Chengchi University, in addition to an MA in International Relations at the National Taiwan University.

Linda J. Yarr has been the director of Partnerships for Strategies in Asia (PISA), the Elliott School of International Affairs at The George Washington University, since 1996. PISA promotes international affairs education training and research in partnership with leading agencies and institutions in Asia. As a member of the research team of The George Washington University Women's Leadership Institute, she is engaged in a project on leadership in the global environment. In 1994–1995, Yarr taught a research course on conflict resolution at American University. From 1987 to 1994, she was assistant professor of global political economy at Friends World College. In addition to teaching global political economy and experiential education, Yarr guided students in active learning projects and designed a program in global women's studies. From 1983 to 1987, she offered courses in comparative politics, development, gender, and the Vietnam War at the University of Colorado Boulder, and the University of Denver. Yarr has held visiting scholar appointments at the Helsinki Collegium for Advanced Studies, the Institute for Malaysian and International Studies of the National University of Malaysia, the School of International Service of American University, and the Rocky Mountain Women's Institute.

Hui Zhang is a senior research associate at the Project on Managing the Atom in the Belfer Center for Science and International Affairs at Harvard University's John F. Kennedy School of Government. His research includes verification techniques of nuclear arms control, the control of fissile material, nuclear terrorism, China's nuclear policy, nuclear safeguards and nonproliferation, nuclear fuel cycle policies, and reprocessing. Before coming to the Kennedy School in September 1999, from 1997–1999 he was a postdoctoral fellow at the Andlinger Center for Energy and Environmental Studies at Princeton University, and in 1998–1999 he received a postdoctoral fellowship from the Social Science Research Council, a MacArthur Foundation program on International Peace and Security. From 2002 to 2003, he received a grant for Research and Writing from the John D. and Catherine T. MacArthur Foundation. Zhang is the author of several technical reports and book chapters and dozens of articles in academic journals and the print media, including *Science and Global Security*, *Arms Control Today*, *The Bulletin of the Atomic Scientists*, *Disarmament Diplomacy*, *Disarmament Forum*, *Nonproliferation Review*, *The Washington Quarterly*, and *China Security*. Hui Zhang received a PhD in nuclear physics from the China Academy of Engineering Physics in Beijing.